T0181874

Springer Tracts in Modern Physics

Volume 270

Springer Tracts in Modern Physics provides comprehensive and critical reviews of topics of current interest in physics. The following fields are emphasized:

- Elementary Particle Physics
- Condensed Matter Physics
- Light Matter Interaction
- Atomic and Molecular Physics
- Complex Systems
- Fundamental Astrophysics

Suitable reviews of other fields can also be accepted. The Editors encourage prospective authors to correspond with them in advance of submitting a manuscript. For reviews of topics belonging to the above mentioned fields, they should address the responsible Editor as listed in "Contact the Editors".

More information about this series at http://www.springer.com/series/426

Shang Yuan Ren

Electronic States in Crystals of Finite Size

Quantum Confinement of Bloch Waves

Second Edition

Shang Yuan Ren
School of Physics
Peking University
Beijing
P. R. China

ISSN 0081-3869 ISSN 1615-0430 (electronic)
Springer Tracts in Modern Physics
ISBN 978-981-13-5210-2 ISBN 978-981-10-4718-3 (eBook)
DOI 10.1007/978-981-10-4718-3

Printed on acid-free paper

This Springer imprint is published by Springer Nature
The registered company is Springer Nature Singapore Pte Ltd.
The registered company address is: 152 Beach Road, #21-01/04 Gateway East, Singapore 189721, Singapore

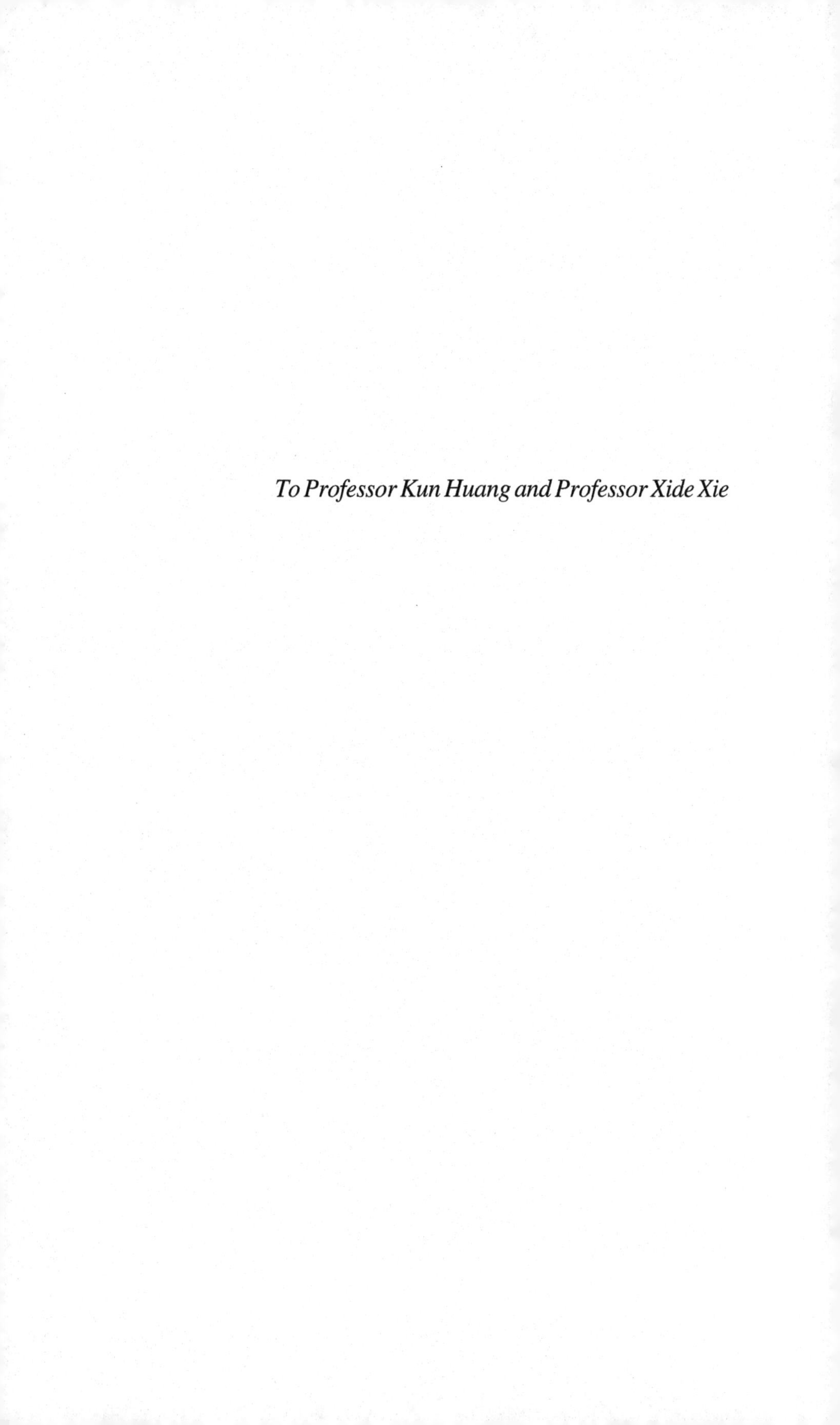

To Professor Kun Huang and Professor Xide Xie

Preface

Based on a theory of differential equations approach, the first edition presented an analytical one-electron non-spin theory of electronic states in some simple and interesting low-dimensional systems and finite crystals. The essential understanding obtained is that different from that all electronic states are Bloch waves in a crystal with translational invariance as in the conventional solid state physics, a two boundary truncation of the translational invariance may produce two different types of electronic states—boundary dependent or size dependent—in some simple and interesting cases. The size dependent states are stationary Bloch waves due to the finite size of the truncated system, which properties and numbers are determined by the size. The boundary dependent states are a different type of electronic states which properties are determined by the very existence of the boundary. Such an understanding was first learned in one-dimensional crystals and later archived in multi-dimensional crystals as well. Since it was first published more than ten years ago, the existence of such two types of confined states or modes has been confirmed by subsequent investigations of other authors in various low-dimensional systems. The very presence of the boundary dependent confined states is a unique distinction of the quantum confinement of Bloch waves. Whether this is the origin of many unusual properties of low-dimensional systems remains to be understood.

The current Chap. 2 is a summary of basic theory of periodic Sturm–Liouville equations, as a part of the mathematical basis of the book. The author recently learned that this mathematical theory could treat more general one-dimensional systems in physics, such as one-dimensional layered photonic crystals and phononic crystals. Therefore, the wave equations of those crystals can be treated in the same way as the Schrödinger differential equation for a one-dimensional electronic crystal. By the theory, eigenmodes of those different one-dimensional crystals have similar properties. As a result, eigenmodes of two types will be obtained if the modes are completely confined to the finite size. The general theory of periodic Sturm-Liouville equations straightforwardly gives many equations or theoretical results previously obtained on layered crystals in the literature. It can further treat more complicated thus previously untreated cases without much extra effort and provide theoretical understandings on results observed in previous numerical calculations but may not be explained. These problems are discussed in Appendices C–F.

The primary focus of this book is still on the electronic states in crystals of finite size. Various substantial revisions, improvements, and corrections are made, in particular, in Chaps. 3, 5, and 8. All main scientific conclusions presented in the first edition remain valid or receive improved understandings, such as the differences between multi-dimensional cases and one-dimensional cases. *In general*, a surface state in a multi-dimensional crystal does not have to be in a band gap, since that in the physical origin, a surface state is closely related to a permitted band rather than a band gap. That a surface state is always in a band gap and decays most at the mid-gap is merely a unique distinction of one-dimensional crystals.

Some readers feel that examples are helpful for understanding the mathematical theory in the book. An advantage of the Kronig–Penney model is that its solutions—in both permitted and forbidden energy ranges—can be analytically obtained and simply expressed. The model provides a concrete example for illustrating the theory presented in Part II. The newly added Appendix A serves this purpose.

Periodicity is one of the most fundamental and extensively investigated mathematical concepts. Many periodic systems can be truncated. The truncations of periodicity do present new issues. In comparison with the periodicity, the investigations and general understandings of the truncated periodicity are much less. The existence of two types of confined states is a preliminary understanding of the simplest cases and is only the beginning. This new edition of the book is a continuing and further effort of the author trying to understand the truncated periodicity in physical systems based on the mathematical theory of periodic differential equations.

This edition is devoted to Professor Kun Huang and Professor Xide Xie, two leading founders of modern solid-state physics in China. The author was fortunate to be personally educated by them, as a human being and as a physicist.

It is a pleasure of the author to take this opportunity to thank Prof. Walter A. Harrison for his guidance and help, leading the author into modern solid-state physics again after he had to stop doing physics for more than ten years during the disastrous Cultural Revolution in China.

The author thanks Profs. Yia-Chung Chang, Zhongqi Ma, Shangfen Ren, Sihong Shao, Huaiyu Wang, Anton Zettl, Han Zhang, Pingwen Zhang, Shengbai Zhang, Yong Zhang, Guangda Zhao for their help or discussions in his working on the new edition.

Last but not least, the author is grateful to his family members. The work presented in the book could not have been completed without their continuing understandings, selfless love and supports in many ways and many years.

Carmel Valley, San Diego, USA Shang Yuan Ren
May 2017

Preface to the First Edition

The theory of electronic states in crystals is the very basis of modern solid state physics. In traditional solid state physics – based on the Bloch theorem – the theory of electronic states in crystals is essentially a theory of electronic states in crystals of infinite size. However, that any real crystal always has a finite size is a physical reality one has to face. The difference between the electronic structure of a real crystal of finite size and the electronic structure obtained based on the Bloch theorem becomes more significant as the crystal size decreases. A clear understanding of the properties of electronic states in real crystals of finite size has both theoretical and practical significance. Many years ago when the author was a student learning solid state physics at Peking University, he was bothered by a feeling that the general use of the periodic boundary conditions seemed unconvincing. At least the effects of such a significant simplification should be clearly understood. Afterward, he learned that many of his schoolmates had the same feeling. Among many solid state physics books, the author found that only in the classic book *Dynamic Theory of Crystal Lattices* by Born and Huang was there a more detailed discussion on the effects of such a simplification in an Appendix.

In the present book, a theory of electronic states in ideal crystals of finite size is developed by trying to understand the quantum confinement effects of Bloch waves. The lack of translational invariance had been a major obstacle in developing a general theory on the electronic states in crystals of finite size. In this book, it was found that on the basis of relevant theorems in the theory of second-order differential equations with periodic coefficients, this major obstacle or difficulty, actually, could be circumvented: Exact and general understanding on the electronic states in some simple and interesting ideal low-dimensional systems and finite crystals could be analytically obtained. Some of the results obtained in the book are quite different from what is traditionally believed in the solid state physics community.

This book consists of five parts. The first part gives a brief introduction to why a theory of electronic states in crystals of finite size is needed. The second part treats one-dimensional semi-infinite crystals and finite crystals; most results in this part can be rigorously proven. The third part treats low-dimensional systems or finite crystals in three-dimensional crystals. The basis is rigorous according to the author's understanding; however, much of the reasoning in this part had to be based on physical intuition due to the lack of enough available mathematical understanding. The fourth part is devoted to concluding remarks. In the fifth part are two appendices. The contents of each chapter in Parts II and III are rather closely related; therefore, readers are expected to read these chapters in the given order. Without appropriate preparation from earlier chapters, readers may find the later chapters difficult to understand. Although the purpose of this book is to present a theory of electronic states in crystals of *finite size*, it is the clear understanding of the electronic states in *crystals with translational invariance* – as obtained in traditional solid state physics – that provided a basis for such a new theory.

One of the feelings the author had frequently while working on the problems in this book is that the mathematicians and the solid state physicists are rather unfamiliar with each other's problems and their respective results. The major mathematical basis of the work presented in this book, Eastham's *The Spectral Theory of Periodic Differential Equations*, was published more than 30 years ago; however, it seems that many of the important results obtained in his book are not yet well known in the solid state physics community. Although the Bloch function is the most fundamental function in the theory of electronic states in modern solid state physics, little is widely known in the community on the general properties of the function except that it can be expressed as the product of a plane wave function and a periodic function. For quite a long time, the author also knew only this about the Bloch function and had many hard working days on some problems without making substantial progress. By mere chance, he saw Eastham's book. He was discouraged by the seemingly difficult mathematics at the beginning but made an effort to understand the book and to apply the new mathematical results learned to relevant physics problems. The book presented here is essentially the result of such effort.

In addition to Eastham's book, the author has also greatly benefited from two classic books: Courant and Hilbert's *Methods of Mathematical Physics* and Titchmarsh's *Eigenfunction Expansions Associated with Second-Order Differential Equations*. The theorems presented in these two books are so powerful that some misconceptions on the electronic states in low-dimensional systems actually could have been clarified much earlier if some theorems in those books published many years ago were clearly and widely understood in the solid state physics community. Unfortunately, these excellent books are out of print now. The wide use of more and more powerful computer-based approaches has unquestionably made great contributions to our understanding of the low-dimensional systems. Nevertheless, the author hopes that the publishing of this book could stimulate more general interest in the use of analytical approaches in understanding these very interesting and challenging systems, which, at least, could be a substantial complement. After all, a really comprehensive and in-depth understanding of a physical problem can usually be obtained from an analytical theory based on a simplified model correctly containing the most essential physics.

It is a pleasure of the author to take this opportunity to thank Professor Kun Huang for his many years of guidance, help, and discussions. It was he who led the author into the field of solid state physics. The author is very grateful to Ms. Avril Rhys (i.e., Mrs. Huang); her concern and help is one of the most appreciated experiences the author had in the process of writing the book. He also wishes to thank Professor Huan-Wu Peng for sharing his experience in the early stage of the solid state physics in the mid-1940s and many interesting discussions. The author was fortunate to have had opportunities to listen to Professor Huang' and Professor Peng's experiences when they worked with Max Born.

The author is grateful to Professors John D. Dow, Hanying Guo, Rushan Han, Walter A. Harrison, Zhongqi Ma, Shangfen Ren, Zhengxing Wang, Sicheng Wu, Shousheng Yan, Lo Yang, Shuxiang Yu, Jinyan Zeng, Ping Zhang, and Pingwen Zhang for their comments and/or discussions. He wishes to thank Miss Yulin Xuan and Miss Zhiling Ruan for much valuable help. He also wishes to thank Dr. Wei Cheng for his help in many computer-related problems.

Last but not least, the author is indebted to his family members, in particular his wife Weimin, his daughters Yujian and Yuhui, his sons-in-law Weidong and Jian, and his grandchildren Nana, Yangyang, and Weiwei. Their love and support not only gave him so much happiness in enjoying family life, but also brought him the strength and courage to fight the sufferings sometimes one had to experience, leading to the birth of this book.

March 2005

Shang Yuan Ren
ZhongGuanYuan, Peking University, Beijing

Contents

Part I
Why a Theory of Electronic States in Crystals of Finite Size Is Needed

Chapter 1
Introduction

Solid-state physics is a field in modern physics in which one is mainly concerned with the physical properties of and physical processes in various solids. Besides its fundamental significance, a clear understanding of different physical properties of and physical processes in solids and their origin may provide insight for possible practical applications of relevant properties and physical processes. Since the middle of the twentieth century, many achievements in the field have made significant contributions to modern science and technology, even resulting in revolutionary developments. We can expect that further achievements in this field will continue to bring tremendous benefits to human beings and society.

A clear understanding of the electronic structure of a solid is always the basis for understanding the physical properties of the solid and the physical processes in the solid. In conventional solid-state physics, the basic theory of electronic states in crystals has been established for more than 80 years. Most further theoretical developments since then have been mainly applications of the basic theory to different physical problems and calculations of detailed electronic structures of various solids. However, this conventional theory also has some fundamental difficulties. Those fundamental difficulties become more significant today when one has to deal with crystals of much smaller size than before.

This chapter is organized as follows. In Sects. 1.1–1.2, we briefly review some of the most basic understandings of the electronic structure of crystals in conventional solid-state physics and how the theory of electronic states in crystals is the very basis for determining the physical properties of and the physical processes in the crystals, by using simple examples. In Sect. 1.3, we point out some fundamental difficulties of the theory of electronic states in conventional solid-state physics. As consequences of these fundamental difficulties, the theory of electronic states in conventional solid-state physics cannot treat the boundary effects and the size effects of crystals, which have substantial significance today when one has to deal with crystals in the submicron and nanometer size range—the low-dimensional systems. In Sects. 1.4–1.5, we briefly review one of the most widely used approaches in theoretical investigations of electronic states in low-dimensional systems—the effective mass approximation

S.Y. Ren, *Electronic States in Crystals of Finite Size*, Springer Tracts in Modern Physics 270, DOI 10.1007/978-981-10-4718-3_1

approach—and some numerical results. In Sect. 1.6 we give a brief introduction to the subject of this book and the main findings.

1.1 Electronic States Based on the Translational Invariance

The very basis of the theory of electronic states in modern solid state physics—the energy band theory—is the Bloch theorem [1]. It is based on the assumption that atoms in a crystal are periodically located—the potential in the crystal has a translational invariance [2–7].

The single-electron Schrödinger differential equation with a periodic potential can be written as

$$-\frac{\hbar^2}{2m}\nabla^2 y(\mathbf{x}) + [V(\mathbf{x}) - E]y(\mathbf{x}) = 0, \tag{1.1}$$

where $V(\mathbf{x})$ is the periodic potential:

$$V(\mathbf{x}+\mathbf{a}_1) = V(\mathbf{x}+\mathbf{a}_2) = V(\mathbf{x}+\mathbf{a}_3) = V(\mathbf{x}). \tag{1.2}$$

Here, $\mathbf{a}_1$, $\mathbf{a}_2$, and $\mathbf{a}_3$ are three primitive lattice vectors of the crystal.

Based on this assumption, the Bloch theorem states that the electronic states in the crystal have the property that

$$\phi(\mathbf{k}, \mathbf{x}+\mathbf{a}_i) = \mathrm{e}^{\mathrm{i}\mathbf{k}\cdot\mathbf{a}_i}\phi(\mathbf{k}, \mathbf{x}), \quad \mathrm{i} = 1, 2, 3; \tag{1.3}$$

this can also be expressed as

$$\phi(\mathbf{k}, \mathbf{x}) = \mathrm{e}^{\mathrm{i}\mathbf{k}\cdot\mathbf{x}}\mathrm{u}(\mathbf{k}, \mathbf{x}), \tag{1.4}$$

where $\mathbf{k}$ is a real wave vector in the $\mathbf{k}$ space and $u(\mathbf{k}, \mathbf{x})$ is a function with the same period as the potential:

$$u(\mathbf{k}, \mathbf{x}+\mathbf{a}_1) = u(\mathbf{k}, \mathbf{x}+\mathbf{a}_2) = u(\mathbf{k}, \mathbf{x}+\mathbf{a}_3) = u(\mathbf{k}, \mathbf{x}). \tag{1.5}$$

The function $\phi(\mathbf{k}, \mathbf{x})$ in (1.3) and (1.4) is called the Bloch function or Bloch wave. This is the most fundamental function in modern solid-state physics.

The range of the wave vector $\mathbf{k}$ in (1.3) or (1.4) can be limited to a specific region in the $\mathbf{k}$ space called the Brillouin zone [8], determined by three primitive vectors of the reciprocal lattice in the $\mathbf{k}$ space $\mathbf{b}_1$, $\mathbf{b}_2$, and $\mathbf{b}_3$:

$$\mathbf{b}_1 = \frac{\mathbf{a}_2 \times \mathbf{a}_3}{\mathbf{a}_1 \cdot \mathbf{a}_2 \times \mathbf{a}_3}, \quad \mathbf{b}_2 = \frac{\mathbf{a}_3 \times \mathbf{a}_1}{\mathbf{a}_1 \cdot \mathbf{a}_2 \times \mathbf{a}_3}, \quad \mathbf{b}_3 = \frac{\mathbf{a}_1 \times \mathbf{a}_2}{\mathbf{a}_1 \cdot \mathbf{a}_2 \times \mathbf{a}_3} \tag{1.6}$$

and thus

$$\mathbf{a}_i \cdot \mathbf{b}_j = \delta_{i,j}; \tag{1.7}$$

here, $\delta_{i,j}$ is the Kronecker symbol.

As the wave vector $\mathbf{k}$ varies in the Brillouin zone, the permitted energy of each Bloch function $\phi(\mathbf{k}, \mathbf{x})$ – the eigenvalue E in (1.1) – also changes. These permitted energy ranges are called energy bands and can be written as $E_n(\mathbf{k})$; here, n is an energy band index. They can be ordered with increasing energy:

$$E_0(\mathbf{k}) \leq E_1(\mathbf{k}) \leq E_2(\mathbf{k}) \leq E_3(\mathbf{k}) \leq E_4(\mathbf{k}) \leq \cdots.$$

The corresponding eigenfunctions are denoted by $\phi_n(\mathbf{k}, x)$ and they can be written as

$$\phi_n(\mathbf{k}, \mathbf{x}) = \mathrm{e}^{\mathrm{i}\mathbf{k}\cdot\mathbf{x}} \mathrm{u}_\mathrm{n}(\mathbf{k}, \mathbf{x}), \tag{1.8}$$

where n is the energy band index, $\mathbf{k}$ is the wave vector, and $u_n(\mathbf{k}, \mathbf{x})$ is a function with the same period as the potential:

$$u_n(\mathbf{k}, \mathbf{x} + \mathbf{a}_1) = u_n(\mathbf{k}, \mathbf{x} + \mathbf{a}_2) = u_n(\mathbf{k}, \mathbf{x} + \mathbf{a}_3) = u_n(\mathbf{k}, \mathbf{x}).$$

The energy band structure formed by the valence electrons of a crystal plays a major role in determining the physical properties of the crystal and which physical processes may happen in the crystal. A crystal with a band-gap between the highest occupied energy band(s) and the lowest unoccupied energy band(s) can only have very few conducting electrons at low temperature thus is either a semiconductor or an insulator, depending on the details of the band structure, such as the size of the band gap. A crystal without a band-gap between the highest occupied energy band(s) and the lowest energy band(s) with unoccupied states usually has net conducting electrons at low temperature, and the crystal is a metal.

Any real crystal always has a finite size and does not have the hypothetical infinite size on which the translational invariance is based. To circumvent this difficulty, in conventional solid state physics the periodic boundary conditions were usually assumed. For a crystal of parallelepiped shape having three edges $N_1\mathbf{a}_1$, $N_2\mathbf{a}_2$, and $N_3\mathbf{a}_3$ meeting at a particular vertex, the periodic boundary conditions require that the wave functions of the electronic states $\phi_n(\mathbf{k}, \mathbf{x})$ in the finite crystal have to satisfy [2–7]:

$$\phi_n(\mathbf{k}, \mathbf{x} + N_1\mathbf{a}_1) = \phi_n(\mathbf{k}, \mathbf{x} + N_2\mathbf{a}_2) = \phi_n(\mathbf{k}, \mathbf{x} + N_3\mathbf{a}_3) = \phi_n(\mathbf{k}, \mathbf{x}). \tag{1.9}$$

The effect of (1.9) is to make the wave vector $\mathbf{k}$ assume discrete values:

$$\mathbf{k} = k_1\mathbf{b}_1 + k_2\mathbf{b}_2 + k_3\mathbf{b}_3, \tag{1.10}$$

where

$$k_i = \frac{j_i}{N_i} 2\pi, \quad j_i = 0, 1, 2, \ldots, N_i - 1, \quad i = 1, 2, 3. \tag{1.11}$$

Thus, in each energy band n, there are, in all,

$$N = N_1 N_2 N_3 \tag{1.12}$$

Bloch states $\phi_n(\mathbf{k}, \mathbf{x})$ for such a finite crystal.

1.2 Energy Band Structure of Several Typical Crystals

The energy band structure of any specific solid is usually described by its energy–wavevector dispersion relation $E_n(\mathbf{k})$. It was first clearly understood by Kramers [9] that the energy band structure of one-dimensional crystals, in general, has some especially simple characteristics (see Chap. 2). The energy band structure of a three-dimensional crystal is usually more complicated [10, 11].

Many different physical properties of a solid can be understood from its specific band structure. For semiconductors (and insulators), the energy bands occupied by valence electrons are called the valence bands, and there is a band gap between the highest occupied valence bands and the lowest unoccupied energy bands called the conduction bands. For crystals with translational invariance, only the energies in permitted energy bands are allowed. No electronic state can have its energy in the band gap.

The most important physics processes in a semiconductor always happen near the band gap. Therefore, the details of the band structure near the band gap - such as, the size of the energy band gap, the locations of the conduction band extreme(s) and the valence band extreme(s), the band structure behaviors near those extremes, and so forth - are almost always technically the most critical and theoretically the most interesting features. It is these details that determine the physical properties of a semiconductor and its possible applications.

Si is the most important semiconductor material today due to the maturity of the processing technology for making devices with it and the abundant source of raw material. It has a band-gap of about 1.2 eV and a valence band maximum (VBM) at the center of the Brillouin zone, with six conduction band minima located on six equivalent [100] axes in the **k** space, near the boundary of the Brillouin zone [10]. Despite its position as the number one semiconductor material in the electronic industry today, a significant shortcoming of Si is that it is an indirect semiconductor—the VBM and the conduction band minima are in different locations in the Brillouin zone so that a direct optical transition between the VBM and any conduction band minimum is forbidden; thus, it is not easy to make optical devices with Si and to integrate optical processing devices with conventional Si electronic integrated circuits.

After Si, GaAs is one of the most important semiconductor materials. It has a band-gap of about 1.5 eV, and both its VBM and its conduction band minimum are located at the center of the Brillouin zone [10]. Therefore, GaAs is a direct-gap semiconductor—a direct optical transition between the VBM and the conduction band minimum is permitted, and this makes GaAs one of the best semiconductor materials for making optical devices and optoelectronic integrated circuits.

For metals, the most important physical processes happen near the Fermi surface; thus, the details of the band structure $E_n(\mathbf{k})$ near the Fermi surface are often of the greatest interest.

Modern solid-state physics is essentially established on the basis of the theory of electronic states in crystals. Many theoretical methods have been developed to study various physical properties in different solids, most of them are based on general understandings such as the following: (i) The electronic states in crystals are Bloch waves and (ii) the physical properties of a specific solid are determined by its specific band structure. It has had great success—many of the electronic, electric, optical, magnetic, thermal, and mechanical properties of various solids of macroscopic size are well understood on the basis of this theory. Based on these understandings, many new electronic devices have been invented and developed; some of them—such as transistors and semiconductor integrated circuits—have brought revolutionary changes into modern science and technology.

1.3 Fundamental Difficulties of the Theory of Electronic States in Conventional Solid-State Physics

The theory of electronic states based on the translational invariance of the potential has been the basis of our current understanding of the electronic states in solids for more than 80 years and has achieved great success. Nevertheless, this conventional theory also has some fundamental difficulties. This is because the translational invariance of the potential can only exist in crystals of infinite size; thus, that the electronic states in crystals can be well described by Bloch waves (1.4) or (1.8) is correct only for crystals of infinite size. According to the Bloch theorem, Bloch waves (1.8) are progressive waves. In general, the flux density of Bloch waves is nonzero,

$$\phi_n^*(\mathbf{k}, \mathbf{x})\nabla\phi_n(\mathbf{k}, \mathbf{x}) - \phi_n(\mathbf{k}, \mathbf{x})\nabla\phi_n^*(\mathbf{k}, \mathbf{x}) \neq 0.$$

These progressive waves travel in all directions; only in the case of a crystal of infinite size will they always remain inside the crystal. Any real crystal has a finite size with a border. If the electronic states in a crystal of finite size are Bloch waves, these progressive waves can move beyond the border, and the electrons in the crystal will flow away from the crystal so that the crystal will continuously lose electrons. Consequently, the electronic states in a crystal of finite size cannot be progressive Bloch waves. To overcome this difficulty, the assumption of the periodic boundary conditions (1.9) implies that if an electron goes out from one boundary face of the crystal, it simultaneously comes back in from the opposite boundary face; obviously, this is not true and not physically possible.

Any real crystal has boundaries. The existence of boundaries—the termination of the periodic potential—may introduce the appearance of new types of electronic states. In 1932, Tamm [12] showed that in a one-dimensional Kronig–Penney [13] crystal, a termination of the periodic potential—a potential barrier outside the boundary of a semi-infinite crystal—may introduce an additional type of electronic states in band gaps of the Bloch wave below the potential barrier. Electronic states of this new type—with energy inside the band gaps—are not permitted in crystals of infinite size or with periodic boundary conditions. They are called surface states because they are located near the surface of crystals. Since then, investigations on the surface states and relevant problems have become a rapidly developing and very productive field in solid state physics and chemistry [14, 15]. It is now well understood that the existence and properties of the surface states can play a significant role in affecting the physical properties of solids and physical processes in solids. The assumption of periodic boundary conditions (1.9) is a simplification that removes any possible boundary effects of the crystal; it does not correspond to the physical reality of any real crystal. For a finite crystal, it gives $N_1N_2N_3$ Bloch states for each energy band. Consequently, the conventional theory of electronic states in solids—based on the translational invariance of the potential—cannot account for the existence of surface states. The very presence of these non-Bloch states has to be built on a separate and different theoretical consideration. This point is another fundamental difficulty of the conventional theory of the electronic states in solids.

Since the theory of the electronic states in crystals in conventional solid-state physics is essentially a theory of electronic states in crystals of infinite size, even some simple but also fundamental problems, such as how many different types of electronic states there are in a simple finite crystal of rectangular cuboid shape such as shown in Fig. 1.1 and how these electronic states are different from each other have not been well understood.

In conventional solid-state physics, all electronic states are considered as Bloch waves. In the early days, when people mainly dealt with solids of macroscopic size in which the bulk properties of the solid were the primary interest, this was acceptable. Because, in a crystal of macroscopic size the number of bulk-like states is much larger than the number of surface-like states, edge-like states, vertex-like state, and so forth.

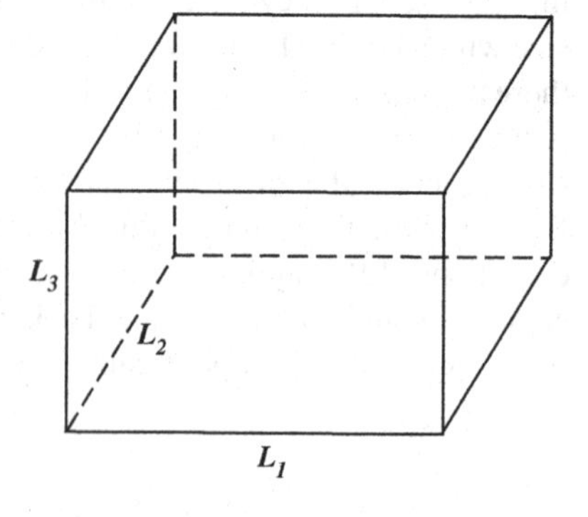

Fig. 1.1 A simple finite crystal of rectangular cuboid shape with sides of length L_1, L_2, and L_3

The difference between the electronic structure of a real crystal of finite size and the electronic structure obtained based on the translational invariance becomes more significant as the crystal size decreases. A clear understanding of the properties of electronic states in real crystals of finite size has both theoretical and practical significance. Since the early 1970s, investigations on the properties of low-dimensional systems such as quantum wells, wires, and dots in the sub-micron and nanometer size range have proceeded rapidly. It was found that in these low-dimensional systems, the properties of a semiconductor crystal change dramatically as the system size decreases: The measured optical band-gap increases as the system size decreases; some indirect semiconductors such as Si may become luminescent, (e.g., [16]), and direct semiconductors such as GaAs may develop into an indirect one, (e.g., [17]). These very interesting size-dependent properties of semiconductors provide both an attractive potential for possible practical applications and a significant theoretical challenge for a clear understanding of the fundamental physics, since the previous theory of electronic states in solids based on the translational invariance can by no means account for these size-dependent effects.

Therefore, a clear understanding of the electronic states in low-dimensional systems and finite crystals can be both interesting theoretically and important practically. However, to develop a general analytical theory of the electronic states in low-dimensional systems or finite crystals with boundaries has been considered as a rather difficult problem: The lack of translational invariance in low-dimensional systems or finite crystals is a major obstacle. It is the use of the translational invariance—the Bloch theorem—that provides both a theoretical frame and a significant mathematical simplification in solving the Schrödinger equation with a periodic potential. Without such a theoretical frame based on the Bloch theorem and the mathematical simplification, the corresponding problem for a finite crystal with border seems to become more difficult. Thus, most previous theoretical investigations on the electronic states in low-dimensional systems were based on approximate or numerical approaches and were usually on a specific material or a specific model (e.g., [18–25]). One of the most widely used approximate methods or approaches is the effective mass approximation.

1.4 The Effective Mass Approximation

The effective mass approximation (EMA) is a widely used approximation in semiconductor physics. It has many different forms; nevertheless, basically the electrons in a semiconductor are treated as electrons with an "effective mass" instead of the free-electron mass. It is a very successful approach to investigating the behavior of electrons in a semiconductor under a weak and slowly varying external field—such as an applied electric and/or magnetic field or the field introduced by a shallow impurity [26].

The theory of electronic states in low-dimensional systems and finite crystals can also be considered as a theory on the quantum confinement of Bloch waves.

The quantum confinement of plane waves—the simplest case is the well-known square potential well problem—is a subject treated in almost all standard quantum mechanics textbooks and is well understood [27]. In the simplest case, when an electron is completely confined in a one-dimensional square potential well of width L, all permitted electronic states are stationary wave states and the energy of the electron may only take discrete values:

$$E_j = \frac{j^2\hbar^2}{2mL^2}, \quad j = 1, 2, 3, \ldots. \tag{1.13}$$

Here, m is the electron mass.[1] Therefore, the lowest possible energy $\hbar^2/2mL^2$ of the electron in the well increases as the well width L decreases. Equation (1.13) can be easily extended to the case where the confinement is in two or three directions. If the barrier heights outside the well are finite rather than infinite, the confinement will not be complete and, consequently, the energy levels inside the well will be somewhat lower. Therefore, the quantum confinement always raises the lowest possible energy level inside the well: The smaller the well width L and/or the higher the barrier outside the well, the higher the lowest possible energy level inside the well.

A well-known experimental fact is that the measured optical energy gap in a semiconductor low-dimensional system increases as the system size decreases. The well-understood concept of the quantum confinement effect of plane waves such as indicated in (1.13) was naturally borrowed to explain this notable fact, (e.g., [29]). According to EMA, the "effective mass" of Bloch electrons should be used instead of the free-electron mass m in (1.13) or related formulas. In a semiconductor crystal, the Bloch electrons have a positive effective mass near the conduction band minimum and a negative effective mass near the VBM. Therefore, as the system size decreases, the consequence of EMA is that the lowest possible energy level in the conduction bands of a semiconductor crystal will go up, and the highest possible energy level in the valence bands will go down as shown in Fig. 1.2—a consequence of the quantum confinement effect of plane waves.

Various forms of EMA have been widely used in investigating the quantum confinements of Bloch electrons (e.g., [18–21, 30]). It turns out that in comparison with the experimental results, the theoretical predictions from the various forms of EMA, in general, overestimate the gap increase as the system size decreases. These general overestimations were usually explained by some factors not included in the EMA, such as the non-parabolicity of the energy band structure and so forth. It is widely believed in the solid state physics community that for the quantum confinements of Bloch waves in semiconductor low-dimensional systems, the physics picture such as

[1]By including the relativistic effect, (1.13) can be further extended to

$$E_j^2 = m^2c^4 + j^2\hbar^2c^2\left(\frac{\pi}{L}\right)^2,$$

where c is the speed of light [28].

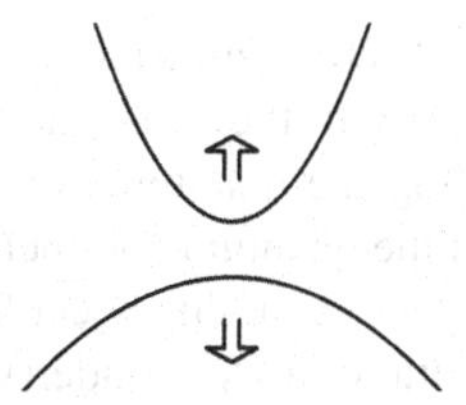

Fig. 1.2 According to EMA, the energy of the lowest unoccupied state will go up, and the energy of the highest occupied state will go down as the size of the semiconductor crystal decreases

shown in Fig. 1.2 is conceptually and qualitatively correct although the EMA might not be able to give quantitatively accurate numerical results.

A natural comment on the use of various forms of EMA in investigating the quantum confinements of Bloch waves is that, originally, EMA was developed for treating the electronic states near band edges in the presence of a slowly varying and weak external perturbation, whereas in a quantum confinement problem, the perturbation is neither weak nor slowly varying near the confinement boundary. Thus, the conditions for justifying the use of EMA are completely violated. Much work has been done on this interesting puzzle, mainly using the envelope function approach [31].

1.5 Some Numerical Results

A very interesting work by Zhang and Zunger [32] on the energy spectrum of confined electrons in Si quantum films obtained results qualitatively different from what one would expect from EMA. In contrast to the prediction of the EMA, in their numerical investigation on Si (001) quantum films Zhang and Zunger [32] observed a band edge state whose energy was approximately equal to the energy of the VBM and hardly changes as the film thickness changes, as shown in Fig. 1.3. Such states have also been observed in numerical investigations on (110) free-standing Si and GaAs quantum

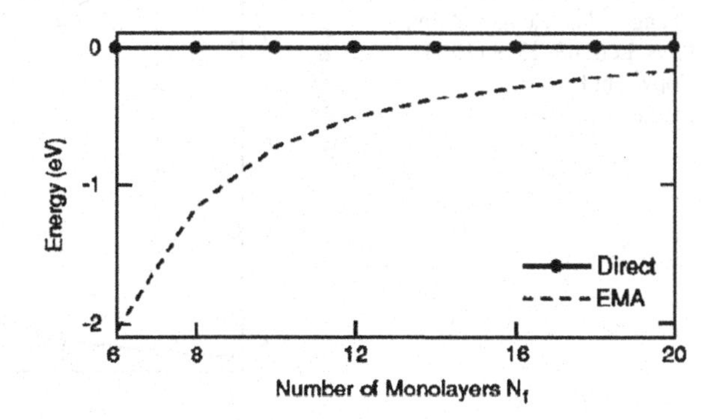

Fig. 1.3 Size dependence of the energy of the "zero-confinement" band edge state in Si (001) films. Reprinted with permission from S. B. Zhang and A. Zunger: Appl. Phys. Lett. **63**, 1399 (1993). Copyright by the American Institute of Physics

film [32–34]. They were called "zero-confinement states". The very existence of such states is directly contradictory to the consequence of the EMA. The obvious failure of EMA for understanding the quantum confinement effect of these band edge states clearly indicates that the quantum confinement of Bloch waves might be fundamentally different from what one might expect from the well-known quantum confinement of plane waves. Without a clear understanding of the physical origin of the existence of such states, the essential physics on the quantum confinement of Bloch waves is not well understood. One may also naturally doubt whether EMA is suitable for describing the quantum confinement of Bloch waves even *conceptually* and one has to be careful when using EMA or EMA-derived ideas or approaches for treating the quantum confinement of Bloch waves. Otherwise, some essential physics might be missed.

The "central observation" of the investigation of Zhang and Zunger in [32] is that the energy spectrum of confined electrons in Si (001) quantum films maps the energy band structure of Si approximately, as shown in Fig. 1.4. Similar maps of the energy spectra of confined electron states were observed in Si (110) and GaAs (110) quantum films [33]. Much previous work also indicates that the eigenvalues of confined Bloch states map the dispersion relations of the unconfined Bloch waves *closely*, (e.g., [35]).

All of these are *observations* obtained by numerical calculations. On the other hand, an analytical result by Pedersen and Hemmer [36] found that the energy spectrum of the majority of the confined electronic states in a finite one-dimensional Kronig–Penney crystal maps the energy bands exactly and does not depend on the boundary location. However, they treated neither the quantum confinement of the band-edge states nor the lowest energy band in their model. Kalotas and Lee [37, 38] investigated bound states in one-dimensional finite N-periodic structures bounded

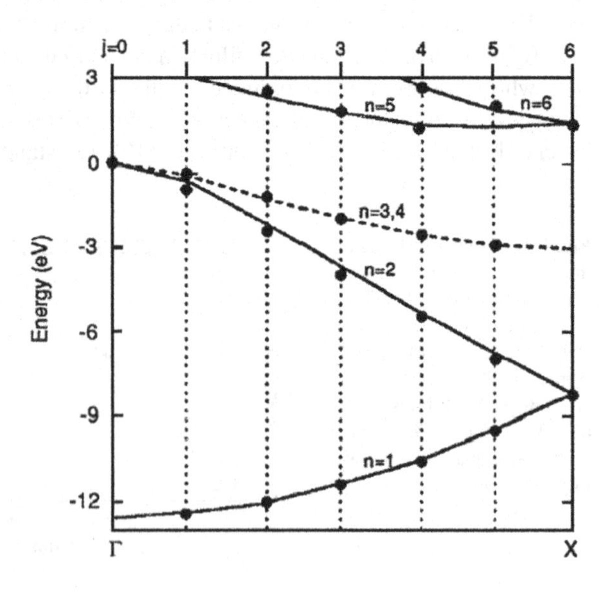

Fig. 1.4 A comparison between the energy bands $E_n(\mathbf{k})$ (*lines*) and the directly calculated eigenvalues in a Si (001) film of 12 monolayers. Reprinted with permission from S. B. Zhang and A. Zunger: Appl. Phys. Lett. **63**, 1399 (1993). Copyright by the American Institute of Physics

by infinite barriers and found analytically that the energy levels of these bound states come from two distinct groups - the permitted bands and the forbidden ranges. Corresponding to each permitted band, there are $N - 1$ energy levels which are dependent on N; The energy levels in the other group corresponding to the forbidden ranges are independent of N.

A correct theory of the electronic states in low-dimensional systems and finite crystals should give clear predictions on the properties of electronic states in low-dimensional systems. It should answer the simple but fundamental questions on how many types of electronic states there are in a simple finite crystal and how those electronic states are different from each other, such as asked in Sect. 1.3. It should clearly indicate the similarities and differences between the quantum confinement of Bloch waves and the well-known quantum confinement of plane waves. It should also give clear explanations of the numerical results such as those shown in Figs. 1.3 and 1.4.

1.6 Subject of the Book and Main Findings

In this book, we try to present a simple theory to obtain some of the most fundamental understandings of the electronic states in low-dimensional systems and finite crystals. We use the Born–Oppenheimer approximation; thus, we can consider the electronic states in a fixed atomic background. Further, we consider only a single-electron and non-spin theory.

Ideal low-dimensional systems and finite crystals are the simplest low-dimensional systems and finite crystals. By "ideal", it is assumed that (i) the potential $v(\mathbf{x})$ inside the low-dimensional system or finite crystal is the same as in a crystal with translational invariance and (ii) the electronic states are completely confined in the limited size of the low-dimensional system or finite crystal. These two simplifying assumptions facilitate the development of an analytical theory of the electronic states in ideal low-dimensional systems and finite crystals and allow us to try to explore and understand some of the most general physics related to the quantum confinement of Bloch waves and the electronic states in crystals of *finite* size.

We proceed by trying to understand two issues: (i) We try to understand the similarities and differences between the complete quantum confinement of plane waves and the complete quantum confinement of Bloch waves in one-dimensional space. It is found that now this problem can be well understood, and it provides new and fundamental understandings on relevant physics problems. (ii) We try to understand the similarities and differences between the complete quantum confinement of Bloch waves in one-dimensional space and the complete quantum confinement of Bloch waves in three-dimensional space. This is a more difficult problem, and now we can only understand some simple cases. We can anticipate that such differences are closely related to the differences between the band structure of the unconfined one-dimensional Bloch waves and that of the unconfined three-dimensional Bloch waves.

An ideal low-dimensional system or finite crystal is a simplified model of a real low-dimensional system or finite crystal. A clear understanding of the electronic states in ideal low-dimensional systems and finite crystals is a first step and the basis for understanding electronic states—and relevant physical properties—of any real low-dimensional system or finite crystal. As Anderson pointed out in his Nobel Prize Lecture [39],

> Very often such a simplified model throws more light on the real workings of nature than any number of *ab initio* calculations of individual situations, which even where correct often contain so much detail as to conceal rather than reveal reality. It can be a disadvantage rather than an advantage to be able to compute or to measure too accurately, since often what one measures or computes is irrelevant in terms of mechanism. After all, the perfect computation simply reproduces Nature, it does not explain her.

Real low-dimensional systems and finite crystals are more complicated than the ideal low-dimensional systems and finite crystals treated in this book. However, in comparison to a crystal of infinite size, an ideal low-dimensional system or a finite crystal is much closer to the physical reality of a real low-dimensional system or a finite crystal. We hope we have reason to expect that the understandings obtained for the ideal low-dimensional systems and finite crystals will be a significant step toward a clearer understanding of the electronic states in the real low-dimensional systems and finite crystals.

The book is organized as follows. In Part II, we treat one-dimensional semi-infinite crystals and finite crystals. These are the simplest systems that can manifest the effects of the existence of boundary and, in most cases, the relevant conclusions can be rigorously proven. As a preparation for the mathematical basis, in Chap. 2 we present a brief introduction of the relevant theory of periodic Sturm–Liouville equations. In Chap. 3, we present a general qualitative analysis and a theoretical formalism for quantitative investigations of the existence and properties of surface states in one-dimensional semi-finite crystals. It is found that, in general, the presence of a boundary does not always introduce a surface state in a specific band gap; only when the boundary in the semi-infinite crystal is located in some specific separated subintervals can a surface state exist in the specific band gap.

In Chap. 4, we present a general analytical theory of the electronic states in one-dimensional crystals of finite length. We prove that in a one-dimensional finite crystal bounded at τ and $\tau + L$, where $L = Na$, a is the potential period and N is a positive integer, there are two different types of electronic states. There are $N - 1$ stationary Bloch states in each energy band, whose energies depend on the crystal length L but not the boundary location τ and map the energy band exactly. There is always one and only one electronic state corresponding to each band gap, whose energy depends on τ but not L, this state can be either a surface state in the band gap or a confined band edge state. It is well known that if one-dimensional plane waves are completely confined, all permitted states are stationary waves. Therefore, the very existence of the boundary-dependent states is a fundamental distinction of the quantum confinement of Bloch waves.

A clear understanding of electronic states in one-dimensional crystals of finite length establishes a sound basis and starting point for the further understanding of the electronic states in low-dimensional systems and three-dimensional crystals of finite size. Part III consists of three chapters devoted to these subjects.

There are similarities and differences between the quantum confinements of three-dimensional Bloch waves and one-dimensional Bloch waves. The problem on the electronic states in a free-standing quantum film can be considered as three-dimensional Bloch waves confined in one specific direction and is treated in Chap. 5. It is found that in some simple and interesting cases where an ideal quantum film is bounded at $x_3 = \tau_3$[2] and $x_3 = (\tau_3 + N_3)$—here τ_3 defines the bottom boundary, and N_3 is a positive integer indicating the thickness of the film—there are two different types of electronic states. For each bulk energy band n and each wave vector $\hat{\mathbf{k}}$ in the film plane, there are $N_3 - 1$ stationary Bloch electronic states in the film whose energies depend on the film thickness N_3. There is one electronic state whose energy depends on τ_3. The energies of these stationary Bloch electronic states map the energy band of the bulk exactly and thus are bulk-like states, whereas the energy of the τ_3-dependent state is usually above or occasionally equal to the highest energy in that energy band with that $\hat{\mathbf{k}}$. This τ_3-dependent state is a surface-like state since it is, in general, a surface state, though, occasionally, it can also be a confined band edge state. A significant difference of the quantum confinement of the three-dimensional Bloch waves is that unlike in one-dimensional cases discussed in Chap. 4, *a surface state in such a film does not have to be in a band gap*. Instead, for the electronic states with the same energy band index n and the same $\hat{\mathbf{k}}$, the following general relation exists:

The energy of the surface-like state
> The energy of every bulk-like state.

The electronic states in some simple quantum wires can be seen as merely the two-dimensional Bloch waves in a quantum film further confined in one more direction. The electronic states in some simple finite crystals or quantum dots can be considered as the one-dimensional Bloch waves in a quantum wire further confined in the third direction. By investigating the effects of the quantum confinement of the Bloch waves step-by- step, we can obtain a general understanding of the electronic states in some simple ideal low-dimensional systems and finite crystals.

The electronic states in a simple rectangular quantum wire having a thickness of N_3 layers and a boundary τ_3 in the $\mathbf{a}_3$ direction, and a width of N_2 layers and a boundary τ_2 in the $\mathbf{a}_2$ direction can be considered as that the two-dimensional Bloch waves in a quantum film are further confined in the $\mathbf{a}_2$ direction. This is the subject treated in Chap. 6. It is found in some simple and interesting cases that the electronic states in such an ideal rectangular quantum wire can generally and exactly be obtained. For such a quantum wire of crystals with a simple cubic, tetragonal, or orthorhombic Bravais lattice, for each bulk energy band n and each wave vector $\bar{\mathbf{k}}$ in the wire direction $\mathbf{a}_1$, there are three types of one-dimensional Bloch waves. There are $(N_2 - 1)(N_3 - 1)$ bulk-like states whose energies depend on N_2 and N_3 and map the bulk energy band exactly. There are $(N_2 - 1) + (N_3 - 1)$ surface-like states whose

[2]In this book, a position vector is usually written as $\mathbf{x} = x_1\mathbf{a}_1 + x_2\mathbf{a}_2 + x_3\mathbf{a}_3$. For quantum films, it is usually assumed that $\mathbf{a}_3$ is the only primitive lattice vector out of the film plane.

energies depend either on N_2 and τ_3 or on N_3 and τ_2. There is one edge-like state[3] whose energy depends on τ_2 and τ_3. For a rectangular quantum wire of crystals with a face-centered-cubic or body-centered-cubic Bravais lattice, the numbers of each type of electronic states are somewhat different. For the electronic states with the same bulk energy band index n and the same $\bar{\mathbf{k}}$, the following general relations exist:

The energy of the edge-like state
> The energy of every surface-like state
> The energy of every relevant bulk-like state.

The electronic states in a simple finite crystal or quantum dot of rectangular cuboid shape having a thickness of N_3 layers and a boundary τ_3 in the $\mathbf{a}_3$ direction, a width of N_2 layers and a boundary τ_2 in the $\mathbf{a}_2$ direction, and a length of N_1 layers and a boundary τ_1 in the $\mathbf{a}_1$ direction can be seen as that the one-dimensional Bloch waves in a rectangular quantum wire are further confined in the $\mathbf{a}_1$ direction. This is the subject treated in Chap. 7. It is found in some simple and interesting cases that the electronic states in such an ideal finite crystal or quantum dot can generally and exactly be obtained. For such a finite crystal or quantum dot of crystals with a simple cubic, tetragonal, or orthorhombic Bravais lattice, for each bulk energy band there are $(N_1-1)(N_2-1)(N_3-1)$ bulk-like states whose energies depend on N_1, N_2 and N_3, $(N_1-1)(N_2-1)+(N_2-1)(N_3-1)+(N_3-1)(N_1-1)$ surface-like states whose energies depend on its sizes in two-dimensions and the boundary in the remaining direction, $(N_1-1)+(N_2-1)+(N_3-1)$ edge-like states whose energies depend on its size in one-dimension and the boundaries in two other directions, and one vertex-like state whose energy depends on τ_1, τ_2 and τ_3.[4] For a finite crystal or quantum dot of rectangular cuboid shape with a face-centered-cubic or body-centered-cubic Bravais lattice, the numbers of each type of electronic states are somewhat different. For the electronic states with the same energy band index n, the following general relations exist:

The energy of the vertex-like state
> The energy of every edge-like state
> The energy of every relevant surface-like state
> The energy of every relevant bulk-like state.

Due to the existence of the boundary-dependent electronic states, the properties of electronic states in a low-dimensional system and finite crystal may be substantially different from the properties of electronic states in a crystal with translational invariance as understood in the conventional solid-state physics. They may also be significantly different from what is widely believed on the electronic states in a low-dimensional system in the solid state physics community, such as those originating from the EMA-derived ideas. The real band-gap in an ideal low-dimensional system of a cubic semiconductor is, in fact, smaller than the band-gap of the bulk semiconductor with translational invariance. It may even be possible that a low-dimensional system of a cubic semiconductor crystal could have the electrical conductivity of metal. *A basic distinction between a macroscopic metal and a macroscopic semiconductor*

[3] The edge-like states were named side-like states in the first edition.

[4] The vertex-like states were named corner-like states in the first edition.

may become blurred when the size of crystal becomes much smaller thus the effects of the existence of the boundary-dependent electronic states in the low-dimensional system or the finite crystal need to be considered.

Chapter 8 in Part IV is devoted to the concluding remarks. The understandings we have had, in fact, are only the very beginning: For the little we have just understood, there is so much more we do not understand. In particular, a natural question is: Are those interesting results merely individual behaviors of this specific problem on the electronic states in low-dimensional systems or finite crystals, or might they originally be one of the consequences of *a whole class* of more general relevant problems concerning the truncated periodicity?

References

1. F. Bloch, Zeit. Phys. **52**, 555 (1928)
2. F. Seitz, *The Modern Theory of Solids* (McGraw-Hill, New York, 1940)
3. H. Jones, *The Theory of Brillouin Zones and Electronic States in Crystals* (North-Holland, Amsterdam, 1960)
4. W.A. Harrison, *Solid State Theory* (McGraw-Hill, New York, 1970)
5. N.W. Ashcroft, N.D. Mermin, *Solid State Physics* (Holt, Rinehart and Winston, New York, 1976)
6. J. Callaway, *Quantum Theory of the Solid State*, 2nd edn. (Academic Press, London, 1991)
7. C. Kittel, *Introduction to Solid State Physics*, 7th edn. (Wiley, New York, 1996)
8. L. Brillouin, J. Phys. Radium **1**, 377 (1930)
9. H.A. Kramers, Physica **2**, 483 (1935)
10. J.R. Chelikowsky, M.L. Cohen, *Electronic Structure and Optical Properties of Semiconductors*, 2nd edn. (Springer, Berlin, 1989)
11. J.R. Chelikowsky, M.L. Cohen, Phys. Rev. B **14**, 556 (1976)
12. I. Tamm, Phys. Z. Sowj. **1**, 733 (1932)
13. R.L. Kronig, W.G. Penney, Proc. R. Soc. Lond. Ser. A **130**, 499 (1931)
14. S.G. Davison, M. Stęślicka, *Basic Theory of Surface States* (Clarendon Press, Oxford, 1992)
15. M.-C. Desjonquéres, D. Spanjaard, *Concepts in Surface Physics* (Springer, Berlin, 1993)
16. M. Nirmal, L. Brus, Acc. Chem. Res. **32**, 417 (1999); S. Ossicini, L. Pavesi, F. Priolo, *Light Emitting Silicon for Microphotonics* (Springer, Berlin, 2003) (and the references therein)
17. A. Franceschetti, A. Zunger, Phys. Rev. **B52**, 14664 (1995); A. Franceschetti, A. Zunger. J. Chem. Phys. **104**, 5572 (1996)
18. A.D. Yoffe, Adv. Phys. **42**, 173 (1993); A.D. Yoffe, Adv. Phys. **50**, 1 (2001); A.D. Yoffe, Adv. Phys. **51**, 799 (2002). (and references therein)
19. J.H. Davies, *The Physics of Low Dimensional Semiconductors* (Cambridge University Press, Cambridge, 1998). (and references therein)
20. P. Harrison, *Quantum Wells, Wires and Dots: Theoretical and Computational Physics* (Wiley, New York, 2000). (and references therein)
21. J.B. Xia, Phys. Rev. B **40**, 8500 (1989); T. Takagahara, K. Takeda, Phys. Rev. B **46**, 15578 (1992); T. Takagahara, Phys. Rev. B **47**, 4569 (1993); Al.L. Efros, M. Rosen, M. Kuno, M. Nirmal, D. J. Norris, M. Bawendi, Phys. Rev. B **54**, 4843 (1996); Al.L. Efros, M. Rosen, Ann. Rev. Mater. Sci. **30**, 475 (2000); A.V. Rodina, Al.L. Efros, A.Yu. Alekseev, Phys. Rev. B **67**, 155312 (2003); D.H. Feng, Z.Z. Xu, T.Q. Jia, X.X. Li, S.Q. Gong. Phys. Rev. B **68**, 035334 (2003)
22. A. Di Carlo, Semicond. Sci. Technol. **18**, R1 (2003). (and references therein)

23. P.E. Lippens, M. Lannoo, Phys. Rev. B**39**, 10935 (1989); S.Y. Ren, J.D. Dow, Phys. Rev. B **45**, 6492 (1992); G.D. Sanders, Y.C. Chang, Phys. Rev. B **45**, 9202 (1992); G. Allan, C. Delurue, M. Lannoo, Phys. Rev. Lett. **76**, 2961 (1996); S.Y. Ren, Phys. Rev. B **55**, 4665 (1997); S.Y. Ren, Solid State Commun. **102**, 479 (1997); Y.M. Niquet, G. Allan, C. Delurue, M. Lannoo, Appl. Phys. Lett. **77**, 1182 (2000); Y.M. Niquet, C. Delurue, G. Allan, M. Lannoo, Phys. Rev. B **62**, 5109 (2000); J. Sée, P. Dollfus, S. Galdin, Phys. Rev. B **66**, 193307 (2002); S. Sapra, D.D. Sarma, Phys. Rev. B **69**, 125304 (2004); P. Chen, K.B. Whaley, Phys. Rev. B **70**, 045311 (2004); G. Allan, C. Delurue. Phys. Rev. B **70**, 245321 (2004)
24. L.W. Wang, A. Zunger, J. Chem. Phys. **100**, 2394 (1994); L.W. Wang, A. Zunger, J. Phys. Chem. **98**, 2158 (1994); L.W. Wang, A. Zunger, Phys. Rev. Lett. **73**, 1039 (1994); C.Y. Yeh, S.B. Zhang, A. Zunger, Phys. Rev. B **50**, 14405 (1994); A. Tomasulo, M.V. Ramakrishna, J. Chem. Phys. **105**, 3612 (1996); L.W. Wang, A. Zunger, Phys. Rev. B **53**, 9579 (1996); H.X. Fu, A. Zunger, Phys. Rev. B **55**, 1642 (1997); H.X. Fu, A. Zunger, Phys. Rev. B **56**, 1496 (1997); L.W. Wang, J.N. Kim, A. Zunger, Phys. Rev. B **59**, 5678 (1999); F.A. Reboredo, A. Franceschetti, A. Zunger, Appl. Phys. Lett. **75**, 2972 (1999); A. Franceschetti, H.X. Fu, L.W. Wang, A. Zunger, Phys. Rev. B **60**, 1919 (1999)
25. B. Delley, E.F. Steigmeier, Appl. Phys. Lett. **67**, 2370 (1995); S. Ogut, J.R. Chelikowsky, S.G. Louie, Phys. Rev. Lett. **79**, 1770 (1997); C.S. Garoufalis, A.D. Zdetsis, S. Grimme, Phys. Rev. Lett. **87**, 276402 (2001); H.-Ch. Weissker, J. Furthmüller, F. Bechstedt, Phys. Rev. B **67**, 245304 (2003); A.S. Barnard, S.P. Russo, I.K. Snook, Phys. Rev. B **68**, 235407 (2003); X. Zhao, C.M. Wei, L. Yang, M.Y. Chou, Phys. Rev. Lett. **92**, 236805 (2004); R. Rurali, N. Lorenti, Phys. Rev. Lett. **94**, 026805 (2005); G. Nesher, L. Kronik, J.R. Chelikowsky. Phys. Rev. B **71**, 035344 (2005)
26. J.M. Luttinger, W. Kohn, Phys. Rev. **97**, 869 (1957); W. Kohn, *Solid State Physics*, vol. 5, ed. by F. Seitz, D. Turnbull (Academic Press, New York, 1955), pp. 257–320
27. L.I. Schiff, *Quantum Mechanics*, 3rd edn. (McGraw-Hill, New York, 1968)
28. S.Y. Ren, Chin. Phys. Lett. **19**, 617 (2002)
29. Al.L. Efros, A.L. Efros, Sov. Phys. Semicond. **16**, 772 (1982); L.E. Brus, J. Phys. Chem. **80**, 4403 (1984); Y. Wang, N. Herron, J. Phys. Chem. **95**, 525 (1991)
30. M.J. Kelly, *Low-Dimensional Semiconductors: Materials, Physics, Technology, Devices*, 3rd ed. (Oxford University Press, Oxford, 1995); B.K. Ridley, *Electrons and Phonons in Semiconductor Multilayers*, 3rd ed. (Cambridge University Press, Cambridge, 1997); L. Bányai, S.W. Koch, *Semiconductor Quantum Dots* (World Scientific, Singapore, 1993); C. Delerue, M. Lannoo, *Nanostructures: Theory and Modeling* (Springer, Berlin, 2004); S. Glutch, *Excitons in Low-Dimensional Semiconductors: Theory, Numerical Methods, Applications* (Springer, Berlin, 2004)
31. M.G. Burt, J. Phys. Condens. Matter **4**, 6651 (1992). (and references therein)
32. S.B. Zhang, A. Zunger, Appl. Phys. Lett. **63**, 1399 (1993)
33. S.B. Zhang, C.-Y. Yeh, A. Zunger, Phys. Rev. B **48**, 11204 (1993)
34. A. Franceschetti, A. Zunger, Appl. Phys. Lett. **68**, 3455 (1996)
35. Z.V. Popovic, M. Cardona, E. Richter, D. Strauch, L. Tapfer, K. Ploog, Phys. Rev. **B40**, 1207 (1989); Z.V. Popovic, M. Cardona, E. Richter, D. Strauch, L. Tapfer, K. Ploog, Phys. Rev. **B40**, 3040 (1989); Z.V. Popovic, M. Cardona, E. Richter, D. Strauch, L. Tapfer, K. Ploog. Phys. Rev. B **41**, 5904 (1990)
36. F.B. Pedersen, P.C. Hemmer, Phys. Rev. B **50**, 7724 (1994)
37. T.M. Kalotas, A.R. Lee, Eur. J. Phys. **16**, 119 (1995)
38. D.W.L. Sprung, J.D. Sigetich, H. Wu, J. Martorell, Am. J. Phys. **68**, 715 (2000)
39. P.W. Anderson, Rev. Mod. Phys. **50**, 191 (1978)

Part II
One-Dimensional Semi-infinite Crystals and Finite Crystals

Chapter 2
The Periodic Sturm–Liouville Equations

One-dimensional crystals are the simplest crystals. Historically, much of our current fundamental understanding of electronic structures of crystals was obtained through the analysis of one-dimensional crystals [1–3]. Among the most well-known examples are the Kronig–Penney model [4], Kramers' general analysis of the band structure of one-dimensional crystals [5], Tamm's surface states [6], and so forth. A clear understanding of electronic states in one-dimensional finite crystals is the basis for further understandings of the electronic states in low-dimensional systems and finite crystals. For this purpose, we need to have a clear understanding of solutions of the Schrödinger equations for one-dimensional crystals.

The Schrödinger differential equation for a one-dimensional crystal can be written as[1]:

$$-y''(x) + [v(x) - \lambda]y(x) = 0, \tag{2.1}$$

where $v(x + a) = v(x)$ is the periodic potential. Eastham's book [7] provided a comprehensive and in-depth mathematical theory of a general class of periodic differential equations, where Eq. (2.1) is a specific and simple form. The theory in his book provided the major mathematics basis of the first edition of this book.

The relevant mathematical theory has made significant progress [8–10] since Eastham's book was published in 1975. Zettl [9] pointed out that in the Eastham's book "strong smoothness and positivity restrictions are placed on the coefficients. However, many, but not all, of the proofs given there are valid under much less severe restrictions on the coefficients." The modern theory of periodic Sturm–Liouville equations [8–10] with less severe restrictions can treat more general problems.

In this chapter, we study the basic theory of the periodic Sturm–Liouville equations [8–10], to prepare for investigations of electronic states in semi-infinite and finite

[1]In this book, a prime on a function denotes differentiation with respect to the variable of the function. If the function has two or more variables, a prime on the function denotes differentiation with respect to the variable x.

S.Y. Ren, *Electronic States in Crystals of Finite Size*, Springer Tracts in Modern Physics 270, DOI 10.1007/978-981-10-4718-3_2

one-dimensional crystals in next two chapters and relevant problems in Appendices. We begin with a brief review of some elementary knowledge of the theory of the second-order linear homogeneous ordinary differential equations. Then, we present two basic Sturm Theorems on zeros of solutions of these equations. In the theory of boundary value problems for ordinary differential equations, the existence and locations of zeros of solutions of such equations are often of central importance. In the major part of this chapter, we learn the basic theory of the periodic Sturm–Liouville equations and the zeros of their solutions. Based on the mathematical theory and theorems in this chapter, a general theoretical formalism for investigations of the existence and properties of surface states in a semi-infinite one-dimensional crystal can be developed, and general results on electronic states in ideal one-dimensional finite crystals can be rigorously proven. For our purpose, we work on to understand how the mathematical results in Eastham's book [7] can be extended by the modern theory of periodic Sturm–Liouville equations [8–10], by following the steps in Chap. 2 of the first edition.

We are interested in the Sturm–Liouville equations with periodic coefficients [8–10]:

$$[p(x)y'(x)]' + [\lambda w(x) - q(x)]y(x) = 0, \tag{2.2}$$

where $p(x) > 0,\ w(x) > 0$ and $p(x),\ q(x),\ w(x)$ are *piecewise continuous* real periodic functions with period a:

$$p(x+a) = p(x), \quad q(x+a) = q(x), \quad w(x+a) = w(x).$$

The one-dimensional Schrödinger equation (2.1) corresponds to a specific and simple form of Eq. (2.2) where $p(x) = w(x) = 1$ and $q(x) = v(x)$. The basic theory of periodic Sturm–Liouville equations summarized in this chapter is somewhat more general and advanced than what we need in treating Eq. (2.1) for electronic states in one-dimensional crystals.[2] However, such a more general and up-to-date theory can treat more general one-dimensional problems in physics, including the one-dimensional photonic crystals and phononic crystals in Appendices. Readers who are interested in a more complete and general mathematical theory are recommended to read original books [7–10]. Readers who are not interested in the proofs of relevant theorems may skip those parts of this chapter.

2.1 Elementary Theory and Two Basic Sturm Theorems

We begin with a class of second-order linear homogeneous ordinary differential equations:

$$[(p(x)y'(x)]' + q(x)y(x) = 0, \quad -\infty < x < +\infty. \tag{2.3}$$

[2] As the mathematical basis of the first edition of the book, in [7] it was assumed that $p(x)$ is real-valued, *continuous* and nowhere zero, and $p'(x)$ is *piecewise continuous*.

Here, $p(x)$ and $q(x)$ are piecewise continuous real finite functions.

$p(x)y'(x)$ is called the quasi-derivative of $y(x)$ [8–11], distinguished from the classical derivative $y'(x)$. In this chapter, the quasi-derivative plays the roles of the classic derivative in the first edition of the book. This is the essential difference between the modern theory of periodic Sturm–Liouville equations [8–10] summarized in this chapter and the theory presented in the Eastham's book [7]. In equations on many physical problems, the classical derivative $y'(x)$ may not exist in some cases, but the quasi-derivative $p(x)y'(x)$ exists and is continuous.[3] The applications of quasi-derivatives [8–11] significantly extends the ranges of the problems treatable by the modern Sturm–Liouville theory.

Equation (2.3) can be written in a matrix form [8–11]:

$$\begin{pmatrix} y(x) \\ p(x)y'(x) \end{pmatrix}' = \begin{pmatrix} 0 & \frac{1}{p(x)} \\ -q(x) & 0 \end{pmatrix} \begin{pmatrix} y(x) \\ p(x)y'(x) \end{pmatrix}, \qquad -\infty < x < +\infty. \tag{2.4}$$

Equation (2.4) is a simple special case of a more general first-order linear homogeneous differential equation of matrices:

$$Y' = A\,Y \tag{2.5}$$

with

$$Y = \begin{pmatrix} y(x) \\ p(x)y'(x) \end{pmatrix}, \qquad A = \begin{pmatrix} 0 & \frac{1}{p(x)} \\ -q(x) & 0 \end{pmatrix}, \qquad -\infty < x < +\infty. \tag{2.6}$$

For physical problems interested in this book, the functions $p(x)$ and $q(x)$ are as stated at the beginning of this section. Mathematicians might be more interested in properties of a more general Eq. (2.5) and relevant equations. The elements of the matrix A might often be more general.

---

[3] Suppose x_i is an isolated point where the function $p(x)$ is not continuous. By integrating (2.3) from $x_i - \delta$ to $x_i + \delta$, where δ is an infinitesimal positive number, we obtain

$$\int_{x_i-\delta}^{x_i+\delta} [p(x)y'(x)]'\mathrm{d}x = -\int_{x_i-\delta}^{x_i+\delta} q(x)y(x)\mathrm{d}x.$$

Since δ is an infinitesimal positive number, we have

$$\int_{x_i-\delta}^{x_i+\delta} q(x)y(x)\mathrm{d}x = 0. \qquad \delta \to 0$$

Thus

$$[p(x)y'(x)]_{x_i-0} = [p(x)y'(x)]_{x_i+0}.$$

That is, the quasi-derivative $p(x)y'(x)$ is continuous at x_i despite the fact that the classic derivative y' does not exist at x_i: $y'_{(x_i-0)} \neq y'_{(x_i+0)}$.

Properties of solutions of Eq. (2.3) or Eq. (2.4) can be obtained from the properties of solutions of the more general Eq. (2.5) [8–12].
1. Two linear-independent solutions.

Any nontrivial solution Y of Eq. (2.4) can be written as a linear combination of two linearly independent solutions Y_1 and Y_2 of Eq. (2.4):

$$Y = c_1 Y_1 + c_2 Y_2, \tag{2.7}$$

or alternatively,

$$\begin{pmatrix} y(x) \\ p(x)y'(x) \end{pmatrix} = c_1 \begin{pmatrix} y_1(x) \\ p(x)y_1'(x) \end{pmatrix} + c_2 \begin{pmatrix} y_2(x) \\ p(x)y_2'(x) \end{pmatrix}. \tag{2.8}$$

Here, c_1 and c_2 are two independent constants.
2. The fundamental matrix and the Wronskian.

Let Y_1 and Y_2 be two linearly independent solutions of Eq. (2.4). The following matrix Φ is called a fundamental matrix of Eq. (2.4):

$$\Phi = \begin{pmatrix} y_1(x) & y_2(x) \\ p(x)y_1'(x) & p(x)y_2'(x) \end{pmatrix}. \tag{2.9}$$

The Wronskian $W(y_1, y_2)$ of two solutions $y_1(x)$ and $y_2(x)$ of Eq. (2.3) is defined as

$$W(y_1, y_2) = y_1(x)\, p(x)y_2'(x) - p(x)y_1'(x)\, y_2(x). \tag{2.10}$$

The Wronskian $W(y_1, y_2)$ of two distinct solutions $y_1(x)$ and $y_2(x)$ of Eq. (2.3) is a constant:

$$[W(y_1, y_2)]' = [y_1(x)\, p(x)y_2'(x) - p(x)y_1'(x)\, y_2(x)]' = 0. \tag{2.11}$$

A necessary and sufficient condition that two solutions y_1 and y_2 of Eq. (2.3) are linearly independent is that

$$W(y_1, y_2) = y_1(x)\, p(x)y_2'(x) - p(x)y_1'(x)\, y_2(x) \neq 0.$$

3. The variation of parameters formula.

The nonhomogeneous differential equation

$$[p(x)z'(x)]' + q(x)z(x) = F \tag{2.12}$$

can be written in a matrix form:

$$\begin{pmatrix} z(x) \\ p(x)z'(x) \end{pmatrix}' = \begin{pmatrix} 0 & \frac{1}{p(x)} \\ -q(x) & 0 \end{pmatrix} \begin{pmatrix} z(x) \\ p(x)z'(x) \end{pmatrix} + \begin{pmatrix} 0 \\ F \end{pmatrix}. \tag{2.13}$$

The nonhomogeneous differential equation (2.13) can be solved as

$$\begin{pmatrix} z(x) \\ p(x)z'(x) \end{pmatrix} = \Phi(x) \int^x \Phi^{-1}(t) \begin{pmatrix} 0 \\ F(t) \end{pmatrix} \mathrm{d}t. \tag{2.14}$$

Here Φ is a fundamental matrix of Eq. (2.4).

Let $y_1(x)$ and $y_2(x)$ be two linearly independent solutions of Eqs. (2.3), (2.14) can be more explicitly written as

$$\begin{pmatrix} z(x) \\ p(x)z'(x) \end{pmatrix} = \begin{pmatrix} y_1(x) & y_2(x) \\ p(x)y_1'(x) & p(x)y_2'(x) \end{pmatrix} \begin{pmatrix} -\int^x \frac{1}{W(t)} y_2(t)F(t)\mathrm{d}t \\ \int^x \frac{1}{W(t)} y_1(t)F(t)\mathrm{d}t \end{pmatrix}, \tag{2.15}$$

or

$$z(x) = -\int^x \frac{F(t)y_2(t)}{W(t)} \mathrm{d}t \; y_1(x) + \int^x \frac{F(t)y_1(t)}{W(t)} \mathrm{d}t \; y_2(x), \tag{2.16}$$

and

$$p(x)z'(x) = -\int^x \frac{F(t)y_2(t)}{W(t)} \mathrm{d}t \; p(x)y_1'(x) + \int^x \frac{F(t)y_1(t)}{W(t)} \mathrm{d}t \; p(x)y_2'(x), \tag{2.17}$$

here $W(t) = [W(y_1, y_2)]_t = y_1(t)\, p(t)y_2'(t) \; - \; p(t)y_1'(t)\, y_2(t)$ is defined in (2.11).

There are two basic theorems on zeros of solutions of the differential equation (2.3).

Theorem 2.1 (Sturm Separation Theorem [Theorem 2.6.2 in [9]]) *Let y_1 and y_2 be two linearly independent real solutions of (2.3); then there is exact one zero of y_2 between two consecutive zeros of y_1.*

Proof Suppose α and β are two consecutive zeros of y_1,

$$y_1(\alpha) = y_1(\beta) = 0; \tag{2.18}$$

then it can be proven that there is, at least, one zero of y_2 in (α, β).

Without losing generality, we may assume that $y_1(x) > 0$ in (α, β). Then we have

$$y_1(\alpha + \delta) - y_1(\alpha) = \int_\alpha^{\alpha+\delta} \frac{p(x)y_1'(x)}{p(x)} \mathrm{d}x > 0 \tag{2.19}$$

for a small $\delta > 0$. Since $p(\alpha)y_1'(\alpha) \neq 0$ and $p(x)y_1'(x)$ is continuous, (2.19) indicates that

$$p(\alpha)y_1'(\alpha) > 0. \tag{2.20}$$

Similarly, we have

$$p(\beta)y_1'(\beta) < 0. \tag{2.21}$$

From (2.11) we have $[W(y_1, y_2)]_\alpha = [W(y_1, y_2)]_\beta \neq 0$, thus

$$y_1(\alpha)\; p(\alpha)y_2'(\alpha) - p(\alpha)y_1'(\alpha)\; y_2(\alpha) - y_1(\beta)\; p(\beta)y_2'(\beta) + p(\beta)y_1'(\beta)\; y_2(\beta) = 0.$$

By (2.18) we further obtain

$$- p(\alpha)y_1'(\alpha)y_2(\alpha) + p(\beta)y_1'(\beta)y_2(\beta) = 0. \tag{2.22}$$

Since y_2 and y_1 are linearly independent, neither $y_2(\alpha)$ nor $y_2(\beta)$ is zero. Equation (2.22) can be true only when $y_2(\alpha)$ and $y_2(\beta)$ have different signs. Since $y_2(x)$ is continuous, there must be at least one zero of $y_2(x)$ in (α, β).

However, if there are more than one zeros of y_2 in (α, β), then according to what we have just proven, there is at least one extra zero of y_1 between two zeros of y_2 in (α, β). This is contradictory to the supposition that α and β are two consecutive zeros of y_1. Therefore, there is always one and only one zero of y_2 between two consecutive zeros of y_1. Similarly, there is always one and only one zero of y_1 between two consecutive zeros of y_2. The zeros of two linearly independent real solutions y_1 and y_2 of (2.3) are distributed alternatively and separated from each other. □

Theorem 2.2 (Sturm Comparison Theorem [Theorem 2.6.3 in [9]]) *Suppose in two differential equations*

$$(p_1 y_1')' + q_1 y_1 = 0, \tag{2.23}$$

and

$$(p_2 y_2')' + q_2 y_2 = 0, \tag{2.24}$$

where

$$0 < p_2 \leq p_1 \text{ and } q_2 \geq q_1 \tag{2.25}$$

are true, and α and β are two zeros of a nontrivial real solution y_1 of the first equation (2.23), then there is, at least, one zero of any nontrivial real solution y_2 of (2.24) in $[\alpha, \beta]$.

Proof Suppose that this is not true — that y_2 is not zero anywhere in $[\alpha, \beta]$, we may assume $y_2 > 0$ in $[\alpha, \beta]$. Without loss of generality, we may assume that α and β are two consecutive zeros of y_1:

$$y_1(\alpha) = y_1(\beta) = 0, \tag{2.26}$$

and $y_1 > 0$ in (α, β), then we can have

$$\left[\frac{y_1}{y_2}(p_1 y_1' y_2 - p_2 y_2' y_1)\right]' = (q_2 - q_1)y_1^2 + (p_1 - p_2){y_1'}^2 + p_2 \frac{(y_1 y_2' - y_2 y_1')^2}{y_2^2} \tag{2.27}$$

in $[\alpha, \beta]$. By doing an integration of (2.27) from α to β and note that the integration of the left side is zero due to (2.26), we obtain that

$$\int_{\alpha}^{\beta}(q_2-q_1)y_1^2\mathrm{d}x+\int_{\alpha}^{\beta}(p_1-p_2)y_1'^2\mathrm{d}x=-\int_{\alpha}^{\beta}p_2\frac{(y_1y_2'-y_2y_1')^2}{y_2^2}\mathrm{d}x.$$

This equation can be valid only when $q_1=q_2$, $p_1=p_2$ and y_1, y_2 are linearly dependent. Thus, the theorem is proven. □

2.2 The Floquet Theory

Now we consider the solutions of an Eq. (2.3) where $p(x)$ and $q(x)$ are real periodic functions with the same period a:

$$[p(x)y']'+q(x)y=0,\qquad p(x+a)=p(x),\ \ q(x+a)=q(x). \tag{2.28}$$

Here, a is a nonzero real constant.

Theorem 2.3 (Theorem 2.7.1 in [9]) *There exist at least one nonzero constant ρ and one nontrivial solution $y(x)$ of (2.28) such that*

$$y(x+a)=\rho\ y(x), \tag{2.29}$$

Proof We can choose two linearly independent solutions $\eta_1(x)$ and $\eta_2(x)$ of (2.28) according to

$$\eta_1(0)=1,\ p(0)\eta_1'(0)=0;\ \ \eta_2(0)=0,\ p(0)\eta_2'(0)=1. \tag{2.30}$$

These solutions are usually called normalized solutions of (2.28) [13].

Since the corresponding $\eta_1(x+a)$ and $\eta_2(x+a)$ are also two linearly independent nontrivial solutions of (2.28), we can write $\eta_1(x+a)$ and $\eta_2(x+a)$ as linearly combinations of $\eta_1(x)$ and $\eta_2(x)$:

$$\begin{aligned}\eta_1(x+a)&=A_{11}\eta_1(x)+A_{12}\eta_2(x),\\ \eta_2(x+a)&=A_{21}\eta_1(x)+A_{22}\eta_2(x),\end{aligned} \tag{2.31}$$

where A_{ij} $(1\le i,j\le 2)$ are four constants. From (2.30) and (2.31), we obtain that

$$A_{11}=\eta_1(a),\ A_{21}=\eta_2(a),\ A_{12}=p(a)\eta_1'(a),\ A_{22}=p(a)\eta_2'(a). \tag{2.32}$$

Any nontrivial solution $y(x)$ of (2.28) can be written as

$$y(x)=c_1\eta_1(x)+c_2\eta_2(x),$$

where c_i are constants. If there is a nonzero ρ that makes

$$\begin{aligned}(A_{11} - \rho)c_1 + A_{21}c_2 &= 0,\\ A_{12}c_1 + (A_{22} - \rho)c_2 &= 0\end{aligned} \tag{2.33}$$

true, then (2.28) has a nontrivial solution of the form (2.29). The requirement that c_i in (2.33) are not both zero leads to the condition

$$\rho^2 - [\eta_1(a) + p(a)\eta_2'(a)]\rho + 1 = 0. \tag{2.34}$$

Here, $[W(\eta_1, \eta_2)]_a = [W(\eta_1, \eta_2)]_0$ thus $\eta_1(a)p(a)\eta_2'(a) - p(a)\eta_1'(a)\eta_2(a) = 1$ was used. The quadratic equation (2.34) is called the characteristic equation associated with (2.28) [13]. Equation (2.34) for ρ has, at least, one nonzero root since it has a nonzero constant term. □

Equation (2.28) may have one nontrivial solution of the form (2.29) or two linearly independent nontrivial solutions of the form (2.29), depending on whether the matrix $A = (A_{ij})$ in (2.32) has only one eigenvector or two linearly independent eigenvectors.

Theorem 2.4 (Theorem 2.7.2 in [9]) *There exist linearly independent solutions $y_1(x)$ and $y_2(x)$ of (2.28) such that either*

(i)

$$\begin{aligned}y_1(x) &= e^{h_1 x} p_1(x),\\ y_2(x) &= e^{h_2 x} p_2(x),\end{aligned} \tag{2.35}$$

here h_1 and h_2 are constants, not necessarily distinct, $p_i(x), i = 1, 2$ are periodic with period a, or

(ii)

$$\begin{aligned}y_1(x) &= e^{hx} p_1(x),\\ y_2(x) &= e^{hx}[x\ p_1(x) + p_2(x)],\end{aligned} \tag{2.36}$$

here h is a constant and $p_i(x), i = 1, 2$ are periodic with period a.

Proof The characteristic equation (2.34) may have either two distinct roots or a repeated root.

1. If the characteristic equation (2.34) has two distinct roots ρ_1 and ρ_2, then there are two linearly independent nontrivial solutions of $y_1(x)$ and $y_2(x)$ of (2.28) such that

$$y_i(x + a) = \rho_i y_i(x),\ i = 1, 2.$$

We can define h_1 and h_2 so that

$$e^{ah_i} = \rho_i \tag{2.37}$$

and then two functions $p_i(x)$ by

$$p_i(x) = e^{-h_i x} y_i(x).$$

It is easy to see that $p_1(x)$ and $p_2(x)$ are periodic functions with period a:

$$p_i(x+a) = e^{-h_i(x+a)} \rho_i y_i(x) = p_i(x).$$

Thus, (2.28) has two linearly independent nontrivial solutions:

$$y_1(x) = e^{h_1 x} p_1(x); \quad y_2(x) = e^{h_2 x} p_2(x). \tag{2.38}$$

2. Now, we consider the case that the characteristic equation (2.34) has a repeated root ρ. Define h by

$$e^{ah} = \rho. \tag{2.39}$$

According to Theorem 2.3, (2.28) has a nontrivial solution of the form (2.29):

$$y_1(x+a) = \rho y_1(x).$$

Suppose $Y_2(x)$ is any solution of (2.28) that is linearly independent of $y_1(x)$. Since $Y_2(x+a)$ is also a nontrivial solution of (2.28) we can write

$$Y_2(x+a) = c_1 y_1(x) + c_2 Y_2(x), \tag{2.40}$$

here c_1 and c_2 are constants. Since

$$[W(y_1, Y_2)]_{x+a} = \rho c_2 [W(y_1, Y_2)]_x$$

and $[W(y_1, Y_2)]_x$ does not depend on x, therefore,

$$\rho c_2 = 1 = \rho^2,$$

the second equality is due to that the constant term in (2.34) is equal to 1. Thus,

$$c_2 = \rho.$$

Equation (2.40) can be written as

$$Y_2(x+a) = c_1 y_1(x) + \rho Y_2(x). \tag{2.41}$$

There could be two different cases:

2.1. $c_1 = 0$.

Equation (2.41) becomes

$$Y_2(x+a) = \rho Y_2(x).$$

We can choose $y_2(x) = Y_2(x)$. Thus, (2.28) has two linearly independent solutions $y_1(x)$ and $y_2(x)$ and

$$y_1(x+a) = e^{ah} y_1(x), \qquad y_2(x+a) = e^{ah} y_2(x).$$

The first part of the theorem is proven. This case corresponds to the case that the matrix $A = (A_{ij})$ has one repeated eigenvalue ρ but two linearly independent eigenvectors. Consequently, (2.28) may have two linearly independent nontrivial solutions of the form (2.29).

2.2. $c_1 \neq 0$.

Define

$$p_1(x) = e^{-hx} y_1(x), \quad p_2(x) = (a\rho/c_1)\, e^{-hx} Y_2(x) - x\, p_1(x);$$

then we have

$$p_1(x+a) = e^{-h(x+a)} y_1(x+a) = p_1(x)$$

and

$$\begin{aligned} p_2(x+a) &= (a\rho/c_1)\, e^{-h(x+a)} Y_2(x+a) - (x+a) p_1(x+a) \\ &= (a\rho/c_1)\, e^{-hx} Y_2(x) - x\, p_1(x) = p_2(x). \end{aligned}$$

Thus, $p_1(x)$ and $p_2(x)$ are periodic functions. Since

$$y_1(x) = e^{hx} p_1(x), \quad Y_2(x) = (c_1/a\rho)\, e^{hx} [x\, p_1(x) + p_2(x)],$$

we may choose

$$y_2(x) = (a\rho/c_1)\, Y_2(x).$$

Thus,

$$y_2(x) = e^{hx} [x\, p_1(x) + p_2(x)]$$

and the part (ii) of the theorem is proven.

□

The part (i) of the theorem corresponds to the cases where Eq. (2.28) has two linearly independent nontrivial solutions of the form (2.29); the part (ii) of the theorem corresponds to the cases where Eq. (2.28) has only one nontrivial solution of the form (2.29).

2.3 Discriminant and Linearly Independent Solutions

From the last section, we see that the linearly independent solutions of Eq. (2.28) are determined by the roots ρ of the characteristic equation (2.34), which are determined by a real number

$$D = \eta_1(a) + p(a)\eta_2'(a). \tag{2.42}$$

This real number D determines the forms of roots ρ of the characteristic equation (2.34) and thus the forms of two linearly independent solutions of (2.28). This real number is called the discriminant of Eq. (2.28).[4]

There can be five different cases.

A. $-2 < D < 2$.

In this case, the two roots ρ_1 and ρ_2 of the characteristic equation (2.34) are two distinct nonreal numbers. They are complex conjugates of each other and have moduli equal to unity. h_i in (2.35) can be chosen as imaginary numbers $\pm ik$, where $0 < k < \pi/a$. Equation (2.28) has two linearly independent solutions,

$$\begin{aligned} y_1(x) &= \mathrm{e}^{ikx} p_1(x), \\ y_2(x) &= \mathrm{e}^{-ikx} p_2(x). \end{aligned} \tag{2.43}$$

Here $p_i(x),\ i = 1, 2$ are periodic functions with period a. k in (2.43) is related to the discriminant D by

$$cos\ ka = \frac{1}{2}D. \tag{2.44}$$

B. $D = 2$.

There are two possible subcases:

B.1. $\eta_2(a)$ and $p(a)\eta_1'(a)$ are not both zero.

In this subcase, not all elements of the matrix $(A - I\rho)$ (I is the unit matrix) are zero. The matrix $A = (A_{ij})$ has only one independent eigenvector. Equation (2.28) can have two linearly independent solutions as

$$\begin{aligned} y_1(x) &= p_1(x), \\ y_2(x) &= x\ p_1(x) + p_2(x). \end{aligned} \tag{2.45}$$

Here $p_i(x), i = 1, 2$ are periodic functions with period a.

B.2. $\eta_2(a) = p(a)\eta_1'(a) = 0$.

---

[4]It is easy to see that as the trace of matrix A_{ij} in (2.32), the discriminant D of Eq. (2.28) does not depend on how the origin 0 is chosen in (2.30).

In this subcase, $\eta_1(a) = p(a)\eta_2'(a) = 1$. All elements of the matrix $(A - I\rho)$ are zero. The matrix $A = (A_{ij})$ has two linearly independent eigenvectors.

Equation (2.28) can have two linearly independent solutions:

$$\begin{aligned} y_1(x) &= p_1(x), \\ y_2(x) &= p_2(x). \end{aligned} \tag{2.46}$$

Here $p_i(x), i = 1, 2$ are periodic functions with period a.

C. $D > 2$.

In this case, the roots ρ_1 and ρ_2 of the characteristic equation (2.34) are two distinct positive real numbers that are not equal to unity. h_i in (2.35) can be chosen as real numbers $\pm\beta$. Equation (2.28) can have two linearly independent solutions:

$$\begin{aligned} y_1(x) &= \mathrm{e}^{\beta x} p_1(x), \\ y_2(x) &= \mathrm{e}^{-\beta x} p_2(x), \end{aligned} \tag{2.47}$$

and $p_i(x), i = 1, 2$ are periodic functions with period a. β in (2.47) is a positive real number related to the discriminant D by

$$cosh\ \beta a = \frac{1}{2} D. \tag{2.48}$$

D. $D = -2$.

There are two possible subcases:

D.1. $\eta_2(a)$ and $p(a)\eta_1'(a)$ are not both zero.

Equation (2.28) can have two linearly independent solutions as

$$\begin{aligned} y_1(x) &= s_1(x), \\ y_2(x) &= x\ s_1(x) + s_2(x). \end{aligned} \tag{2.49}$$

Here $s_i(x), i = 1, 2$ are semi-periodic functions with semi-period a: $s_i(x + a) = -s_i(x)$.

D.2. $\eta_2(a) = p(a)\eta_1'(a) = 0$.

Equation (2.28) can have two linearly independent solutions:

$$\begin{aligned} y_1(x) &= s_1(x), \\ y_2(x) &= s_2(x). \end{aligned} \tag{2.50}$$

Here $s_i(x), i = 1, 2$ are semi-periodic functions with semi-period a.

E. $D < -2$.

In this case, the roots ρ_1 and ρ_2 of the characteristic equation (2.34) are two distinct negative real numbers that are not equal to -1. h_i in (2.35) can be chosen as complex numbers $\pm(\beta + i\pi/a)$. Equation (2.28) can have two linearly independent solutions as

$$\begin{aligned} y_1(x) &= e^{\beta x} s_1(x), \\ y_2(x) &= e^{-\beta x} s_2(x), \end{aligned} \tag{2.51}$$

and $s_i(x), i = 1, 2$ are semi-periodic functions with semi-period a. β in (2.51) is a positive real number related to the discriminant D by

$$cosh\ \beta a = -\frac{1}{2} D. \tag{2.52}$$

2.4 The Spectral Theory

Now, we consider a periodic Sturm–Liouville equation (2.2),

$$[p(x)y'(x)]' + [\lambda w(x) - q(x)]y(x) = 0,$$

where $p(x) > 0,\ w(x) > 0$ and $p(x),\ q(x),\ w(x)$ are piecewise continuous real periodic functions with period a:

$$p(x + a) = p(x), \quad q(x + a) = q(x), \quad w(x + a) = w(x).$$

The normalized solutions $\eta_i(x, \lambda), i = 1, 2$ of (2.2) are defined as

$$\eta_1(0, \lambda) = 1,\ p(0)\eta_1'(0, \lambda) = 0;\ \ \eta_2(0, \lambda) = 0,\ p(0)\eta_2'(0, \lambda) = 1, \tag{2.53}$$

and the discriminant of (2.2) is[5]

$$D(\lambda) = \eta_1(a, \lambda) + p(a)\eta_2'(a, \lambda). \tag{2.54}$$

The two linearly independent solutions of (2.2) are determined by $D(\lambda)$ in (2.54). To understand the properties of solutions of Eq. (2.2), we need to know how $D(\lambda)$ changes as λ changes. For this purpose, we first give two definitions.

[5]For the Schrödinger equation (2.1) of a one-dimensional crystal, the discriminant is $D(\lambda) = \eta_1(a, \lambda) + \eta_2'(a, \lambda)$ since $p(x) = 1$.

2.4.1 Two Eigenvalue Problems

We consider the solutions of Eq. (2.2) under the conditions

$$y(a) = y(0), \quad p(a)y'(a) = p(0)y'(0). \tag{2.55}$$

The corresponding eigenvalues are denoted by λ_n and can be ordered according to

$$\lambda_0 \leq \lambda_1 \leq \lambda_2 \leq \cdots,$$

and the eigenfunctions can be chosen as to be real-valued and denoted as $\zeta_n(x)$. $\zeta_n(x)$ can be further required to form an orthonormal set over $[0, a]$:

$$\int_0^a \zeta_m(x)\zeta_n(x)\,\mathrm{d}x = \delta_{m,n}.$$

$\zeta_n(x)$ can be extended by (2.55) to the whole of $(-\infty, +\infty)$ as continuous and piecewise quasi-differentiable functions[6] with period a. Thus, λ_n are the values of λ for which Eq. (2.2) has a nontrivial solution with period a.

Similarly, we can also consider the solutions of (2.2) under the conditions

$$y(a) = -y(0), \quad p(a)y'(a) = -p(0)y'(0). \tag{2.56}$$

The corresponding eigenvalues are denoted by μ_n and can be ordered according to

$$\mu_0 \leq \mu_1 \leq \mu_2 \leq \cdots.$$

The corresponding eigenfunctions can be chosen to be real-valued and denoted as $\xi_n(x)$. $\xi_n(x)$ can be further required to form an orthonormal set over $[0, a]$:

$$\int_0^a \xi_m(x)\xi_n(x)\,\mathrm{d}x = \delta_{m,n}.$$

$\xi_n(x)$ can be extended by (2.56) to the whole of $(-\infty, +\infty)$ as continuous and piecewise quasi-differentiable functions with semi-period a. Thus, μ_n are the values of λ for which Eq. (2.2) has a nontrivial solution with semi-period a.

[6]Each $\zeta_n(x)$ and its quasi-derivative $p(x)\zeta_n'(x)$ are continuous.

2.4.2 The Function $D(\lambda)$

The following theorem describes how $D(\lambda)$ changes as λ changes regarding the eigenvalues λ_n and μ_n defined by the two eigenvalue problems (2.55) and (2.56)[7]:

Theorem 2.5 (Extended Theorem 2.3.1 in [7])

(i) The numbers λ_n and μ_n occur in the order

$$\lambda_0 < \mu_0 \leq \mu_1 < \lambda_1 \leq \lambda_2 < \mu_2 \leq \mu_3 < \lambda_3 \leq \lambda_4 < \cdots . \tag{2.57}$$

As λ increases from $-\infty$ to $+\infty$, $D(\lambda)$ changes as the following, where $m = 0, 1, 2, \cdots$:

(ii) In the interval $(-\infty, \lambda_0)$, $D(\lambda) > 2$.
(iii) In the intervals $[\lambda_{2m}, \mu_{2m}]$, $D(\lambda)$ decreases from $+2$ to -2.
(iv) In the intervals (μ_{2m}, μ_{2m+1}), $D(\lambda) < -2$.
(v) In the intervals $[\mu_{2m+1}, \lambda_{2m+1}]$, $D(\lambda)$ increases from -2 to $+2$.
(vi) In the intervals $(\lambda_{2m+1}, \lambda_{2m+2})$, $D(\lambda) > 2$.

Proof This theorem can be proven in several steps.

(1) *There exists a Λ such that for all $\lambda < \Lambda$, $D(\lambda) > 2$.*

We can choose a Λ so that for all x in $(-\infty, +\infty)$,

$$[q(x) - \Lambda w(x)] > 0$$

is true.

Suppose $y(x)$ is *any* nontrivial solution of (2.2) for which $y(0) \geq 0$ and $p(0)y'(0) \geq 0$; then there is always an interval $(0, \Delta)$ in which $y(x) > 0$.

For all $\lambda \leq \Lambda$, in any interval $(0, X)$ for which $y(x) > 0$ we have

$$[p(x)y'(x)]' = [q(x) - \lambda w(x)]y(x) > 0;$$

thus, in the interval $(0, X)$, we have $p(x)y'(x) > 0$ and $y(x)$ is increasing in $(0, X)$. Therefore, $y(x)$ has no zero $x = x_0$ in $(0, +\infty)$ and both $y(x)$ and $p(x)y'(x)$ are increasing.

Since both $\eta_1(x, \lambda)$ and $\eta_2(x, \lambda)$ defined in (2.53) satisfy

$$\eta_1(0, \lambda) \geq 0, \;\; p(0)\eta_1'(0, \lambda) \geq 0; \;\; \eta_2(0, \lambda) \geq 0, \;\; p(0)\eta_2'(0, \lambda) \geq 0,$$

[7]Relevant contents can also be found in Theorem 12.7 in [8], Theorem 4.8.1 in [9], Theorems 1.6.1 and 2.4.2 in [10], Theorem 5.33 in [11], and also in References [14, 15].

both $\eta_1(x,\lambda)$, $p(x)\eta_1'(x,\lambda)$ and $\eta_2(x,\lambda)$, $p(x)\eta_2'(x,\lambda)$ are increasing in $(0,+\infty)$ for all $\lambda \le \Lambda$. In particular, we have

$$\eta_1(a,\lambda) > \eta_1(0,\lambda) = 1; \quad p(a)\eta_2'(a,\lambda) > p(0)\eta_2'(0,\lambda) = 1.$$

Thus, for all $\lambda \le \Lambda$, we have $D(\lambda) > 2$.

However, as λ increases, $[p(x)y'(x)]'/y(x) = [q(x) - \lambda w(x)]$ will become negative and, consequently, $D(\lambda)$ will decrease as λ increases.

(2) *For any* λ *such that* $|D(\lambda)| < 2$, $D'(\lambda)$ *is not zero.*

Differentiating (2.2) with respect to λ with $y(x) = \eta_1(x,\lambda)$, we obtain

$$\frac{\mathrm{d}}{\mathrm{d}x}\left[p(x)\frac{\mathrm{d}}{\mathrm{d}x}\left[\frac{\partial\eta_1(x,\lambda)}{\partial\lambda}\right]\right] + [\lambda w(x) - q(x)]\frac{\partial\eta_1(x,\lambda)}{\partial\lambda} = -w(x)\eta_1(x,\lambda). \tag{2.58}$$

By using the variation of parameters formula (2.16), we solve $\partial\eta_1(x,\lambda)/\partial\lambda$ from (2.58) with the initial condition $\frac{\partial\eta_1(0,\lambda)}{\partial\lambda} = 0$ (from (2.53)) and obtain that

$$\frac{\partial\eta_1(x,\lambda)}{\partial\lambda} = \int_0^x [\eta_1(x,\lambda)\eta_2(t,\lambda) - \eta_2(x,\lambda)\eta_1(t,\lambda)]w(t)\eta_1(t,\lambda)\,\mathrm{d}t, \tag{2.59}$$

by noting that $W[\eta_1(t,\lambda), \eta_2(t,\lambda)] = 1$.

Similarly by differentiating (2.2) with respect to λ with $y(x) = \eta_2(x,\lambda)$, we obtain

$$\frac{\mathrm{d}}{\mathrm{d}x}\left[p(x)\frac{\mathrm{d}}{\mathrm{d}x}\left[\frac{\partial\eta_2(x,\lambda)}{\partial\lambda}\right]\right] + [\lambda w(x) - q(x)]\frac{\partial\eta_2(x,\lambda)}{\partial\lambda} = -w(x)\eta_2(x,\lambda). \tag{2.60}$$

By using the variation of parameters formula (2.16), we solve $\frac{\partial\eta_2(x,\lambda)}{\partial\lambda}$ from (2.60) with the initial condition that $\frac{\partial\eta_2(0,\lambda)}{\partial\lambda} = 0$ (from (2.53)) and obtain that

$$\frac{\partial\eta_2(x,\lambda)}{\partial\lambda} = \int_0^x [\eta_1(x,\lambda)\eta_2(t,\lambda) - \eta_2(x,\lambda)\eta_1(t,\lambda)]w(t)\eta_2(t,\lambda)\,\mathrm{d}t, \tag{2.61}$$

By using the variation of parameters formula (2.17), we solve $\frac{\partial p(x)\eta_2'(x,\lambda)}{\partial\lambda}$ from (2.60) with the initial condition $\frac{\partial p(0)\eta_2'(0,\lambda)}{\partial\lambda} = 0$ (from (2.53)) and obtain that

$$\begin{aligned}\frac{\partial p(x)\eta_2'(x,\lambda)}{\partial\lambda} = \int_0^x &[p(x)\eta_1'(x,\lambda)\eta_2(t,\lambda) \\ &- p(x)\eta_2'(x,\lambda)\eta_1(t,\lambda)]w(t)\eta_2(t,\lambda)\mathrm{d}t.\end{aligned} \tag{2.62}$$

Combining (2.62) and (2.59) and putting $x = a$, we obtain that

$$D'(\lambda) = \int_0^a [p\eta_1'\eta_2^2(t,\lambda) + (\eta_1 - p\eta_2')\eta_1(t,\lambda)\eta_2(t,\lambda) - \eta_2\eta_1^2(t,\lambda)]\, w(t)\, dt, \tag{2.63}$$

and thus

$$\begin{aligned} 4\eta_2 D'(\lambda) = &- \int_0^a [2\eta_2\eta_1(t,\lambda) - (\eta_1 - p\eta_2')\eta_2(t,\lambda)]^2\, w(t)\, dt \\ &- [4 - D^2(\lambda)] \int_0^a \eta_2^2(t,\lambda)\, w(t)\, dt. \end{aligned} \tag{2.64}$$

In (2.63) and (2.64) we have written $\eta_i = \eta_i(a,\lambda)$ and $p\eta_i' = p(a)\eta_i'(a,\lambda)$ for brevity.

If $|D(\lambda)| < 2$, from (2.64) we have $\eta_2 D'(\lambda) < 0$; thus, $D'(\lambda) \neq 0$. Therefore, only in the regions of λ in which $|D(\lambda)| \geq 2$ can $D'(\lambda) = 0$ be true.

(3) *At a zero λ_n of $D(\lambda) - 2 = 0$, if and only if*

$$\eta_2(a,\lambda_n) = p(a)\eta_1'(a,\lambda_n) = 0, \tag{2.65}$$

$D'(\lambda_n) = 0$ is true. Further, if $D'(\lambda_n) = 0$, then $D''(\lambda_n) < 0$.

(3a) Equation (2.65) gives that $\eta_1(a,\lambda_n)p(a)\eta_2'(a,\lambda_n) = 1$ by $W[\eta_1(a,\lambda_n), \eta_2(a,\lambda_n)] = 1$. It further gives

$$\eta_1(a,\lambda_n) = p(a)\eta_2'(a,\lambda_n) = 1 \tag{2.66}$$

by $D(\lambda_n) = 2$. Equation (2.63) then gives that $D'(\lambda_n) = 0$. According to **B.2** in Sect. 2.3, this case corresponds to the case that $D(\lambda) - 2$ has a double zero at $\lambda = \lambda_n$.

(3b) On the other hand, if $D'(\lambda_n) = 0$, then the first integrand on the right of (2.64) must be identically zero since $D(\lambda_n) = 2$. Since $\eta_1(t,\lambda)$ and $\eta_2(t,\lambda)$ are linearly independent, $\eta_2(a,\lambda_n) = 0$ and $\eta_1(a,\lambda_n) = p(a)\eta_2'(a,\lambda_n)$ must be true. From (2.63), we obtain that $p(a)\eta_1'(a,\lambda_n) = 0$.

(3c) To further prove $D''(\lambda_n) < 0$ when $D'(\lambda_n) = 0$, we differentiate (2.58) with respect to λ and obtain

$$\begin{aligned} \frac{d}{dx}\left[\frac{\partial^2 p(x)\eta_1'(x,\lambda)}{\partial\lambda^2}\right] + [\lambda w(x) - q(x)]\frac{\partial^2\eta_1(x,\lambda)}{\partial\lambda^2} \\ = -2w(x)\frac{\partial}{\partial\lambda}\eta_1(x,\lambda). \end{aligned} \tag{2.67}$$

Applying the variation of parameters formula (2.16) to solve $\partial^2\eta_1(x,\lambda)/\partial\lambda^2$ from (2.67) with the initial condition $\frac{\partial^2\eta_1(0,\lambda)}{\partial\lambda^2} = 0$ (from (2.53)) we obtain that

$$\frac{\partial^2 \eta_1(x,\lambda)}{\partial \lambda^2} = 2\int_0^x [\eta_1(x,\lambda)\eta_2(t,\lambda) - \eta_2(x,\lambda)\eta_1(t,\lambda)]w(t)\frac{\partial}{\partial \lambda}\eta_1(t,\lambda)\, dt \tag{2.68}$$

by noting that $W[\eta_1(t,\lambda), \eta_2(t,\lambda)] = 1$.

We differentiate (2.60) with respect to λ and obtain

$$\frac{d}{dx}\left[\frac{\partial^2 p(x)\eta_2'(x,\lambda)}{\partial \lambda^2}\right] + [\lambda w(x) - q(x)]\frac{\partial^2 \eta_2(x,\lambda)}{\partial \lambda^2} = -2w(x)\frac{\partial}{\partial \lambda}\eta_2(x,\lambda). \tag{2.69}$$

By applying the variation of parameters formula (2.17) again to solve $\frac{\partial^2 p(x)\eta_2'(x,\lambda)}{\partial \lambda^2}$ from (2.69) with the initial condition $\frac{\partial^2 p(0)\eta_2'(0,\lambda)}{\partial \lambda^2} = 0$ (from (2.53)), we obtain

$$\frac{\partial^2 p(x)\eta_2'(x,\lambda)}{\partial \lambda^2} = 2\int_0^x [p(x)\eta_1'(x,\lambda)\eta_2(t,\lambda) - p(x)\eta_2'(x,\lambda)\eta_1(t,\lambda)]w(t)\frac{\partial}{\partial \lambda}\eta_2(t,\lambda)\, dt. \tag{2.70}$$

Therefore, by combining (2.68) and (2.70) and noting that when $D'(\lambda_n) = 0$, (2.65) and (2.66) are true, we obtain that

$$\begin{aligned} D''(\lambda_n) &= 2\int_0^a \left[\eta_2(t,\lambda_n)\frac{\partial \eta_1(t,\lambda)}{\partial \lambda} - \eta_1(t,\lambda_n)\frac{\partial \eta_2(t,\lambda)}{\partial \lambda}\right]_{\lambda=\lambda_n} w(t)dt \\ &= -2\int_0^a w(t)dt \int_0^t [\eta_1(t,\lambda_n)\eta_2(\tau,\lambda_n) - \eta_2(t,\lambda_n)\eta_1(\tau,\lambda_n)]^2 w(\tau)d\tau. \end{aligned} \tag{2.71}$$

Equations (2.59) and (2.61) were used in obtaining the second equality. The right side of (2.71) is less than zero since the integrand in the double integral is positive.

(4) It can be similarly proven that there is a corresponding result to (3) for the zeros μ_n of $D(\lambda) + 2$: If and only if

$$\eta_2(a,\mu_n) = p(a)\eta_1'(a,\mu_n) = 0, \tag{2.72}$$

$D'(\mu_n) = 0$ is true. Further, $D''(\mu_n) > 0$ when $D'(\mu_n) = 0$. This case corresponds to that $D(\lambda) + 2$ has a double zero at $\lambda = \mu_n$.

(5) Therefore, except cases in (3) or (4), only in the regions of λ in which $D(\lambda) < -2$ or $D(\lambda) > 2$ can $D'(\lambda) = 0$ be true. The $D(\lambda)$–λ curve can change direction only in such regions.

(6) From above results of (1)–(5), we can discuss the behavior of $D(\lambda)$ as λ increases from $-\infty$ to $+\infty$. When λ is a large negative real number, $D(\lambda) > 2$ by (1). As λ increases from $-\infty$, we have $D(\lambda) > 2$ until λ reaches the first zero λ_0 of $D(\lambda) - 2$. Since λ_0 is not a maximum of $D(\lambda)$, $D''(\lambda_0) \not< 0$; thus, $D'(\lambda_0) \neq 0$ by (3). The $D(\lambda)$–λ curve intersects the line $D = 2$ at $\lambda = \lambda_0$; thus, to the immediate right of λ_0, we have $D(\lambda) < 2$. Then by (2), as λ increases from λ_0, $D(\lambda)$ decreases until λ reaches the first zero μ_0 of $D(\lambda) + 2$. Thus, in the interval $(-\infty, \lambda_0)$, $D(\lambda) > 2$, and in the interval $[\lambda_0, \mu_0]$, $D(\lambda)$ decreases from $+2$ to -2.
In general, μ_0 will be a simple zero of $D(\lambda) + 2$, so the $D(\lambda)$–λ curve intersects the line $D = -2$ at $\lambda = \mu_0$, and to the immediate right of μ_0 $D(\lambda) < -2$. As λ increase, $D(\lambda) < -2$ will remain true until λ reaches the second zero μ_1 of $D(\lambda) + 2$, since, by (5), the $D(\lambda)$–λ curve can change direction in a region where $D(\lambda) < -2$. Since μ_1 is not a minimum of $D(\lambda)$, μ_1 is a simple zero of $D(\lambda) + 2$. The $D(\lambda)$–λ curve intersects the line $D = -2$ again at $\lambda = \mu_1$. To the immediate right of μ_1, we have $D(\lambda) > -2$; then according to (2), as λ increases from μ_1, $D(\lambda)$ increases until λ reaches the next zero λ_1 of $D(\lambda) - 2$. Thus, in the interval (μ_0, μ_1), $D(\lambda) < -2$, and in the interval $[\mu_1, \lambda_1]$, $D(\lambda)$ increases from -2 to $+2$.
In general, λ_1 will be a simple zero of $D(\lambda) - 2$, so the $D(\lambda)$–λ curve intersects the line $D = 2$ at $\lambda = \lambda_1$, and to the immediate right of λ_1, we have $D(\lambda) > 2$. As λ increase, $D(\lambda) > 2$ will remain to be true until λ reaches the third zero λ_2 of $D(\lambda) - 2$, since, by (5), the $D(\lambda)$–λ curve can change direction in a region where $D(\lambda) > 2$. The argument we used starting from $\lambda = \lambda_0$ can be repeated starting from $\lambda = \lambda_2$, and can be repeated again and again as λ increases to $+\infty$.

Now all parts of the theorem have been proven except when $D(\lambda) \pm 2$ has double zeros. If, for example, $D(\lambda) - 2$ has a double zero at a specific $\lambda = \lambda_{2m+1}$ (i.e., $\lambda_{2m+2} = \lambda_{2m+1}$). From **B.2** in Sect. 2.3, this can happen only when $\eta_2(a, \lambda_{2m+1}) = \eta_1'(a, \lambda_{2m+1}) = 0$; therefore, $D'(\lambda_{2m+1}) = 0$ and $D''(\lambda_{2m+1}) < 0$ is true by (3). Consequently, to the immediate right of $\lambda = \lambda_{2m+1} = \lambda_{2m+2}$ we have $D(\lambda) < 2$. In such a case, the $D(\lambda)$–λ curve merely *touches* the line $D = 2$ at $\lambda = \lambda_{2m+1} = \lambda_{2m+2}$ rather than intersects the line $D = 2$ twice at $\lambda = \lambda_{2m+1}$ and at $\lambda = \lambda_{2m+2}$. The previous analysis of $D(\lambda)$ can repeatedly continue again. The cases where $D(\lambda) + 2$ has double zeros can be similarly analyzed by using (4). □

Therefore, in general, when λ increases from $-\infty$ to $+\infty$, the discriminant $D(\lambda)$ of Eq. (2.2) as defined in (2.54) changes, as shown typically in Fig. 2.1 [1, 5, 8, 9, 11, 14, 15]. The permitted eigenvalue bands of (2.2) are in the ranges of λ for which $-2 \leq D(\lambda) \leq 2$ (solid lines). No eigenmode exists in the ranges of λ for which $D(\lambda) > 2$ or $D(\lambda) < -2$ (dashed lines).

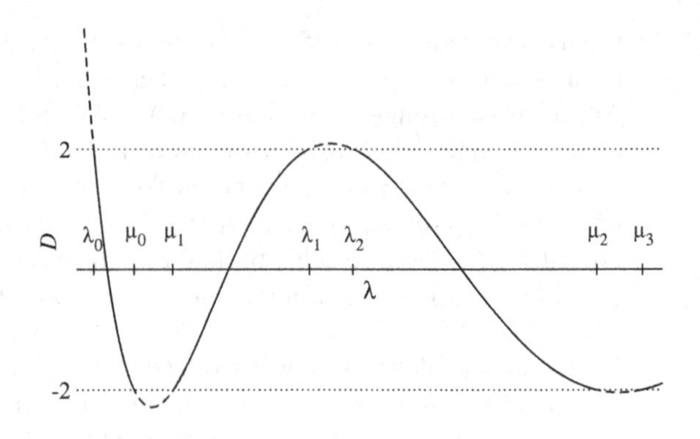

Fig. 2.1 A typical $D(\lambda)$–λ curve. The permitted eigenvalue bands of Eq. (2.2) are in the ranges of λ for which $-2 \leq D(\lambda) \leq 2$ (*solid lines*). No eigenvalue exists in the ranges of λ for which $D(\lambda) > 2$ or $D(\lambda) < -2$ (*dashed lines*)

2.5 Band Structure of Eigenvalues

The permitted band structure of a periodic Sturm–Liouville equation (2.2) has some especially simple and general properties [1, 2, 5, 8, 9, 11, 15].

Only for λ in the range $[\lambda_0, +\infty)$ can eigenmodes exist in a crystal with translational invariance. There are five different cases:

A. λ in an interval (λ_{2m}, μ_{2m}) or $(\mu_{2m+1}, \lambda_{2m+1})$, where $m = 0, 1, 2, \ldots, -2 < D(\lambda) < 2$. Two linearly independent solutions can be chosen as

$$\begin{aligned} y_1(x, \lambda) &= \mathrm{e}^{ik(\lambda)x} p_1(x, \lambda), \\ y_2(x, \lambda) &= \mathrm{e}^{-ik(\lambda)x} p_2(x, \lambda), \end{aligned} \tag{2.73}$$

and $p_i(x, \lambda)$ are periodic functions depending on λ. They are the two well-known one-dimensional Bloch states $\phi_n(k, x)$ and $\phi_n(-k, x)$ with wave vector k or $-k$. Their corresponding energies can be written as $\varepsilon_n(k)$ and $\varepsilon_n(-k)$, with $\varepsilon_n(-k) = \varepsilon_n(k)$. Therefore, the intervals (λ_{2m}, μ_{2m}) and $(\mu_{2m+1}, \lambda_{2m+1})$ correspond to the inside of permitted energy bands of Eq. (2.2).
In each permitted band, where $0 < k(\lambda) < \pi/a$ is determined by (2.51):

$$cos\ ka = \frac{1}{2} D(\lambda). \tag{2.74}$$

It has been shown that $\varepsilon_n(k)$ is always a monotonic function of k inside each permitted energy band [2]. Here we give another simple proof.
Suppose this is not true, there is an energy band in which $\varepsilon_n(k)$ is not a monotonic function of k. Then there must be at least one λ inside the energy band for which there are at least two distinct k_1 and k_2 in $(0, \frac{\pi}{a})$ for which $\varepsilon_n(k_1) = \varepsilon_n(k_2) = \lambda$. That means (2.2) has at least four linearly independent solutions for such a λ: two with $k = k_1$ and two with $k = k_2$ in (2.73). This is contradictory to that a periodic Sturm–Liouville equation such as (2.2) can only have two linearly independent solutions.

B. At $\lambda = \lambda_n$, $D(\lambda) = 2$.

B.1. In most cases, λ_n is a simple zero of $D(\lambda) - 2$. Equation (2.2) has two linearly independent solutions with forms as

$$\begin{aligned} y_1(x, \lambda) &= p_1(x, \lambda_n), \\ y_2(x, \lambda) &= x\, p_1(x, \lambda_n) + p_2(x, \lambda_n), \end{aligned} \tag{2.75}$$

and $p_i(x, \lambda)$ are periodic functions depending on λ. In crystals with translational invariance, only the periodic function solution y_1 is permitted. λ_n corresponding to a band-edge eigenvalue at $k = 0$, $\varepsilon_n(0)$. For $n > 0$, Case **B.1** corresponds to the cases where there is a nonzero band gap between $\varepsilon_{2m+1}(0)$ and $\varepsilon_{2m+2}(0)$.

B.2. In some special cases, λ_n $(n > 0)$ is a double zero of $D(\lambda) - 2$: $\lambda_{2m+1} = \lambda_{2m+2}$. Equation (2.2) has two linearly independent solutions with forms as

$$\begin{aligned} y_1(x, \lambda) &= p_1(x, \lambda_{2m+1}), \\ y_2(x, \lambda) &= p_2(x, \lambda_{2m+1}). \end{aligned} \tag{2.76}$$

Their corresponding eigenvalues are $\varepsilon_{2m+1}(0) = \varepsilon_{2m+2}(0) = \lambda_{2m+1}$, and there is a zero band gap between $\varepsilon_{2m+1}(0)$ and $\varepsilon_{2m+2}(0)$.

C. λ in an interval $(\lambda_{2m+1}, \lambda_{2m+2})$, $D(\lambda) > 2$. In such a case λ is inside a band gap at $k = 0$ of Eq. (2.2). The two linearly independent solutions of Eq. (2.2) can be written as

$$\begin{aligned} y_1(x, \lambda) &= e^{\beta(\lambda)x} p_1(x, \lambda), \\ y_2(x, \lambda) &= e^{-\beta(\lambda)x} p_2(x, \lambda). \end{aligned} \tag{2.77}$$

Here, $\beta(\lambda) > 0$ is determined by (2.48):

$$cosh\ \beta a = \frac{1}{2} D(\lambda) \tag{2.78}$$

and $p_i(x, \lambda)$, $i = 1, 2$ are periodic functions. These forbidden solutions in crystals with translational invariance might play a significant role in the physics of one-dimensional semi-infinite crystals and crystals of finite length.

D. At $\lambda = \mu_n$, $D(\lambda) = -2$.

D.1. In most cases, μ_n is a simple zero of $D(\lambda) + 2$. Equation (2.2) has two linearly independent solutions with forms as

$$\begin{aligned} y_1(x, \lambda) &= s_1(x, \mu_n), \\ y_2(x, \lambda) &= x\, s_1(x, \mu_n) + s_2(x, \mu_n), \end{aligned} \tag{2.79}$$

and $s_i(x, \lambda)$ are semi-periodic functions depending on λ. In crystals of infinite size, only the semi-periodic function solution y_1 is permitted. μ_n

corresponds to a band-edge eigenvalue $\varepsilon_n(\frac{\pi}{a})$ at $k = \frac{\pi}{a}$. Case **D.1** corresponds to the cases that there is a nonzero band gap between $\varepsilon_{2m}(\frac{\pi}{a})$ and $\varepsilon_{2m+1}(\frac{\pi}{a})$.

D.2. In some special cases where μ_n is a double zero of $D(\lambda) + 2$: $\mu_{2m} = \mu_{2m+1}$. Equation (2.2) has two linearly independent solutions as

$$\begin{aligned} y_1(x, \lambda) &= s_1(x, \mu_{2m}), \\ y_2(x, \lambda) &= s_2(x, \mu_{2m}). \end{aligned} \tag{2.80}$$

Their corresponding eigenvalues are $\varepsilon_{2m}(\frac{\pi}{a}) = \varepsilon_{2m+1}(\frac{\pi}{a}) = \mu_{2m}$, and there is a zero band gap between $\varepsilon_{2m}(\frac{\pi}{a})$ and $\varepsilon_{2m+1}(\frac{\pi}{a})$.

E. λ in an interval (μ_{2m}, μ_{2m+1}), $D(\lambda) < -2$. In such a case λ is inside a band gap at $k = \pi/a$ of Eq. (2.2). The two linearly independent solutions can be written as

$$\begin{aligned} y_1(x, \lambda) &= e^{\beta(\lambda)x} s_1(x, \lambda), \\ y_2(x, \lambda) &= e^{-\beta(\lambda)x} s_2(x, \lambda). \end{aligned} \tag{2.81}$$

Here, $\beta(\lambda) > 0$ is determined by (2.52):

$$cosh\ \beta a = -\frac{1}{2} D(\lambda), \tag{2.82}$$

and $s_i(x, \lambda)$ are semi-periodic functions. These forbidden solutions in crystals with translational invariance might play a significant role in the physics of one-dimensional semi-infinite crystals and crystals of finite length.

Therefore, in above cases **A, B**, and **D**, the permitted eigen solutions can exist as solutions of the periodic Sturm–Liouville equation (2.2). By combining our discussions in these three cases, we see that in the permitted band $\varepsilon_n(k)$ and permitted eigen solutions $\phi_n(k, x)$, the wave vector k in $\varepsilon_n(k)$ and $\phi_n(k, x)$ is limited in the Brillouin zone,

$$-\frac{\pi}{a} < k \leq \frac{\pi}{a}, \tag{2.83}$$

and that λ_n, μ_n, and $\zeta_n(x)$, $\xi_n(x)$ defined in Sect. 2.4.1 are the band-edge energies and band-edge wave functions:

$$\lambda_n = \varepsilon_n(0), \quad \zeta_n(x) = \phi_n(0, x), \tag{2.84}$$

and

$$\mu_n = \varepsilon_n\left(\frac{\pi}{a}\right), \quad \xi_n(x) = \phi_n\left(\frac{\pi}{a}, x\right). \tag{2.85}$$

Therefore, (2.57) can be rewritten as

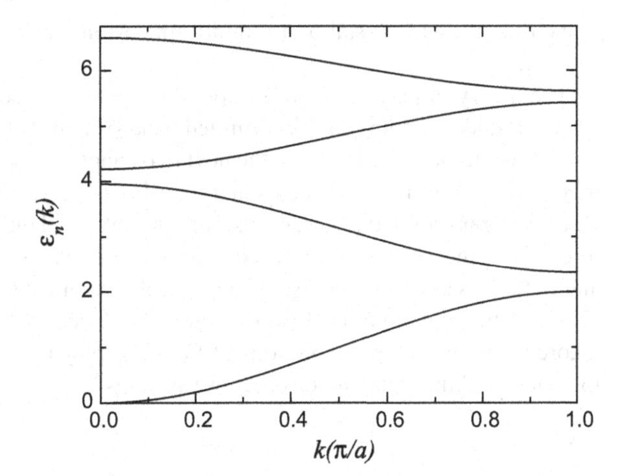

Fig. 2.2 A typical permitted band structure determined by (2.74) as solutions of a periodic Sturm–Liouville equation (2.2)

$$\varepsilon_0(0) < \varepsilon_0\left(\frac{\pi}{a}\right) \leq \varepsilon_1\left(\frac{\pi}{a}\right) < \varepsilon_1(0) \leq \varepsilon_2(0)$$
$$< \varepsilon_2\left(\frac{\pi}{a}\right) \leq \varepsilon_3\left(\frac{\pi}{a}\right) < \varepsilon_3(0) \leq \varepsilon_4(0) < \cdots . \quad (2.86)$$

A typical permitted band structure determined by (2.74) as solutions of a Sturm–Liouville equation Eq. (2.2) is shown in Fig. 2.2.

From the figure, we can see and understand that the permitted band structure as solutions of a periodic Sturm–Liouville equation has some very simple and general properties, see, for example [2, 15]:

1. In each permitted band, $\varepsilon_n(-k) = \varepsilon_n(k)$ and $\varepsilon_n(k)$ $(k > 0)$ is a monotonic function of k.
2. The minimum and the maximum of each permitted band are always located either at $k = 0$ or $k = \frac{\pi}{a}$.
3. There is no permitted band crossing or permitted band overlap.
4. Each permitted band and each band gap occur alternatively.
5. The band gaps are between $\varepsilon_{2m}(\frac{\pi}{a})$ and $\varepsilon_{2m+1}(\frac{\pi}{a})$ or between $\varepsilon_{2m+1}(0)$ and $\varepsilon_{2m+2}(0)$.

A crystal described by a one-dimensional Schrödinger equation (2.1) can also be considered as a one-dimensional free space—a crystal with a zero potential (empty lattice)—reconstructed by a perturbation of the crystal potential $v(x)$. The crystal with an empty lattice has a free-electron-like energy spectrum $\lambda(k) = ck^2$ (c is a proportional constant) without any band gap in an extended Brillouin zone scheme. In the reduced Brillouin zone scheme, the energy spectrum $\lambda(k) = ck^2$ is folded at the Brillouin zone boundary, *continuously and simply touches at* $k = 0$ *and* $k = \frac{\pi}{a}$ *with no band crossings or band overlaps. Finite Fourier components of the crystal potential* $v(x)$—no matter how small it is—*will open band gaps at* $k = 0$ *or* $k = \frac{\pi}{a}$, *with no band crossing or band overlap*, see, for example [2, 15]. Such a picture would be helpful for future understanding of the significant differences between a one-dimensional crystal and a multi-dimensional crystal that we will discuss in Chap. 5.[8]

The theory of periodic Sturm–Liouville equations discussed so far has provided a general understanding of the permitted band structure of a one-dimensional crystal with translational invariance. In the next two chapters and in the appendices, we will treat states/modes in one-dimensional semi-infinite crystals or ideal finite crystals, in which the translational invariance is truncated. We will meet the complex band structure of periodic Sturm–Liouville equations. Similarly, the complex band structure of an Eq. (2.2) is also completely and analytically obtained from the discriminant $D(\lambda)$ (2.54) of the equation, as shown in Eqs. (2.78) and (2.82). The following several theorems on the zeros of solutions of Eq. (2.2) play a significant role in helping us to understand the states/modes in those systems.[9]

2.6 Zeros of Solutions

Now, we consider the solutions of (2.2) under the condition

$$y(a) = y(0) = 0. \tag{2.87}$$

The eigenvalues are denoted by Λ_n, and the corresponding eigenfunctions are denoted by $\Psi_n(x)$.

For physics problems in one-dimensional phononic crystals or photonic crystals, correspondingly there is an eigenvalue problem of (2.2) under the condition

$$p(a)y'(a) = p(0)y'(0) = 0, \tag{2.88}$$

the eigenvalues can be denoted by ν_n, and the corresponding eigenfunction can be written as $\Phi_n(x)$.

Theorem 2.6 (Extended Theorem 3.1.1 in [7], Theorem 13.10 in [8]) *For* $m = 0, 1, 2, \ldots$, *we have*

$$\varepsilon_{2m}\left(\tfrac{\pi}{a}\right) \leq \Lambda_{2m} \leq \varepsilon_{2m+1}\left(\tfrac{\pi}{a}\right); \quad \varepsilon_{2m+1}(0) \leq \Lambda_{2m+1} \leq \varepsilon_{2m+2}(0). \tag{2.89}$$

and

$$\begin{gathered} \varepsilon_{2m}\left(\tfrac{\pi}{a}\right) \leq \nu_{2m+1} \leq \varepsilon_{2m+1}\left(\tfrac{\pi}{a}\right); \quad \varepsilon_{2m+1}(0) \leq \nu_{2m+2} \leq \varepsilon_{2m+2}(0), \\ \nu_0 \leq \varepsilon_0(0). \end{gathered} \tag{2.90}$$

---

[8]These properties of the permitted bands are closely related to that a periodic Sturm–Liouville equation can have no more than two independent solutions.

[9]Relevant contents can be found in [8, 10].

Proof Since $\Psi_n(x)$ is the eigenfunction corresponding to the nth eigenvalue under the condition (2.87), it has exactly n zeros in $(0, a)$. According to (2.53), $\eta_2(0, \lambda) = 0$; thus, each Λ_n is the solution of the equation

$$\eta_2(a, \Lambda_n) = 0, \tag{2.91}$$

and the corresponding eigenfunction

$$\Psi_n(x) = \eta_2(x, \Lambda_n).$$

Therefore, $\eta_2(x, \Lambda_n)$ has exactly n zeros in the interval $(0, a)$. According to (2.53), $p(0)\ \eta_2'(0, \Lambda_n) > 0$; thus,

$$p(a)\eta_2'(a, \Lambda_n) < 0\ (n = \text{even}); \quad p(a)\eta_2'(a, \Lambda_n) > 0\ (n = \text{odd}). \tag{2.92}$$

Since we have

$$\eta_1(a, \Lambda_n)p(a)\eta_2'(a, \Lambda_n) = 1,$$

from

$$\eta_1(a, \Lambda_n)p(a)\eta_2'(a, \Lambda_n) - p(a)\eta_1'(a, \Lambda_n)\eta_2(a, \Lambda_n) = 1$$

and (2.91), therefore

$$D(\Lambda_n) = \eta_1(a, \Lambda_n) + p(a)\eta_2'(a, \Lambda_n) = \frac{1}{p(a)\eta_2'(a, \Lambda_n)} + p(a)\eta_2'(a, \Lambda_n).$$

For $n = even$, we have

$$-D(\Lambda_{n=even}) = [|p(a)\eta_2'(a, \Lambda_{n=even})|^{-1/2} - |p(a)\eta_2'(a, \Lambda_{n=even})|^{1/2}]^2 + 2 \geq 2$$

and thus

$$D(\Lambda_{n=even}) \leq -2$$

from (2.92). Similarly

$$D(\Lambda_{n=odd}) \geq 2$$

from (2.92). Therefore, Λ_n and Λ_{n+1} are always in different band gaps, if we consider the special cases in which $\varepsilon_{2m}(\frac{\pi}{a}) = \varepsilon_{2m+1}(\frac{\pi}{a})$ or $\varepsilon_{2m+1}(0) = \varepsilon_{2m+2}(0)$ as a band gap with the gap size being zero.

Now, we consider two consecutive zeros $\varepsilon_{2m+1}(0)$ and $\varepsilon_{2m+2}(0)$ of $D(\lambda) - 2$, with either $D(\lambda) > 2$ between them, or $D'(\lambda) = 0$ thus $D(\lambda) - 2$ has a double zero at $\lambda = \varepsilon_{2m+1}(0) = \varepsilon_{2m+2}(0)$ (see Fig. 2.1).

In the special cases where $D'(\varepsilon_{2m+1}(0)) = 0$, $D(\lambda) - 2$ has repeated solutions $\varepsilon_{2m+1}(0) = \varepsilon_{2m+2}(0)$; then by (2.65) we always have $\eta_2(a, \varepsilon_{2m+1}(0)) = 0$ and thus (2.91) has one solution $\Lambda_n = \varepsilon_{2m+1}(0)$.

In most cases, $D(\lambda) > 2$ in $(\varepsilon_{2m+1}(0), \varepsilon_{2m+2}(0))$. According to (2.64), we have $\eta_2(a, \lambda)D'(\lambda) \leq 0$ at both $\varepsilon_{2m+1}(0)$ and $\varepsilon_{2m+2}(0)$. Since $D'(\varepsilon_{2m+1}(0)) > 0$ and $D'(\varepsilon_{2m+2}(0)) < 0$, we have $\eta_2(a, \varepsilon_{2m+1}(0)) \leq 0$ and $\eta_2(a, \varepsilon_{2m+2}(0)) \geq 0$. Thus, $\eta_2(a, \lambda)$ has at least one zero in $[\varepsilon_{2m+1}(0), \varepsilon_{2m+2}(0)]$. Since Λ_n and Λ_{n+1} must be in different band gaps, there is no more than one Λ_n in $[\varepsilon_{2m+1}(0), \varepsilon_{2m+2}(0)]$. Thus, there is one Λ_n ($n =$ odd) in $[\varepsilon_{2m+1}(0), \varepsilon_{2m+2}(0)]$.

The cases of two consecutive zeros $\varepsilon_{2m}(\frac{\pi}{a})$ and $\varepsilon_{2m+1}(\frac{\pi}{a})$ of $D(\lambda) + 2$ can be similarly considered; we will obtain the conclusion that there is one and only one Λ_n ($n =$ even) in $[\varepsilon_{2m}(\frac{\pi}{a}), \varepsilon_{2m+1}(\frac{\pi}{a})]$.

Therefore, Λ_n starts occurring in $[\varepsilon_0(\frac{\pi}{a}), \varepsilon_1(\frac{\pi}{a})]$ and always occurs alternatively between $[\varepsilon_{2m}(\frac{\pi}{a}), \varepsilon_{2m+1}(\frac{\pi}{a})]$ or $[\varepsilon_{2m+1}(0), \varepsilon_{2m+2}(0)]$.

Equation (2.90) can be similarly proven. Each ν_n is the solution of the equation

$$p(a)\eta_1'(a, \nu_n) = 0, \tag{2.93}$$

and

$$\Phi_n(x) = \eta_1(x, \nu_n),$$

where $\eta_i(x, \lambda)$ and $p(x)\eta_i'(x, \lambda) (i = 1, 2)$ are defined in (2.53). Therefore, $\eta_1(x, \nu_n)$ has exactly n zeros in the interval $[0, a)$ and exactly n zeros in the interval $(0, a)$. That means $\eta_1(a, \nu_0) > 0$ and $\eta_1(a, \nu_{n=even}) > 0$, $\eta_1(a, \nu_{n=odd}) < 0$.

We have

$$D(\nu_n) = \eta_1(a, \nu_n) + p(a)\eta_2'(a, \nu_n) = \eta_1(a, \nu_n) + [\eta_1(a, \nu_n)]^{-1},$$

since $\eta_1(a, \nu_n)p(a)\eta_2'(a, \nu_n) = 1$ by (2.93). Therefore if $n =$ even, $\eta_1(a, \nu_n) > 0$, $D(\nu_n) > 2$; $n =$ odd, $\eta_1(a, \nu_n) < 0$, $D(\nu_n) < -2$;

Therefore, ν_n and ν_{n+1} are always in different forbidden ranges, if we consider the special cases in which $\varepsilon_{2m}(\frac{\pi}{a}) = \varepsilon_{2m+1}(\frac{\pi}{a})$ or $\varepsilon_{2m+1}(0) = \varepsilon_{2m+2}(0)$ as a band gap with the gap size being zero. ν_n starts to occur in $(-\infty, \varepsilon_0(0)]$ and then always occurs alternatively between $[\varepsilon_{2m}(\frac{\pi}{a}), \varepsilon_{2m+1}(\frac{\pi}{a})]$ or $[\varepsilon_{2m+1}(0), \varepsilon_{2m+2}(0)]$. □

Theorem 2.7 (Extended Theorem 3.1.2 in [7], Theorem 2.5.1 in [10], Theorems 13.7 and 13.8 in [8]) *As solutions of Eq. (2.2),*

(i) $\phi_0(0, x)$ *has no zero in* $[0, a]$.

(ii) $\phi_{2m+1}(0, x)$ *and* $\phi_{2m+2}(0, x)$ *have exactly* $2m + 2$ *zeros in* $[0, a)$.

(iii) $\phi_{2m}(\frac{\pi}{a}, x)$ *and* $\phi_{2m+1}(\frac{\pi}{a}, x)$ *have exactly* $2m + 1$ *zeros in* $[0, a)$.

Proof The Theorem can be proven by Theorem 2.2 and Eqs. (2.86) and (2.89).

(1) Since $\Psi_0(x)$ (i.e., $\eta_2(x, \Lambda_0)$) has no zero in $(0, a)$ and by (2.86) and (2.89) $\varepsilon_0(0) < \varepsilon_0(\frac{\pi}{a}) \leq \Lambda_0$, from Theorem 2.2, both $\phi_0(0, x)$ and $\phi_0(\frac{\pi}{a}, x)$ have no more than one zero in $[0, a)$. $\phi_0(0, x)$ as a periodic function can only have an even numbers of zeros in $[0, a)$, and $\phi_0(\frac{\pi}{a}, x)$ as a semi-periodic function can only have an odd number of zeros in $[0, a)$. Therefore, $\phi_0(0, x)$ must have no zero in $[0, a)$, and $\phi_0(\frac{\pi}{a}, x)$ must have exactly one zero in $[0, a)$.

(2) By (2.89) and (2.86) $\Lambda_{2m} \leq \varepsilon_{2m+1}(\frac{\pi}{a}) < \varepsilon_{2m+1}(0) \leq \Lambda_{2m+1}$, from Theorem 2.2 both $\phi_{2m+1}(\frac{\pi}{a}, x)$ and $\phi_{2m+1}(0, x)$ has at least $2m+1$ but no more than $2m+2$ zeros in $(0, a)$. $\phi_{2m+1}(\frac{\pi}{a}, x)$ as a semi-periodic function can only have an odd number of zeros in $[0, a)$, and $\phi_{2m+1}(0, x)$ as a periodic function can only have an even number of zeros in $[0, a)$. Thus $\phi_{2m+1}(\frac{\pi}{a}, x)$ must have exactly $2m+1$ zeros in $[0, a)$ and $\phi_{2m+1}(0, x)$ must have exactly $2m+2$ zeros in $[0, a)$.

(3) By (2.89) and (2.86) $\Lambda_{2m+1} \leq \varepsilon_{2m+2}(0) < \varepsilon_{2m+2}(\frac{\pi}{a}) \leq \Lambda_{2m+2}$, from Theorem 2.2 both $\phi_{2m+2}(0, x)$ and $\phi_{2m+2}(\frac{\pi}{a}, x)$ has at least $2m+1$ but no more than $2m+2$ zeros in $(0, a)$. $\phi_{2m+2}(0, x)$ as a periodic function can only have an even number of zeros in $[0, a)$, $\phi_{2m+2}(\frac{\pi}{a}, x)$ as a semi-periodic function can only have an odd number of zeros in $[0, a)$. Thus $\phi_{2m+2}(0, x)$ must have exactly $2m+2$ zeros in $[0, a)$, $\phi_{2m+2}(\frac{\pi}{a}, x)$ must have exactly $2m+1$ zeros in $[0, a)$.

The theorem is proven by combining items (1)–(3). □

Now, we consider an eigenvalue problem of (2.2) in $[\tau, \tau + a]$ for a real number τ under the boundary condition

$$y(\tau) = y(\tau + a) = 0. \tag{2.94}$$

The corresponding eigenvalues can be written as $\Lambda_{\tau,n}$.

Correspondingly there is an eigenvalue problem of (2.2) under the condition

$$p(\tau + a)y'(\tau + a) = p(\tau)y'(\tau) = 0, \tag{2.95}$$

the eigenvalues can be denoted by $\nu_{\tau,n}$.

Theorem 2.8 (Extended Theorem 3.1.3 in [7], Theorem 13.10 in [8]) *As functions of τ, the ranges of $\Lambda_{\tau,2m}$ are $[\varepsilon_{2m}(\frac{\pi}{a}), \varepsilon_{2m+1}(\frac{\pi}{a})]$ and the ranges of $\Lambda_{\tau,2m+1}$ are $[\varepsilon_{2m+1}(0), \varepsilon_{2m+2}(0)]$;*

As functions of τ, the ranges of $\nu_{\tau,2m+1}$ are $[\varepsilon_{2m}(\frac{\pi}{a}), \varepsilon_{2m+1}(\frac{\pi}{a})]$ and the ranges of $\nu_{\tau,2m+2}$ are $[\varepsilon_{2m+1}(0), \varepsilon_{2m+2}(0)]$.

Proof Since $\varepsilon_n(0)$ are the values of λ for which the corresponding solutions $\phi_n(0, x)$ of (2.2) are periodic, and $\varepsilon_n(\frac{\pi}{a})$ are the values of λ for which the corresponding solutions $\phi_n(\frac{\pi}{a}, x)$ of (2.2) are semi-periodic, $\varepsilon_n(0)$ and $\varepsilon_n(\frac{\pi}{a})$ will remain unchanged if the basic interval in (2.55) and (2.56) is changed from $[0, a]$ to $[\tau, \tau + a]$. Consequently, the conclusions of Theorem 2.6 will remain unchanged if the basic interval is changed from $[0, a]$ to $[\tau, \tau + a]$. Therefore, from Theorem 2.6, we have

$$\varepsilon_{2m}\left(\frac{\pi}{a}\right) \leq \Lambda_{\tau,2m} \leq \varepsilon_{2m+1}\left(\frac{\pi}{a}\right), \quad \varepsilon_{2m+1}(0) \leq \Lambda_{\tau,2m+1} \leq \varepsilon_{2m+2}(0). \tag{2.96}$$

From part (iii) of Theorem 2.7, both $\phi_{2m}(\frac{\pi}{a}, x)$ and $\phi_{2m+1}(\frac{\pi}{a}, x)$ have exactly $2m+1$ zeros in $[0, a)$. According to Theorem 2.2, the zeros of $\phi_{2m}(\frac{\pi}{a}, x)$ and $\phi_{2m+1}(\frac{\pi}{a}, x)$ are distributed alternatively. There is always one and only one zero of $\phi_{2m+1}(\frac{\pi}{a}, x)$ between two consecutive zeros of $\phi_{2m}(\frac{\pi}{a}, x)$, and there is always one and only one zero of $\phi_{2m}(\frac{\pi}{a}, x)$ between two consecutive zeros of $\phi_{2m+1}(\frac{\pi}{a}, x)$.

Suppose x_0 is any zero of $\phi_{2m}(\frac{\pi}{a}, x)$. Let $\tau = x_0$; then $\phi_{2m}(\frac{\pi}{a}, x)$ satisfies (2.94):

$$\phi_{2m}\left(\frac{\pi}{a}, \tau\right) = \phi_{2m}\left(\frac{\pi}{a}, \tau + a\right) = 0.$$

Again, from part (iii) of Theorem 2.7, $\phi_{2m}(\frac{\pi}{a}, x)$ has $2m$ zeros in the open interval $(x_0, x_0 + a)$; thus, $\phi_{2m}(\frac{\pi}{a}, x)$ is an eigenfunction of (2.2) under the boundary condition (2.94) corresponding to the eigenvalue $\Lambda_{x_0,2m}$: $\Lambda_{x_0,2m} = \varepsilon_{2m}(\frac{\pi}{a})$. Similarly, if x_1 is any zero of $\phi_{2m+1}(\frac{\pi}{a}, x)$, then $\phi_{2m+1}(\frac{\pi}{a}, x)$ is an eigenfunction of (2.2) under the boundary condition (2.94) corresponding to the eigenvalue $\Lambda_{x_1,2m}$: $\Lambda_{x_1,2m} = \varepsilon_{2m+1}(\frac{\pi}{a})$. Hence, as τ as the variable changes from $\tau = x_0$ to $\tau = x_1$, a zero of $\phi_{2m+1}(\frac{\pi}{a}, x)$ next to x_0, as a function of τ, $\Lambda_{\tau,2m}$ correspondingly and continuously changes from $\varepsilon_{2m}(\frac{\pi}{a})$ to $\varepsilon_{2m+1}(\frac{\pi}{a})$. Similarly, as τ as the variable changes from $\tau = x_1$ to $\tau = x_2$, the other zero of $\phi_{2m}(\frac{\pi}{a}, x)$ next to x_1, as a function of τ, $\Lambda_{\tau,2m}$ correspondingly and continuously changes back from $\varepsilon_{2m+1}(\frac{\pi}{a})$ to $\varepsilon_{2m}(\frac{\pi}{a})$. Therefore, as functions of τ, the ranges of $\Lambda_{\tau,2m}$ are $[\varepsilon_{2m}(\frac{\pi}{a}), \varepsilon_{2m+1}(\frac{\pi}{a})]$.

Similarly, we can obtain that as functions of τ, the ranges of $\Lambda_{\tau,2m+1}$ are $[\varepsilon_{2m+1}(0), \varepsilon_{2m+2}(0)]$. The first part of the theorem is proven.

Similarly, we have

$$\varepsilon_{2m}\left(\frac{\pi}{a}\right) \leq \nu_{\tau,2m+1} \leq \varepsilon_{2m+1}\left(\frac{\pi}{a}\right), \quad \varepsilon_{2m+1}(0) \leq \nu_{\tau,2m+2} \leq \varepsilon_{2m+2}(0), \quad \nu_{\tau,0} \leq \varepsilon_0(0). \tag{2.97}$$

If τ is not an isolated discontinuous point of $p(x)$, then the results for $\nu_{\tau,n}$ can be similarly obtained. Since, when $x = \tau$ is a turning point of the periodic or semi-periodic eigenfunction $\phi_n(k_g, x)$, $p(\tau + a)\phi_n'(k_g, \tau + a) = p(\tau)\phi_n'(k_g, \tau) = 0$ thus the equalities in Eq. (2.95) are attained.

In case if τ is an isolated discontinuous point of $p(x)$, then $\tau - \delta$ and $\tau + \delta$—the real number δ can be as small as needed—are not an isolated discontinuous point of $p(x)$, hence the argument in the previous paragraph works.

Since $p(x)y'(x)$ is continuous at any isolated discontinuous point of $p(x)$, $\nu_{\tau,n}$ is also continuous at any isolated discontinuous point τ of $p(x)$. Thus the second part of the theorem is proven. □

This theorem indicates that: *There is always one and only one eigenvalue* $\Lambda_{\tau,n}$ *of (2.2) under the boundary condition (2.94) in each gap* $[\varepsilon_{2m}(\frac{\pi}{a}), \varepsilon_{2m+1}(\frac{\pi}{a})]$ *if* $\varepsilon_{2m}(\frac{\pi}{a}) < \varepsilon_{2m+1}(\frac{\pi}{a})$ *or* $[\varepsilon_{2m+1}(0), \varepsilon_{2m+2}(0)]$ *if* $\varepsilon_{2m+1}(0) < \varepsilon_{2m+2}(0)$. *In some special cases when* $\varepsilon_{2m}(\frac{\pi}{a}) = \varepsilon_{2m+1}(\frac{\pi}{a})$ *or* $\varepsilon_{2m+1}(0) = \varepsilon_{2m+2}(0)$, *then we have* $\Lambda_{\tau,2m} = \varepsilon_{2m}(\frac{\pi}{a})$ *or* $\Lambda_{\tau,2m+1} = \varepsilon_{2m+1}(0)$.

A consequence of Theorem 2.8 is that *in general a one-dimensional Bloch function* $\phi_n(k, x)$ *does not have a zero except* $k = 0$ *or* $k = \frac{\pi}{a}$. Since if $\phi_n(k, x)$ has a zero at $x = x_0$, $\phi_n(k, x_0) = 0$, then we must have $\phi_n(k, x_0 + a) = 0$. According to Theorem 2.8, the corresponding eigenvalue $\Lambda_{\tau,n}$ must be in either $[\varepsilon_{2m}(\frac{\pi}{a}), \varepsilon_{2m+1}(\frac{\pi}{a})]$ or $[\varepsilon_{2m+1}(0), \varepsilon_{2m+2}(0)]$. Since $(\varepsilon_{2m}(\frac{\pi}{a}), \varepsilon_{2m+1}(\frac{\pi}{a}))$ and $(\varepsilon_{2m+1}(0), \varepsilon_{2m+2}(0))$ are band gaps, only the Bloch functions at a band-edge $\phi_{n\neq 0}(0, x)$ or $\phi_n(\frac{\pi}{a}, x)$ may have a zero.

Theorem 2.9 (Extended Theorem 3.2.2 in [7], Theorem 2.5.2 in [10]) *Any nontrivial solution of (2.2) with $\lambda \le \varepsilon_0(0)$ has at most one zero in $-\infty < x < +\infty$.*

Proof This theorem can be proven in two steps.

(1) From part (i) of Theorem 2.7, we know that $\phi_0(0, x)$, which is a nontrivial solution of (2.2) with $\lambda = \varepsilon_0(0)$, has no zero in $[0, a]$ and thus has no zero in $(-\infty, +\infty)$.

(2) If any nontrivial solution $y(x, \lambda)$ of (2.2) with $\lambda \le \varepsilon_0(0)$ had more than one zeros in $(-\infty, +\infty)$, $\phi_0(0, x)$ would have at least one zero between two zeros of $y(x, \lambda)$ by Theorem 2.2. This is contradictory to (1). □

The theory in this chapter could play a fundamental role in investigations of many physical problems on general one-dimensional crystals, including infinite crystals, semi-infinite crystals, and finite crystals. In the next two chapters, we will investigate electronic states in one-dimensional semi-infinite crystals and finite crystals. Theorems 2.7–2.9, especially the Theorem 2.8, play an essential role in the theory of electronic states in ideal one-dimensional crystals of finite length.

Based on the theory in this chapter, a corresponding theory for treating other one-dimensional crystals, including one-dimensional phononic crystals or photonic crystals will be presented in the Appendices.

References

1. F. Seitz, *The Modern Theory of Solids* (McGraw-Hill, New York, 1940)
2. H. Jones, *The Theory of Brillouin Zones and Electronic States in Crystals* (North-Holland, Amsterdam, 1960)
3. C. Kittel, *Introduction to Solid State Physics*, 7th edn. (Wiley, New York, 1996)
4. R.L. Kronig, W.G. Penney, Proc. Roy. Soc. Lond. Ser. A. **130**, 499 (1931)
5. H.A. Kramers, Physica **2**, 483 (1935)
6. I. Tamm, Phys. Z. Sowj. **1**, 733 (1932)
7. M.S.P. Eastham, *The Spectral Theory of Periodic Differential Equations* (Scottish Academic Press, Edinburgh, 1973). (and references therein)
8. J. Weidmann, *Spectral Theory of Ordinary Differential Operators* (Springer, Berlin, 1987)
9. A. Zettl, *Sturm-Liouville Theory* (American Mathematical Society, Providence, 2005) (In particular, p. 39)
10. B. Malcolm Brown, M.S.P. Eastham, K.M. Schmidt, *Periodic Differential Operators*. Operator Theory: Advances and Applications, vol. 230 (Springer, Heidelberg, 2013) (and references therein)
11. G. Teschl, *Ordinary Differential Equations and Dynamical Systems* (American Mathematical Society, Providence, 2012)
12. V.A. Yakubovich, V.M. Starzhinskii, *Linear Differential Equations with Periodic Coefficients*, vol. 1 (Chapter II) (Wiley, New York, 1975)
13. W. Magnus, S. Winkler, *Hill's Equation* (Interscience, New York, 1966)
14. W. Kohn, Phys. Rev. **115**, 809 (1959)
15. P.A. Kuchment, Bull. (New Ser.) Am. Math. Soc. **53**, 343 (2016)

Chapter 3
Surface States in One-Dimensional Semi-infinite Crystals

A one-dimensional semi-infinite crystal is the simplest periodic system with a boundary. Based on a Kronig–Penney model, Tamm was the first to find that the termination of the periodic potential due to the existence of a barrier at the boundary in a one-dimensional semi-infinite crystal can cause localized surface states to exist in band gaps below the barrier height [1]. Now after more than 70 years, the investigations of the properties of surface states and relevant physical and chemical processes have become an important field in solid state physics and chemistry [2–6]. Among the many surface states of various origins, the surface states caused purely by the termination of the crystal periodic potential are not only the simplest but also the most fundamental surface states. In this chapter, we present a general single-electron analysis on the existence and properties of surface states caused purely by the termination of the crystal periodic potential in one-dimensional semi-infinite crystals.

There are, basically, two different approaches for theoretical investigations of surface states: the potential approach and the atomic orbital or tight-binding approach [2, 7]. The potential approach is mainly developed by physicists. The crystal potential models treated in the investigations of surface states in one-dimensional semi-infinite crystals include the Kronig–Penney model [1], the nearly free electron model [8–10], the square potential model [10], the sinusoidal potential model [2, 7, 10, 11], and so forth. Much significant progress has been made. In particular, by using a sinusoidal crystal potential model, Levine [11] systematically investigated the surface states caused by a step barrier of variable barrier height at a variable location and obtained many interesting results. In this chapter, we investigate the properties of surface states in general one-dimensional semi-infinite crystals without using a specific crystal potential model and try to obtain a more general understanding of the problem. This chapter is organized as follows: In Sect. 3.1, we present the problem in a general way. In Sect. 3.2, two qualitative relations on the properties of surface states in one-dimensional semi-infinite crystals are presented. In Sect. 3.3, we consider the simplest cases where the barrier height outside the crystal is infinite. In Sect. 3.4 we consider cases where the barrier height is finite. In Sect. 3.5 we present a general theoretical

S.Y. Ren, *Electronic States in Crystals of Finite Size*, Springer Tracts in Modern Physics 270, DOI 10.1007/978-981-10-4718-3_3

approach to quantitatively investigate the existence and properties of surface states in semi-infinite one-dimensional crystals. In Sect. 3.6 are comparisons with previous work and discussions.

3.1 Basic Considerations

The Schrödinger differential equation for a one-dimensional periodic potential can be written as

$$- y''(x) + [v(x) - \lambda]y(x) = 0, \quad - \infty < x < +\infty, \tag{3.1}$$

where

$$v(x + a) = v(x)$$

is the periodic crystal potential. Equation (3.1) is a specific form of Eq. (2.2). The theory in Chap. 2 could be applied to treat (3.1). We assume that Eq. (3.1) is solved and all solutions are known. The eigenvalues are energy bands $\varepsilon_n(k)$, and the corresponding eigenfunctions are Bloch functions $\phi_n(k, x)$, where $n = 0, 1, 2, \ldots,$ and $-\frac{\pi}{a} < k \leq \frac{\pi}{a}$. We are mainly interested in cases where there is always a band gap between two consecutive energy bands. The band gaps of (3.1) are always located either at the center of the Brillouin zone $k = 0$ or the boundary of the Brillouin zone $k = \frac{\pi}{a}$. They can be ordered: The band gap $n = 0$ is the lowest band gap at $k = \frac{\pi}{a}$, the band gap $n = 1$ is the lowest band gap at $k = 0$, and so on.

Till now, most theoretical investigations of the basic physics of surface states were based on a semi-infinite crystal approach. It is usually assumed that the potential inside the crystal is the same as that in an infinite crystal. Based on this assumption, in general, a termination of the crystal periodic potential at the boundary of a semi-infinite one-dimensional crystal has two variables: the position of the termination τ and the potential outside the crystal $V_{out}(x)$.

For a one-dimensional semi-infinite crystal with a left boundary at τ, the potential can be written as

$$v(x, \tau) = \begin{cases} V_{out}(x), & x \leq \tau, \\ v(x), & x > \tau. \end{cases} \tag{3.2}$$

We are only interested in cases where outside the semi-infinite crystal there is a barrier; that is, $V_{out}(x)$ is always above $v(x)$ and the energy of the surface state Λ. We call a semi-infinite crystal given by (3.2) a right semi-infinite crystal, whereas a left semi-infinite crystal is defined by a periodic potential in $(-\infty, \tau)$. The energies Λ and eigenfunctions $\psi(x)$ of electronic states in a right semi-infinite crystal can be obtained as solutions of the Schrödinger differential equation

$$- \psi''(x) + [v(x) - \Lambda]\psi(x) = 0, \quad \tau < x < +\infty, \tag{3.3}$$

with a certain boundary condition determined by $V_{out}(x)$ at the boundary τ. A finite $V_{out}(x)$ will allow a small part of the electronic state to spill out of the semi-infinite crystal and thus make the boundary condition to become [12]

$$(\psi'/\psi)_{x=\tau} = \sigma, \tag{3.4}$$

where σ is a positive real number depending on $V_{out}(x)$ for a right semi-infinite one-dimensional crystal. σ will decrease monotonically as $V_{out}(x)$ decreases. Although $V_{out}(x)$ may have different forms, the effect of different $V_{out}(x)$ on the problem treated here can be simplified to be given by the effect of σ.

In general, there may be two different types of solutions for (3.3) and (3.4).

For any τ and V_{out}, inside each energy band of the infinite crystal, that is, for any specific Λ in $(\varepsilon_{2m}(0), \varepsilon_{2m}(\frac{\pi}{a}))$ or $(\varepsilon_{2m+1}(\frac{\pi}{a}), \varepsilon_{2m+1}(0))$ where $m = 0, 1, 2, \ldots$, there is always a solution of (3.3) and (3.4) inside the semi-infinite crystal that is

$$\psi_{n,k}(x) = c_1\phi_n(k, x) + c_2\phi_n(-k, x), \tag{3.5}$$

where $0 < k < \frac{\pi}{a}$ can be uniquely determined by $\varepsilon_n(k) = \Lambda$. This is because both $\phi_n(k, x)$ and $\phi_n(-k, x)$ are two linearly independent non-divergent solutions of (3.3) and, thus, one of their linear combinations can always satisfy the boundary condition (3.4). In this chapter, we are not interested in those states.

In contrast, for a specific Λ not inside a permitted energy band of the infinite crystal, there is only *one* non-divergent solution of (3.3), and this solution usually cannot also satisfy the boundary condition (3.4). Only when the non-divergent solution also satisfies (3.4) can we have a solution of both (3.3) and (3.4). If it exists, the energy Λ of such a solution is dependent on τ and V_{out}. The existence and properties of such a state are the primary interest of this chapter.

In general, a solution of (3.3) and (3.4) with an energy Λ not inside a permitted band is different from (3.5) if it exists. Its wave function inside a right semi-infinite one-dimensional crystal always has the form[1]

$$\psi(x, \Lambda) = e^{-\beta(\Lambda)x} f(x, \Lambda), \tag{3.6}$$

---

[1] Any solution $y(x)$ of (3.3) can always be expressed as a linear combination of two linearly independent solutions of (3.1):

$$y(x, \Lambda) = c_1 y_1(x, \Lambda) + c_2 y_2(x, \Lambda).$$

If Λ is *inside* a band gap at $k = 0$, the two linearly independent solutions y_1 and y_2 can be chosen as $y_1(x, \Lambda) = e^{\beta(\Lambda)x} p_1(x, \Lambda)$ and $y_2(x, \Lambda) = e^{-\beta(\Lambda)x} p_2(x, \Lambda)$, where $p_i(x, \Lambda)$ are periodic functions, and $\beta(\Lambda)$ is a positive number, all depending on Λ. For a non-divergent solution in right semi-infinite crystals, $c_1 = 0$ has to be chosen.

If Λ is *at* a band edge at $k = 0$, the two linearly independent solutions y_1 and y_2 can be chosen as $y_1(x, \Lambda) = p_1(x, \Lambda)$ and $y_2(x, \Lambda) = x\, p_1(x, \Lambda) + p_2(x, \Lambda)$, where $p_i(x, \Lambda)$ are periodic functions depending on Λ. For a non-divergent solution in semi-infinite crystals, we have to choose $c_2 = 0$.

The combination of the two requirements leads to (3.6) for a band gap at $k = 0$. Similar arguments can be applied to a band gap at $k = \frac{\pi}{a}$ as well.

where $\beta(\Lambda) \geq 0$, and $f(x, \Lambda)$ is a periodic function $f(x + a, \Lambda) = f(x, \Lambda)$ if Λ is in a band gap or at a band edge at the center of the Brillouin zone $k = 0$, or a semi-periodic function $f(x + a, \Lambda) = -f(x, \Lambda)$ if Λ is in a band gap or at a band edge at the boundary of the Brillouin zone $k = \frac{\pi}{a}$. A surface state located near the boundary τ has $\beta(\Lambda) > 0$. On the other hand, a state with $\beta(\Lambda) = 0$ can also be a solution of (3.3) and (3.4) for some specific τ and V_{out}. It is easy to see that $\psi(x, \Lambda)$ in (3.6) can always be chosen as a real function.[2]

Correspondingly, inside a left semi-infinite one-dimensional crystal, such a state always has the form

$$\psi(x, \Lambda) = e^{\beta(\Lambda)x} f(x, \Lambda). \tag{3.6a}$$

3.2 Two Qualitative Relations

If a localized surface state exists in a specific band gap in a right semi-infinite crystal under a specific boundary τ and a specific $V_{out}(x)$, concerning how its energy depends on τ and V_{out}, the following two Eqs. (3.7) and (3.8) can be proven with the help of the Hellmann–Feynman theorem [13]:

1. For the energy Λ_n of a surface state in a specific nth band gap,

$$\frac{\partial}{\partial \sigma}\Lambda_n > 0. \tag{3.7}$$

The equation can be proven in two steps.

(1) In the simple cases where $V_{out}(x) = V_{out}$ is a step barrier. Define a new potential

$$\tilde{V}(x, \eta) = (1 - \eta)v_0(x, \tau) + \eta\, v_1(x, \tau),$$

where

$$v_i(x, \tau) = \begin{cases} V_i, & x \leq \tau, \\ v(x), & x > \tau; \end{cases}$$

$V_1 = V_0 + \delta V$ and δV is an infinitesimal positive number. According to the Hellmann–Feynman theorem [13], for an eigenvalue $\tilde{\Lambda}_n(\eta)$ of a localized surface state $|\,\rangle_n$ of the Hamiltonian $\tilde{H} = T + \tilde{V}(x, \eta)$, where T is the kinetic energy operator, we have

$$\frac{\partial \tilde{\Lambda}_n(\eta)}{\partial \eta} = \left\langle \frac{\partial \tilde{H}(\eta)}{\partial \eta} \right\rangle_n = \left\langle \frac{\partial \tilde{V}(\eta)}{\partial \eta} \right\rangle_n = \langle v_1(x, \tau) - v_0(x, \tau)\rangle_n > 0,$$

[2] $\psi^*(x, \Lambda)$ is also a solution of (3.3) and (3.4).

since $V_1 > V_0$. Thus, $\tilde{\Lambda}_n(\eta)$ is a monotonic increasing function of η. However, $\tilde{V}(x, 0) = v_0(x, \tau)$; thus, $\tilde{\Lambda}_n(0) = \Lambda_n(\tau, V_{out} = V_0)$, whereas $\tilde{V}(x, 1) = v_1(x, \tau)$ and thus $\tilde{\Lambda}_n(1) = \Lambda_n(\tau, V_{out} = V_1)$. Therefore, $\Lambda_n(\tau, V_{out} = V_1) > \Lambda_n(\tau, V_{out} = V_0)$; that is,

$$\frac{\partial \Lambda_n}{\partial V_{out}} > 0.$$

Obviously $\partial V_{out}/\partial \sigma > 0$ for a right semi-infinite crystal; consequently,

$$\frac{\partial \Lambda_n}{\partial \sigma} = \frac{\partial \Lambda_n}{\partial V_{out}} \frac{\partial V_{out}}{\partial \sigma} > 0$$

for those simple cases where $V_{out}(x)$ is a step barrier. This case is a special case of (3.7).

(2) Since for a specific $v(x)$, Λ_n is a function of only τ and σ, (3.7) should also be true for more $V_{out}(x)$ in general.

A similar discussion for a left semi-infinite crystal can lead to the statement that if a surface state solution of energy Λ_n exists in a band gap in a left semi-infinite crystal with periodic potential in $(-\infty, \tau)$, then

$$\partial \Lambda_n/\partial \sigma < 0, \tag{3.7a}$$

just the opposite of (3.7), because of that for a left semi-infinite crystal, σ is a negative real number depending on V_{out} and thus $\partial V_{out}/\partial \sigma < 0$.

2. The energy Λ_n of a surface state in a right semi-infinite crystal increases as τ increases, or

$$\frac{\partial}{\partial \tau} \Lambda_n > 0. \tag{3.8}$$

Suppose a surface state with an energy $\Lambda_n(\tau_0, V_{out})$ exists in the nth band gap for a specific $\tau = \tau_0$ and V_{out}. Now, consider that $\tau_1 = \tau_0 + \delta\tau$, where $\delta\tau$ is an infinitesimal positive real number. We can define a new potential

$$\tilde{V}(x, \eta) = (1 - \eta) v(x, \tau_0) + \eta\, v(x, \tau_1),$$

where

$$v(x, \tau) = \begin{cases} V_{out}, & x \leq \tau, \\ v(x), & x > \tau, \end{cases}$$

and a new Hamiltonian $\tilde{H}(\eta) = T + \tilde{V}(x, \eta)$, where T is the kinetic energy operator. According to the Hellmann–Feynman theorem [13], for an eigenvalue $\tilde{\Lambda}_n(\eta)$ of a localized surface state $| \rangle_n$ of $\tilde{H}(\eta)$, we have

$$\frac{\partial \tilde{\Lambda}(\eta)}{\partial \eta} = \left\langle \frac{\partial \tilde{H}(\eta)}{\partial \eta} \right\rangle_n = \left\langle \frac{\partial \tilde{V}(\eta)}{\partial \eta} \right\rangle_n = \langle v(x, \tau_1) - v(x, \tau_0) \rangle_n > 0$$

since $V_{out} > v(x)$, thus $v(x, \tau_1) - v(x, \tau_0) > 0$. Hence, $\tilde{\Lambda}_n(\eta)$ is a monotonic increasing function of η. However, $\tilde{V}(x, 0) = v(x, \tau_0)$, thus, $\tilde{\Lambda}_n(0) = \Lambda_n(\tau_0, V_{out})$, whereas $\tilde{V}(x, 1) = v(x, \tau_1)$, thus, $\tilde{\Lambda}_n(1) = \Lambda_n(\tau_1, V_{out})$. Therefore, $\Lambda_n(\tau_1, V_{out}) > \Lambda_n(\tau_0, V_{out})$; that is $\partial \Lambda_n / \partial \tau > 0$.

A similar discussion for a left semi-infinite crystal can lead to the interesting point that if a surface state solution of energy Λ_n exists in the same band gap in a left semi-infinite crystal with periodic potential in $(-\infty, \tau)$, then $\partial \Lambda_n / \partial(-\tau) > 0$ is true. In other words,

$$\frac{\partial}{\partial \tau} \Lambda_n < 0, \tag{3.8a}$$

just the opposite of (3.8).

These two Eqs. (3.7) and (3.8) on the properties of the surface states in one-dimensional semi-infinite crystal should be true in general for any crystal potential $v(x)$, the crystal boundary τ and the barrier potential outside the crystal $V_{out}(\tau) > v(\tau)$.[3]

3.3 Surface States in Ideal Semi-infinite Crystals

The significance of Eq. (3.8) and its consequence can be more clearly seen if we consider the ideal semi-infinite crystals where the potential outside the crystal is $V_{out}(x) = +\infty$ and thus (3.4) becomes

$$\psi(x, \Lambda)|_{x=\tau} = 0. \tag{3.9}$$

The solutions of (3.3) and (3.9) can be investigated with the help of the theorems in Chap. 2 regarding the *zeros* of solutions of one-dimensional Schrödinger differential equations with a periodic potential.

First, as a consequence of (3.9) and Theorem 2.8, there is *at most* one solution[4] of (3.3) and (3.9) in each band gap of (3.1).

---

[3]The investigations in Appendix D indicates that the properties of surface states/modes in more general one-dimensional semi-infinite crystals, including electronic crystals, photonic crystals, and phononic crystals are similar: The dependencies of the energy of a surface state/mode on the boundary location and the boundary condition in a general one-dimensional semi-infinite crystals can be analytically obtained by a periodic Sturm–Liouville theory approach. The results obtained there can be considered as a further quantitation and extension of Eqs. (3.7) and (3.8).

[4]A function (3.6) satisfying (3.9) must satisfy $\psi(\tau + a, \Lambda) = 0$ as well. According to Theorem 2.8, for any real τ there is always one and only one solution of (3.1) in each gap for which $y(\tau, \lambda) = y(\tau + a, \lambda) = 0$. However, such a solution may or may not have the form of (3.6): It may have the form of $y(x, \lambda) = e^{\beta(\lambda)x} f(x, \lambda)$. Only when such a solution has the form of (3.6) is it a solution of (3.3) and (3.9).

Now, we consider a specific nth band gap at $k = k_g$, where either $k_g = 0$ for $n = 1, 3, 5, \ldots$ or $k_g = \frac{\pi}{a}$ for $n = 0, 2, 4, \ldots$. According to Theorem 2.7, the two band edge wave functions $\phi_n(k_g, x)$ and $\phi_{n+1}(k_g, x)$ have exactly $n + 1$ zeros for x in a potential period $[0, a)$. The locations of these zeros are determined by the crystal potential $v(x)$. According to Theorem 2.2, the zeros of $\phi_n(k_g, x)$ and $\phi_{n+1}(k_g, x)$ must be distributed alternatively: There is always one and only one zero of $\phi_{n+1}(k_g, x)$ between two consecutive zeros of $\phi_n(k_g, x)$, and there is always one and only one zero of $\phi_n(k_g, x)$ between two consecutive zeros of $\phi_{n+1}(k_g, x)$.

If τ is at any one of these zeros, then a solution of (3.3) and (3.9) is simple: The corresponding band edge wave function $\phi_n(k_g, x)$ or $\phi_{n+1}(k_g, x)$ satisfies both (3.3) and (3.9) and thus is a solution $\psi(x, \Lambda)$ of (3.3) and (3.9) corresponding to that band gap, with the eigenvalue $\Lambda_n(\tau)$ equal to the band edge energy $\varepsilon_n(k_g)$ or $\varepsilon_{n+1}(k_g)$. The semi-infinite semiconductor has a band edge state solution for this specific band gap n. We can use a label $M(n)$ to express the set of all zeros of $\phi_n(k_g, x)$ and $\phi_{n+1}(k_g, x)$. In the interval $[0, a)$ – where 0 can be chosen to be any specific zero of $\phi_n(k_g, x)$ – the set $M(n)$ contains $n + 1$ zeros of $\phi_n(k_g, x)$ and $n + 1$ zeros of $\phi_{n+1}(k_g, x)$.

If τ is not a zero of either $\phi_n(k_g, x)$ or $\phi_{n+1}(k_g, x)$, it must be between a zero of $\phi_n(k_g, x)$ and a zero of $\phi_{n+1}(k_g, x)$: Suppose $x_{l,n}$ and $x_{r,n}$ are two consecutive zeros of $\phi_n(k_g, x)$, on the left and the right of τ respectively, and suppose $x_{m,n+1}$ is the zero of $\phi_{n+1}(k_g, x)$ in the interval $(x_{l,n}, x_{r,n})$. Then τ must be either in the interval $(x_{l,n}, x_{m,n+1})$ or in the interval $(x_{m,n+1}, x_{r,n})$.

Because $\Lambda_n(x_{l,n}) = \Lambda_n(x_{r,n}) = \varepsilon_n(k_g)$ and $\Lambda_n(x_{m,n+1}) = \varepsilon_{n+1}(k_g)$, when τ increases from $x_{l,n}$ to $x_{m,n+1}$, $\Lambda_n(\tau)$ as a continuous function of τ goes up from $\varepsilon_n(k_g)$ to $\varepsilon_{n+1}(k_g)$. Therefore, when τ is in the interval $(x_{l,n}, x_{m,n+1})$, (3.8) is true, i.e., a surface state solution of (3.3) and (3.9) may exist in the right semi-infinite crystal. We can use a label $L(n)$ to express the set of all such points. In the interval $[0, a)$, the set $L(n)$ contains $n + 1$ sub-open-intervals, since from each of the $n + 1$ zeros of $\phi_n(k_g, x)$ in the interval $[0, a)$ one can obtain such an open interval in which (3.8) is true. As an example, in Fig. 3.1 is shown $\Lambda_1(\tau)$ (the energy of the surface state in the lowest band gap at $k = 0$) as a function of τ in an interval of length a for a right semi-infinite crystal.

We can also use a label $R(n)$ to express the set of all points in the interval $(x_{m,n+1}, x_{r,n})$. In the interval $[0, a)$, the set $R(n)$ also contains $n + 1$ sub-open-intervals. It is easy to see that when τ is in any such interval $(x_{m,n+1}, x_{r,n})$, there is no solution for (3.3) and (3.9). If there were a solution, then $\Lambda_n(\tau)$ as a continuous function of τ would go down from $\varepsilon_{n+1}(k_g)$ to $\varepsilon_n(k_g)$ when τ goes from $x_{m,n+1}$ to $x_{r,n}$. It means that $\Lambda_n(\tau)$ would decrease as τ increases, this behavior is contradictory to (3.8). However, when τ is in the interval $(x_{m,n+1}, x_{r,n})$, a surface state solution can exist in the left semi-infinite crystal $(-\infty, \tau)$ since then (3.8a) is true.[5] Figure 3.2 shows $\Lambda_1(\tau)$ as a function of τ in the interval $[0, a)$ for a left semi-infinite crystal. Note in both Figs. 3.1 and 3.2, there are regions in which there is no corresponding Λ_1 for a τ, indicating that no surface state solution exists in the right semi-infinite

[5]Therefore, $\partial\Lambda_n/\partial\tau > 0$ when τ is in $L(n)$, $\partial\Lambda_n/\partial\tau < 0$ when τ is in $R(n)$, and $\partial\Lambda_n/\partial\tau = 0$ when τ is in $M(n)$.

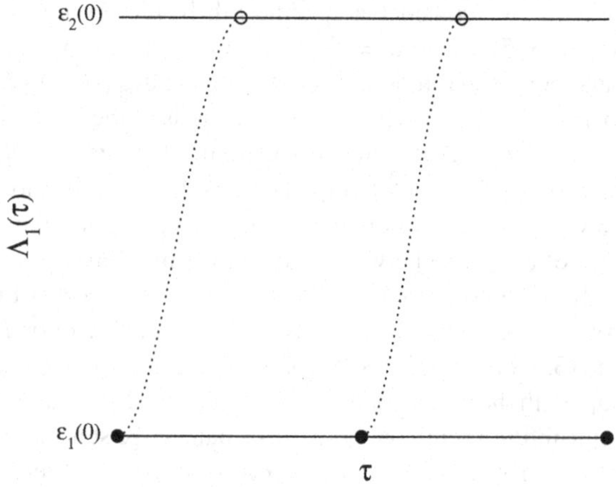

Fig. 3.1 $\Lambda_1(\tau)$ as a function of τ in the interval $[0, a]$ for a right semi-infinite crystal with periodic potential in $(\tau, +\infty)$. Zeros of $\phi_1(0, x)$ are shown as *solid circles*, and zeros of $\phi_2(0, x)$ are shown as *open circles*. The *dotted lines* indicate that a surface state exists in the semi-infinite crystal if τ is in the corresponding regions

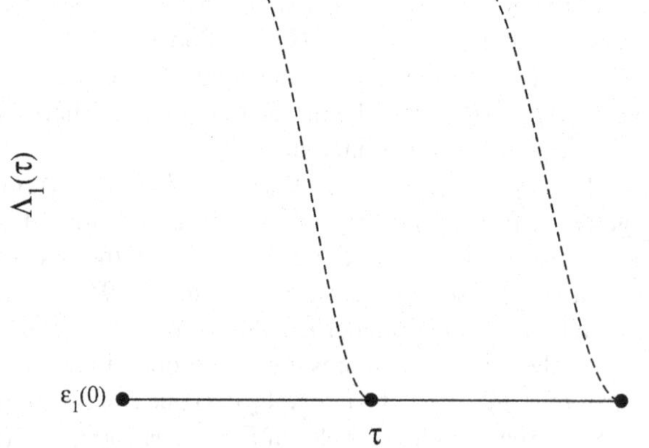

Fig. 3.2 $\Lambda_1(\tau)$ as a function of τ in the interval $[0, a]$ for a left semi-infinite crystal with periodic potential in $(-\infty, \tau)$. Zeros of $\phi_1(0, x)$ are shown as *solid circles*, and zeros of $\phi_2(0, x)$ are shown as *open circles*. The *dashed lines* indicate that a surface state exists in the semi-infinite crystal if τ is in the corresponding regions

crystal or in the left semi-infinite crystal for that τ. Any τ in the interval $[0, a)$ must belong to one set of either $L(n)$, $M(n)$, or $R(n)$ for any specific band gap n. Therefore, for $V_{out} = +\infty$, it is not that a termination of the periodic potential at any τ in a potential period interval $[0, a)$ may cause a surface state existing in a specific band gap in the right semi-infinite one-dimensional crystal or in the left semi-infinite one-dimensional crystal.

Therefore, we have seen that there are two seemingly different types of solutions for (3.3) and (3.9) in a band gap: a band edge state or a surface state.[6] Essentially

[6]Note that due to (3.6) if τ is in either $L(n)$ or $M(n)$, a solution of (3.3) and (3.9) always has

$$\psi(x, \Lambda)|_{x=\tau+Na} = 0$$

if N is a positive integer; that is, a solution in the right semi-infinite crystal is also a solution in a finite crystal of length Na. Since this equation is true for any integer N; therefore, the energy Λ of they are not very different: A band edge state can also be considered merely as a particular surface-like state with its energy equal to a band edge energy and thus its decay factor in (3.6) is $\beta(\Lambda) = 0$.

3.4 Cases Where V_{out} Is Finite

Now, we consider cases where V_{out} is finite. For a finite V_{out}, the boundary condition (3.4) rather than (3.9) should be used for a right semi-infinite one-dimensional crystal. Equation (3.9) corresponds to $\sigma = +\infty$, and σ monotonically decreases as V_{out} decreases.

Since (3.6) is a general form of a solution of (3.3) and (3.4) in a band gap, from (3.4) and (3.6) we have

$$\left.\frac{f'(x, \Lambda)}{f(x, \Lambda)}\right|_{x=\tau} - \beta(\Lambda) = \sigma. \tag{3.10}$$

Unlike the simplest cases discussed in Sect. 3.3, now the intervals in which a termination boundary τ can cause a surface state in a band gap will depend on $V_{out}(x)$ and, consequently, the corresponding $\Lambda_n(\tau) - \tau$ curve in a right semi-infinite crystal such as shown in Fig. 3.1 will move to the right.

We still consider the lowest band gap at $k = 0$ as an example. For a solution of (3.3) and (3.4) with the energy at the lower band edge $\varepsilon_1(0)$, (3.10) becomes

$$\left.\frac{\phi_1'(0, x)}{\phi_1(0, x)}\right|_{x=\tau} = \sigma_{\varepsilon_1(0)} \tag{3.11}$$

by noting that $\beta(\Lambda) = 0$ in (3.6) for a state with an energy of a band edge and that σ depends on the energy of the state for a specific V_{out}. $\frac{\phi_1'(0,x)}{\phi_1(0,x)}$ is determined by the periodic potential $v(x)$. In Fig. 3.3 is shown a typical $\frac{\phi_1'(0,x)}{\phi_1(0,x)}$ as a function of x. Given a specific finite V_{out} and thus a specific positive $\sigma_{\varepsilon_1(0)}$, on the right of an (any) zero $x_{a,1}$ (solid circle) of the lower band edge wave function $\phi_1(0, x)$, there is always a specific point $x_{a,1} + \delta_{a,1}$ where $\delta_{a,1} > 0$, which makes

$$\frac{\phi_1'(0, x_{a,1} + \delta_{a,1})}{\phi_1(0, x_{a,1} + \delta_{a,1})} = \sigma_{\varepsilon_1(0)}$$

true, as is shown by the short-dashed lines in Fig. 3.3. The smaller V_{out} is, the smaller $\sigma_{\varepsilon_1(0)}$ is and the larger $\delta_{a,1}$ is, as can be seen in Fig. 3.3.

Depending on the crystal potential $v(x)$, the details of Fig. 3.3 might be more or less different, such as the shapes of $\frac{\phi_1'(0,x)}{\phi_1(0,x)}$ and the locations of the zeros of it.

(Footnote 6 continued)
such a state is independent of the crystal length. Similarly, if τ is in $R(n)$ or $M(n)$, such a solution in the left semi-infinite crystal is also a solution in a finite crystal of length Na and its energy Λ is independent of the crystal length.

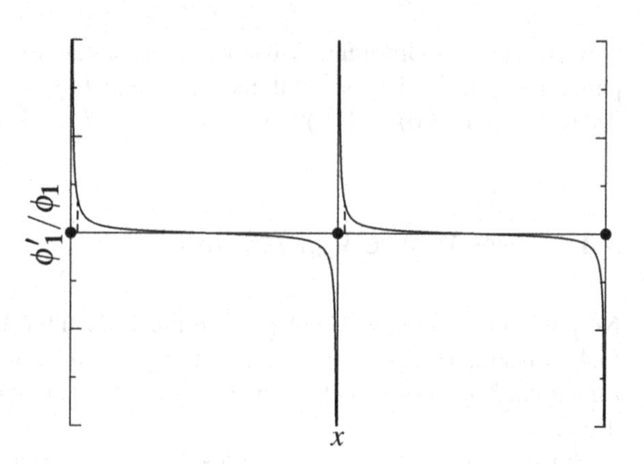

Fig. 3.3 $\frac{\phi_1'(0,x)}{\phi_1(0,x)}$ as function of x in the same interval $[0, a]$. Zeros of $\phi_1(0, x)$ are shown as *solid circles*. The two *short lines* indicate the two τ for which $\frac{\phi_1'(0,\tau)}{\phi_1(0,\tau)} = \sigma$ is satisfied. Each τ is on the right of a zero of $\phi_1(0, x)$ for a positive finite σ

However, there is always, at least, one zero of $\phi_1'(0, x)$ and thus one zero of $\frac{\phi_1'(0,x)}{\phi_1(0,x)}$ between two consecutive zeros of $\phi_1(0, x)$. Therefore, the analysis given here is valid for any $v(x)$ in general: For a finite V_{out} and thus a finite $\sigma_{\varepsilon_1(0)}$, there is always an $x = x_{a,1} + \delta_{a,1}$ for which $\delta_{a,1} > 0$, which makes (3.11) true. Thus, there will be a solution of (3.3) and (3.4) with energy $\Lambda_1 = \varepsilon_1(0)$ in the right semi-infinite crystal with the boundary $\tau = x_{a,1} + \delta_{a,1}$ and the potential barrier V_{out}. Its wave function $\psi(x, \Lambda_1) = \phi_1(0, x)$ in the crystal.

A similar analysis can be applied to the upper band edge as well: When V_{out} decreases from $+\infty$ to a specific finite value, the boundary τ for a solution of (3.3) and (3.4) with $\Lambda_1 = \varepsilon_2(0)$ (the upper band edge energy of the lowest band gap at $k = 0$) will move to the right, from $\tau = x_{a,2}$ to a specific $\tau = x_{a,2} + \delta_{a,2}$, in which $\delta_{a,2} > 0$. Similar analysis can also be applied to each surface state in that band gap; thus, instead of the $\Lambda_1(\tau) - \tau$ curves in Fig. 3.1, we have $\Lambda_1(\tau) - \tau$ curves for a finite V_{out}, as shown in Fig. 3.4: The $\Lambda_1(\tau) - \tau$ curves in Fig. 3.4 are on the right side of the $\Lambda_1(\tau) - \tau$ curves in Fig. 3.1; that how far away it is depends on the barrier potential V_{out} (and the crystal potential $v(x)$).

Corresponding to Fig. 3.2, in Fig. 3.5 are shown the $\Lambda_1(\tau) - \tau$ curves in a left semi-infinite crystal with periodic potential in $(-\infty, \tau)$ for a finite V_{out}, for which $\frac{\psi'}{\psi} = \sigma < 0$ is the boundary condition. Thus, the curves in Fig. 3.5 are always on the left side of the corresponding curves in Fig. 3.2.

Therefore, the effect of a finite V_{out} is to change the positions (and probably, the shapes) of the $\Lambda_1(\tau) - \tau$ curves. This point can be understood by the Eqs. (3.7) and (3.8) in Sect. 3.2: Because for a right semi-infinite crystal, both $\frac{\partial}{\partial\sigma}\Lambda_n > 0$ and $\frac{\partial}{\partial\tau}\Lambda_n > 0$ are true for an existing surface state in the nth gap, a τ increase in the right semi-infinite crystal is needed to compensate for the effect of a V_{out} decrease to keep a fixed Λ_1. On the contrary, a τ decrease in the left semi-infinite crystal is needed to compensate the effect of a V_{out} decrease.

Again, in both Figs. 3.4 and 3.5, there are regions in which there is no Λ_1 for a τ, indicating that no surface state solution exists in the right semi-infinite crystal or in the left semi-infinite crystal for that τ. Therefore, for a specific finite V_{out}, again

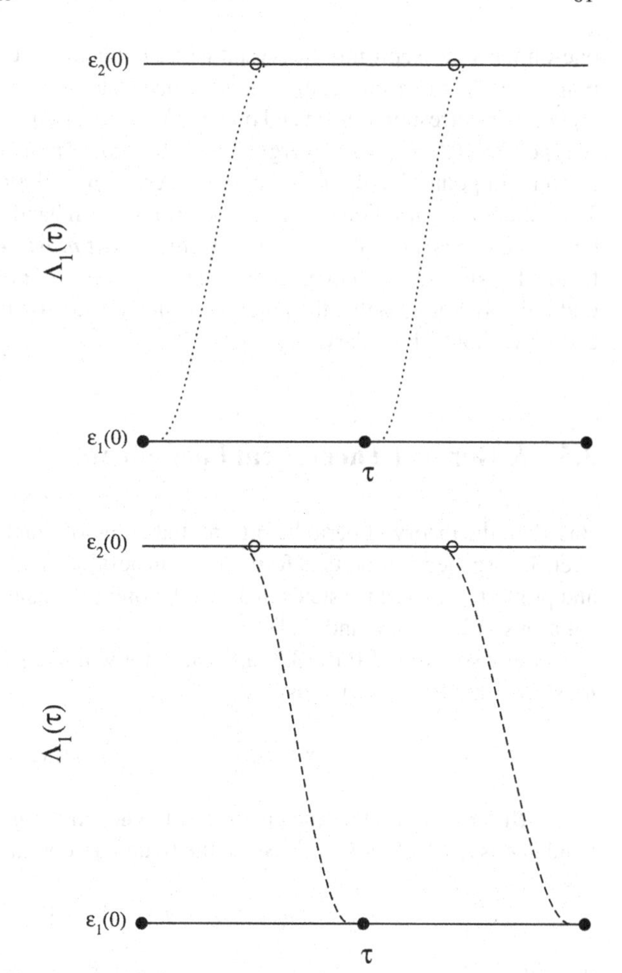

Fig. 3.4 $\Lambda_1(\tau)$ as a function of τ in the same interval $[0, a]$ for a finite $V_{out}(x)$ for a right semi-infinite crystal with periodic potential in $(\tau, +\infty)$. Zeros of $\phi_1(0, x)$ are shown as *solid circles*, and zeros of $\phi_2(0, x)$ are shown as *open circles*. The *dotted lines* indicate that a surface state exists in the semi-infinite crystal if τ is in the corresponding regions. Note that no surface state exists when τ is in the near neighborhood of a *solid circle*

Fig. 3.5 $\Lambda_1(\tau)$ as a function of τ in the same interval $[0, a]$ for a finite $V_{out}(x)$ for a left semi-infinite crystal with periodic potential in $(-\infty, \tau)$. Zeros of $\phi_1(0, x)$ are shown as *solid circles*, and zeros of $\phi_2(0, x)$ are shown as *open circles*. The *dashed lines* indicate that a surface state exists in the semi-infinite crystal if τ is in the corresponding regions. Note that no surface state exists when τ is in the near neighborhood of a *solid circle*

it is not that a termination of the periodic potential at any τ in a potential period interval $[0, a)$ may cause a surface state existing in this band gap in the right semi-infinite one-dimensional crystal or in the left semi-infinite one-dimensional crystal. A consequence of the moving of $\Lambda_1(\tau) - \tau$ curves due to a finite V_{out} is that there is no longer a surface state or a band edge state possible for a τ in the near neighborhood of a zero of the *lower* band edge wave function, as can be seen in Figs. 3.4 and 3.5.

The analysis presented here for the lowest band gap at $k = 0$ is valid for a general crystal potential $v(x)$ and outside potential V_{out}. A similar analysis can also be applied to other band gaps and results corresponding to Figs. 3.1, 3.2, 3.3, 3.4 and 3.5 can be obtained.

The left side of (3.10) is determined by $v(x)$ and τ, and the right side is determined by V_{out}. After $v(x)$ and V_{out} are given, $\Lambda_n(\tau)$ as a function of τ such as shown in Fig. 3.4 is uniquely determined. Therefore, there are three possibilities for a one-dimensional semi-infinite crystal with a boundary τ. If τ corresponds to an inside point of a $\Lambda_n(\tau)$ - τ curve, there is a surface state of type (3.6) ($\beta > 0$) with energy $\Lambda_n(\tau)$. If τ corresponds to an end point of a $\Lambda_n(\tau)$ - τ curve, there is a band edge state of type (3.6) ($\beta = 0$) with energy $\Lambda_n(\tau)$. If τ corresponds to neither an inside point nor an end point of a $\Lambda_n(\tau)$ - τ curve, There is no solution of Eqs. (3.3) and (3.4). Thus, there is *at most* one state of type (3.6) in each band gap in a right semi-infinite one-dimensional crystal. Consequently, there is *at most* one surface state ($\beta > 0$) or band edge state ($\beta = 0$) in each band gap in a semi-infinite one-dimensional crystal under the assumption that the potential inside the crystal is the same as in an infinite crystal without a boundary.

3.5 A General Theoretical Formalism

Based on the theory of periodic differential equations and the basic consideration in Sect. 3.1, a general theoretical formalism can be developed to investigate the existence and properties of surface states in a specific one-dimensional semi-infinite crystal as solutions of Eqs. (3.3) and (3.4).

For any solution of Eqs. (3.3) and (3.4) not within a permitted band, the solution must have the form (3.6), thus

$$\psi(x+a,\Lambda) = \pm e^{-\beta(\Lambda)a}\psi(x,\Lambda),$$

where $\beta(\Lambda) \geq 0$, whether the positive or the negative sign is chosen depends on the band gap is at $k = 0$ or $k = \frac{\pi}{a}$. Since the boundary condition (3.4) can be written as

$$\sigma\psi(\tau,\Lambda) - \psi'(\tau,\Lambda) = 0,$$

the following equation is *a necessary* condition for the existing of such a solution:

$$\sigma\psi(\tau+a,\Lambda) - \psi'(\tau+a,\Lambda) = \sigma\psi(\tau,\Lambda) - \psi'(\tau,\Lambda) = 0. \qquad (3.12)$$

Inside the semi-infinite crystal any solution ψ of Eq. (3.3) can be expressed as a linear combination of two linearly independent normalized solutions η_1 and η_2 of Eq. (3.1):

$$\psi(x,\lambda) = c_1\eta_1(x,\lambda) + c_2\eta_2(x,\lambda), \qquad x \geq \tau, \qquad (3.13)$$

where $\eta_1(x,\lambda)$ and $\eta_2(x,\lambda)$ are two linearly independent normalized solutions of Eq. (3.1) satisfying that

$$\eta_1(\tau,\lambda) = 1,\ \eta_1'(\tau,\lambda) = 0; \quad \eta_2(\tau,\lambda) = 0,\ \eta_2'(\tau,\lambda) = 1. \qquad (3.14)$$

Equation (3.12) becomes

$$\begin{aligned} \sigma[c_1\eta_1(\tau+a,\Lambda)+c_2\eta_2(\tau+a,\Lambda)]-[c_1\eta_1'(\tau+a,\Lambda)+c_2\eta_2'(\tau+a,\Lambda)]=0;\\ \sigma c_1-c_2=0. \end{aligned} \tag{3.15}$$

The condition that c_1 and c_2 in Eq. (3.15) are not both zero is

$$\sigma^2\eta_2(\tau+a,\Lambda)+\sigma[\eta_1(\tau+a,\Lambda)-\eta_2'(\tau+a,\Lambda)]-\eta_1'(\tau+a,\Lambda)=0.$$

It can be solved as

$$\sigma=\frac{-\eta_1(\tau+a,\Lambda)+\eta_2'(\tau+a,\Lambda)\pm\sqrt{D(\Lambda)^2-4}}{2\,\eta_2(\tau+a,\Lambda)}, \tag{3.16}$$

where $D(\Lambda)=\eta_1(\tau+a,\Lambda)+\eta_2'(\tau+a,\Lambda)$ is the discriminant of Eq. (3.1).[7]

Using Eq. (3.16) in Eq. (3.15), we obtain c_2/c_1, and then the ratio P of $\psi(\tau+a,\Lambda)$ over $\psi(\tau,\Lambda)$ is,

$$P=\frac{\psi(\tau+a,\Lambda)}{\psi(\tau,\Lambda)}=\frac{1}{2}[D(\Lambda)\pm\sqrt{D(\Lambda)^2-4}] \tag{3.17}$$

from Eqs. (3.13) and (3.14). There are three possibilities: (a) $0<|P|<1$; (b) $|P|=1$; (c) $|P|>1$. The case (a) corresponds to an oscillatory decreasing solution in the $+x$ direction $\psi(x,\Lambda)$ of Eqs. (3.3) and (3.4) and it is a surface state localized near the boundary τ of the semi-infinite crystal; The case (b) corresponds to that $\psi(x,\Lambda)$ in Eq. (3.13) is a band edge state of Eq. (3.1); The case (c) corresponds to an oscillatory increasing $\psi(x,\Lambda)$ in the $+x$ direction. Such a $\psi(x,\Lambda)$ could not exist as a solution of Eqs. (3.3) and (3.4).

Therefore, to have a solution corresponding to $0<|P|\le 1$ in (3.17), the minus (plus) sign in (3.17) thus the minus (plus) sign in (3.16) should be chosen if $D(\Lambda)\ge 2$ ($D(\Lambda)\le -2$). Consequently from Eqs. (3.16) and (3.17) the existence of a surface state in a band gap is determined by

$$\sigma=\frac{-\eta_1(\tau+a,\Lambda)+\eta_2'(\tau+a,\Lambda)+\sqrt{D(\Lambda)^2-4}}{2\,\eta_2(\tau+a,\Lambda)} \tag{3.18}$$

for a band gap at $k_x=\pi/a$, or

$$\sigma=\frac{-\eta_1(\tau+a,\Lambda)+\eta_2'(\tau+a,\Lambda)-\sqrt{D(\Lambda)^2-4}}{2\,\eta_2(\tau+a,\Lambda)} \tag{3.19}$$

for a band gap at $k_x=0$.

---

[7] $D(\Lambda)$ of Eq. (3.1) does not depend on τ in (3.14).

There are interesting cases in physical problems where $\eta_i(\tau + a)$ and $\eta_i'(\tau + a)$ can be easily calculated. Applying Eqs. (3.18) and (3.19) could be a practically convenient approach to investigate the existence and properties of surface states in one-dimensional semi-infinite crystals, in particular, their dependencies on the boundary location τ and σ. In Appendix A, the existence and properties of surface states in a Kronig–Penney model are investigated by using Eqs. (3.18) and (3.19). The qualitative understandings we learned in last sections can be clearly seen from the quantitatively calculated results. In Appendices E and F, this approach is further extended to treat surface modes in one-dimensional semi-infinite phononic crystals and photonic crystals.

In simple cases on surface states in an ideal semi-infinite crystal where the boundary condition (3.4) is simplified to $\psi(\tau, \Lambda) = 0$, from (3.6) we obtain that

$$\psi(\tau + a, \Lambda) = \psi(\tau, \Lambda) = 0. \tag{3.20}$$

Equations (3.18) and (3.19) now become

$$\eta_2(\tau + a, \Lambda) = 0, \tag{3.21}$$

by (3.13) and (3.14).

3.6 Comparisons with Previous Work and Discussions

In this chapter, we have presented a general analysis of the surface states due to the termination of the crystal periodic potential in one-dimensional semi-infinite crystals. Although to the authors knowledge, the two qualitative relations (3.7) and (3.8) in Sect. 3.2, the general analysis in Sects. 3.3 and 3.4, and the general theoretical formalism in Sect. 3.5 have not been explicitly published by other authors, many results presented here have been more or less known in previous theoretical investigations based on different specific potential models.

Many previous theoretical investigations of the conditions for the existence of surface states in one-dimensional semi-infinite crystals were based on a nearly free electron model and for the lowest band gap at $k = \frac{\pi}{a}$ [2]. By using a nearly free electron model, many authors [8–10] found that the termination of the periodic potential at a potential minimum rather than at a potential maximum can cause a surface state in the lowest band gap at $k = \frac{\pi}{a}$. That is, only a "Tamm" type of surface state rather than a "Shockley" type of surface state could exist in that band gap. More generally, Goodwin [9] found that in a nearly free electron model, a surface state could appear in a specific band gap when the termination is at a minimum rather than at a maximum of the corresponding Fourier component of the crystal potential. These are consistent with our general result that a surface state is not possible in a band gap for a τ equal to a zero of the *lower* band edge wave function obtained in Sect. 3.4. Since in a nearly free electron model, the zeros of the *lower* band edge wave function of a band gap are always at the maxima of the corresponding Fourier component of the crystal potential, and the zeros of the *upper* band edge wave function are always at the minima of the corresponding component of the crystal potential. Our general results (Figs. 3.4 and 3.5) indicate that a surface state could not exist when τ is in the near neighborhood of the zeros of the *lower* band edge wave function. For the lowest band gap at $k = \frac{\pi}{a}$, this means that when τ is in the neighborhood of a maximum of the crystal potential, a surface state cannot exist. Therefore, only a "Tamm" type rather than a "Shockley" type of surface state can exist in such a case.

A particularly interesting work is a systematic investigation by Levine [11] on the existence of surface states in different band gaps in a sinusoidal crystal potential (Mathieu problem) for different boundary positions. By using some further approximations, Levine found that for the nth band gap, a potential period $[0, a)$ can be separated into $2(n+1)$ intervals. Only when the boundary is in one of the $n+1$ specific intervals can there be a surface state in the band gap in the semi-infinite one-dimensional crystal. These $n+1$ surface-state-allowed intervals are separated by the $n+1$ surface-state-unallowed intervals. In each of the surface-state-allowed intervals, the energy of the surface state increases as the boundary goes inside the semi-infinite crystal. The results obtained in Sect. 3.4 are more general yet consistent with Levine's results. Since for a sinusoidal crystal potential, the properties of two band edge wave functions $\phi_n(k_g, x)$ and $\phi_{n+1}(k_g, x)$ of a specific band gap, including their zeros, $\frac{\phi_n'(k_g,x)}{\phi_n(k_g,x)}$, $\frac{\phi_{n+1}'(k_g,x)}{\phi_{n+1}(k_g,x)}$, and so forth, can be precisely known (e.g., [14, 15]), many results in [11] may be directly obtained from the general analysis in Sects. 3.2–3.4 without the use of those further approximations.

Many people believed that the termination of periodic potential in a one-dimensional semi-infinite crystal *always* causes a surface state to exist in each band gap below the potential barrier. From the analysis presented here, we have seen that this is, in fact, a misconception: The termination of the periodic potential $v(x)$ at the boundary of a semi-infinite one-dimensional crystal may or may not cause a state in a specific band gap. If it does cause a state for that specific band gap, this state may be either a surface state located near the boundary of the semi-infinite crystal or a band edge state not decaying in the semi-infinite crystal with a decay factor $\beta = 0$.

References

1. I. Tamm, Phys. Z. Sowj. **1**, 733 (1932)
2. S.G. Davison, M. Stęślicka, *Basic Theory of Surface States* (Clarendon Press, Oxford, 1992)
3. M.-C. Desjonquéres, D. Spanjaard, *Concepts in Surface Physics* (Springer, Berlin, 1993)
4. A. Zangwill, *Physics at Surfaces* (Cambridge University Press, Cambridge, 1988)
5. F. Bechstedt, *Principles of Surface Physics* (Springer, Berlin, 2003)
6. A. Groß, *Theoretical Surface Science: A Microscopic Perspective* (Springer, Berlin, 2003)
7. S.G. Davison, J.D. Levine, *Solid State Physics*, vol. 25, ed. by H. Ehrenreich, F. Seitz, D. Turnbull (Academic Press, New York, 1970), pp. 1–149
8. A.M. Maue, Z. Phys. **94**, 717 (1935); F. Forstmann, Z. Phys. **235**, 69 (1970); F. Forstmann, *Photoemission and the Electronic Properties of Surfaces*, ed. by B. Feuerbacher, B. Fitten, R.F. Willis (Wiley, New York, 1978), pp. 193–226; R.O. Jones, *Surface Physics of Phosphors and*

Semiconductors, ed. by C.G. Scott, C.E. Reed (Academic Press, London 1975), pp. 95–142; J.B. Pendry, S.J. Gurman, Surf. Sci. **49**, 87 (1975)
9. E.T. Goodwin, Proc. Camb. Phils. Soc. **35**, 205 (1939)
10. H. Statz, Z. Naturforsch. **5a**, 534 (1950)
11. J.D. Levine, Phys. Rev. **171**, 701 (1968)
12. W. Shockley, Phys. Rev. **56**, 317 (1939)
13. H. Hellmann, Acta physicochimi. URSS I, **6**, 913 (1935); IV, **2**, 224 (1936); H. Hellmann, *Einführung in die Quantenchemie* (Deuticke, Leipzig, 1937); R.P. Feynman. Phys. Rev. **56**, 340 (1939)
14. S.Y. Ren, Y.-C. Chang, Ann. Phys. (N. Y.) **325**, 937 (2010)
15. F.P. Mechel, *Mathieu Functions: Formulas, Generation, Use* (S. Hirzel Verlag, Stuttgart, 1997)

Chapter 4
Electronic States in Ideal One-Dimensional Crystals of Finite Length

In this chapter, we present a general investigation of the electronic states in ideal one-dimensional crystals of finite length $L = Na$, where a is the potential period and N is a positive integer.[1] On the basis of the theory of the periodic differential equations in Chap. 2, exact and general results on the electronic states in such an ideal finite crystal can be analytically obtained. We will see that in obtaining the results in this chapter, it is the understanding of the *zeros* of the solutions of a one-dimensional Schrödinger differential equation with a periodic potential that plays a fundamental role.

This chapter is organized as follows. After giving a basic consideration of the problem in Sect. 4.1, in Sect. 4.2 we prove the major results of this chapter: In contrast with the well-known understanding that all electronic states are Bloch waves in an infinite one-dimensional crystal, there are two different types of electronic states in an ideal finite one-dimensional crystal. In Sect. 4.3, we give more discussions on the boundary-dependent states, which are a fundamental distinction of the quantum confinement of Bloch waves. In Sect. 4.4, we briefly discuss one-dimensional stationary Bloch states. In Sect. 4.5, we treat one-dimensional symmetric finite crystals in which the energies of all electronic states can be obtained from the bulk energy band structure. Sections 4.6–4.8 are comments on several relevant problems. Section 4.9 is a simple summary.

4.1 Basic Considerations

The Schrödinger differential equation for a one-dimensional periodic potential can be written as

$$- y'' + [v(x) - \lambda]y = 0, \quad -\infty < x < +\infty, \tag{4.1}$$

[1] Part of the results of this chapter was published in [1, 2].

S.Y. Ren, *Electronic States in Crystals of Finite Size*, Springer Tracts in Modern Physics 270, DOI 10.1007/978-981-10-4718-3_4

where $v(x + a) = v(x)$ is the periodic crystal potential.

We assume that (4.1) is solved. The eigenvalues of (4.1) are energy bands $\varepsilon_n(k)$ and the corresponding eigenfunctions are Bloch functions $\phi_n(k, x)$, where $n = 0, 1, 2, \ldots$ and $-\frac{\pi}{a} < k \leq \frac{\pi}{a}$. We are mainly interested in the cases where there is always a band gap between two consecutive energy bands of (4.1). For these cases, the band edges $\varepsilon_n(0)$ and $\varepsilon_n(\frac{\pi}{a})$ occur in the order

$$\varepsilon_0(0) < \varepsilon_0\left(\frac{\pi}{a}\right) < \varepsilon_1\left(\frac{\pi}{a}\right) < \varepsilon_1(0) < \varepsilon_2(0)$$
$$< \varepsilon_2\left(\frac{\pi}{a}\right) < \varepsilon_3\left(\frac{\pi}{a}\right) < \varepsilon_3(0) < \varepsilon_4(0) < \cdots. \tag{4.2}$$

The band gaps are between $\varepsilon_{2m}(\frac{\pi}{a})$ and $\varepsilon_{2m+1}(\frac{\pi}{a})$ or between $\varepsilon_{2m+1}(0)$ and $\varepsilon_{2m+2}(0)$.

For an ideal one-dimensional crystal of finite length $L = Na$, we assume that the potential *inside* the crystal is still $v(x)$, as in (4.1). The two ends of the crystal are denoted as τ and $\tau + L$, where τ is a real number.

The eigenvalues Λ and eigenfunctions $\psi(x, \Lambda)$ of the electronic states in the finite crystal are solutions of the Schrödinger differential equation

$$-\psi''(x) + [v(x) - \Lambda]\psi(x) = 0, \quad \tau < x < \tau + L, \tag{4.3}$$

inside the crystal with certain boundary conditions at the two boundaries τ and $\tau + L$. For an ideal finite crystal, we have the boundary conditions

$$\psi(x) = 0, \quad x \leq \tau \text{ or } x \geq \tau + L. \tag{4.4}$$

Our purpose is to find solutions of (4.3) under the boundary conditions (4.4).

Suppose $y_1(x, \lambda)$ and $y_2(x, \lambda)$ are two linearly independent solutions of (4.1). In general, a nontrivial solution of (4.3) and (4.4), if it exists, can be expressed as

$$\psi(x, \Lambda) = \begin{cases} y(x, \Lambda), & \tau < x < \tau + L, \\ 0, & x \leq \tau \text{ or } x \geq \tau + L. \end{cases}$$

Here,

$$y(x, \lambda) = c_1 y_1(x, \lambda) + c_2 y_2(x, \lambda) \tag{4.5}$$

—in which c_1 and c_2 are not both zero—is a nontrivial solution of (4.1) and satisfies

$$y(\tau, \Lambda) = y(\tau + L, \Lambda) = 0. \tag{4.6}$$

The nontrivial solutions of (4.3) and (4.4) can be found through (4.5) and (4.6) based on the general properties of linearly independent solutions of (4.1) in different energy intervals, as discussed in Sect. 2.5.

4.2 Two Types of Electronic States

We have understood in Chap. 2 that the forms of linearly independent solutions $y_1(x, \lambda)$ and $y_2(x, \lambda)$ in (4.5) can be determined by the discriminant $D(\lambda)$ of (4.1). The existence and the properties of nontrivial solutions Λ and $y(x, \Lambda)$ in (4.6) can be straightforwardly obtained on this basis.

For a finite crystal, both the permitted and the forbidden energy ranges of the infinite crystal should be considered. In principle, we need to consider solutions of (4.6) for λ in $(-\infty, +\infty)$. However, according to Theorem 2.9, any nontrivial solution of (4.1) with $\lambda \leq \varepsilon_0(0)$ can have only at most one zero for x in $(-\infty, +\infty)$; thus, it cannot satisfy (4.6). Consequently, there is not a nontrivial solution of (4.6) for λ in $(-\infty, \varepsilon_0(0)]$; we need only to consider λ in $(\varepsilon_0(0), +\infty)$. Similar to our discussions in Sect. 2.5, depending on λ, there are five different cases:

A. $|D(\lambda)| < 2$.

In this case, λ is inside an energy band of (4.1). According to (2.73), two linearly independent solutions of (4.1) can be chosen as

$$y_1(x, \lambda) = e^{ik(\lambda)x} p_1(x, \lambda), \quad y_2(x, \lambda) = e^{-ik(\lambda)x} p_2(x, \lambda),$$

where $k(\lambda)$ is a real number depending on λ and

$$0 < k(\lambda)a < \pi,$$

and $p_1(x, \lambda)$ and $p_2(x, \lambda)$ have period a: $p_i(x + a, \lambda) = p_i(x, \lambda)$. All $k(\lambda)$ and $p_i(x, \lambda)$ are functions of λ. Simple mathematics leads to that the existence of non-trivial solutions of (4.5) and (4.6) requires[2]

$$e^{ik(\Lambda)L} - e^{-ik(\Lambda)L} = 0. \qquad (A.1)$$

Note $(A.1)$ does not contain τ. The nontrivial solutions can be obtained if

$$k(\Lambda)L = j\pi,$$

where j is a positive integer. Thus in each energy band $\varepsilon_n(k)$, there are $N - 1$ values of Λ_j, for which

$$k(\Lambda_j) = j\pi/L, \quad j = 1, 2, \ldots, N - 1.$$

---

[2]Otherwise

$$c_1 p_1(\tau, \Lambda) = 0, \text{ and } c_2 p_2(\tau, \Lambda) = 0. \qquad (A.2)$$

It was pointed out on p. 48 that, in general, a one-dimensional Bloch function $\phi_n(k, x)$ does not have a zero except $k = 0$ where $D(\lambda) = 2$ or $k = \frac{\pi}{a}$ where $D(\lambda) = -2$. Thus, neither $p_1(\tau, \Lambda)$ nor $p_2(\tau, \Lambda)$ in (A.2) can be zero. $(A.2)$ leads to that $c_1 = c_2 = 0$ and no nontrivial solution of (4.5) and (4.6) from $(A.2)$ exists.

Correspondingly, for each energy band, there are $N - 1$ electronic states $\psi_{n,j}(x;\tau)$ whose energies are given by

$$\Lambda_{n,j} = \varepsilon_n\left(\frac{j\pi}{L}\right), \quad j = 1, 2, \ldots, N-1. \tag{4.7}$$

Each energy for this case is a function of L, the crystal length. However, all do not depend on the location of the crystal boundary τ or $\tau + L$. These states are stationary Bloch states consisting of two Bloch waves with wave vectors $k = j\pi/L$ and $-k = -j\pi/L$ in the finite crystal, formed due to the multiple reflections of Bloch waves at the two ends τ and $\tau + L$ of the finite crystal. For simplicity, we call these states L-dependent states; although only the eigenvalue of such a state depends only on L, the wave function of such a state depends on both τ and L.

The energies $\Lambda_{n,j}$ in (4.7) map the energy band structure $\varepsilon_n(k)$ of the infinite crystal exactly. By using a Kronig–Penney potential, Pedersen and Hemmer found that the energy spectrum of the confined Bloch waves maps the energy bands exactly [3]. The fact that the energy spectra of confined electrons in Si (001), (110) quantum films and in GaAs (110) quantum films approximately map the energy band structure of the bulk was observed in numerical calculations by Zhang and Zunger [4] and Zhang et al. [5]. Much previous work also found that the eigenvalues of confined Bloch states map the dispersion relations of the unconfined Bloch waves *closely* (e.g., [6]). From (4.7), we see that this is, in fact, an exact correspondence between the electronic states in an ideal one-dimensional crystal of finite length and a one-dimensional crystal of infinite length. Furthermore, this exact correspondence does not depend on τ, the location of the crystal boundary. These electronic states can be considered as bulk-like electronic states in a one-dimensional crystal of finite length.

B. $D(\lambda) = 2$.

In this case, λ is at a band-edge at $k = 0$: $\lambda = \varepsilon_{2m+1}(0)$ or $\lambda = \varepsilon_{2m+2}(0)$.

According to (2.75), two linearly independent solutions of (4.1) for this case can be expressed as

$$y_1(x, \lambda) = p_1(x, \lambda), \quad y_2(x, \lambda) = x\, p_1(x, \lambda) + p_2(x, \lambda),$$

where $p_1(x, \lambda)$ and $p_2(x, \lambda)$ are periodic functions with period a.

Due to Theorem 2.1, the zeros of $p_1(x, \lambda)$ are separated from the zeros of $p_2(x, \lambda)$. From (4.5) and (4.6), simple mathematics shows that, in this case, the existence of a nontrivial solution (4.6) requires

$$p_1(\tau, \Lambda) = 0 \text{ and } c_2 = 0. \tag{B.1}$$

$(B.1)$ indicates that if a solution of (4.3) and (4.4) exists at a band edge at $k = 0$, the corresponding wave function $y(x, \Lambda)$ of the confined electronic state must be a periodic function, with a zero at the crystal boundary τ (and also $\tau + L$).

C. $D(\lambda) > 2$.

In this case, λ is *inside* a band gap at $k = 0$: $\varepsilon_{2m+1}(0) < \lambda < \varepsilon_{2m+2}(0)$. According to (2.77), two linearly independent solutions of (4.1) can be expressed as

$$y_1(x, \lambda) = e^{\beta(\lambda)x} p_1(x, \lambda), \quad y_2(x, \lambda) = e^{-\beta(\lambda)x} p_2(x, \lambda),$$

where $\beta(\lambda)$ is a positive real number depending on λ and $p_1(x, \lambda)$ and $p_2(x, \lambda)$ are periodic functions with period a.

Again due to Theorem 2.1, the zeros of $p_1(x, \lambda)$ are separated from the zeros of $p_2(x, \lambda)$. If there is a nontrivial solution $y(x, \Lambda)$ in this case, simple mathematics from (4.5) and (4.6) gives that we must have either

$$p_1(\tau, \Lambda) = 0 \text{ and } c_2 = 0, \tag{C.1}$$

or

$$p_2(\tau, \Lambda) = 0 \text{ and } c_1 = 0. \tag{C.2}$$

$(C.1)$ and $(C.2)$ indicate that if a solution of (4.3) and (4.4) exists inside a band gap at $k = 0$, the corresponding wave function $y(x, \Lambda)$ of the confined electronic state must be a product of an exponential function and a periodic function. The periodic function must have a zero at the crystal boundary τ (and also $\tau + L$). Note that $(C.1)$ and $(C.2)$ cannot be true simultaneously.

From the discussions on Cases B and C, we can see that the existence of a nontrivial solution of (4.6) in a band gap $[\varepsilon_{2m+1}(0), \varepsilon_{2m+2}(0)]$ at $k = 0$ requires that either one of $(B.1)$, $(C.1)$, or $(C.2)$ must be true. Since all functions $p_i(x, \lambda)$ in $(B.1)$, $(C.1)$, and $(C.2)$ are periodic functions, we always have $y(\tau + a, \Lambda) = 0$ if we have $y(\tau, \Lambda) = 0$. Therefore, the following equation is a necessary condition for having a solution Λ in (4.6) for a band gap at $k = 0$:

$$y(\tau + a, \Lambda) = y(\tau, \Lambda) = 0. \tag{4.8}$$

It is easy to see that (4.8) is also a sufficient condition for having a solution (4.6): From (4.8), one can obtain $y(\tau + \ell a, \Lambda) = 0$, where $\ell = 0, 1, 2, \ldots, N$.

D. $D(\lambda) = -2$.

In this case, λ is at a band-edge at $k = \frac{\pi}{a}$: $\lambda = \varepsilon_{2m}(\frac{\pi}{a})$ or $\lambda = \varepsilon_{2m+1}(\frac{\pi}{a})$. According to (2.79), two linearly independent solutions of (4.1) can be expressed as

$$y_1(x, \lambda) = s_1(x, \lambda), \quad y_2(x, \lambda) = x\, s_1(x, \lambda) + s_2(x, \lambda),$$

where $s_1(x, \lambda)$ and $s_2(x, \lambda)$ are semi-periodic functions with semi-period a.

Due to Theorem 2.1, the zeros of $s_1(x, \lambda)$ are separated from the zeros of $s_2(x, \lambda)$. From (4.5) and (4.6), simple mathematics leads to that the existence of a nontrivial solution (4.6) in this case requires

$$s_1(\tau, \Lambda) = 0 \text{ and } c_2 = 0. \tag{D.1}$$

$(D.1)$ indicates that if a solution of (4.3) and (4.4) exists at a band-edge at $k = \frac{\pi}{a}$, the corresponding wave function $y(x, \Lambda)$ of the confined electronic state must be a semi-periodic function with semi-period a, with a zero at the crystal boundary τ (and also $\tau + L$).

E. $D(\lambda) < -2$.

In this case, λ is *inside* a band gap at $k = \frac{\pi}{a}$: $\varepsilon_{2m}(\frac{\pi}{a}) < \lambda < \varepsilon_{2m+1}(\frac{\pi}{a})$. According to (2.81), two linearly independent solutions of (4.1) can be expressed as

$$y_1(x, \lambda) = e^{\beta(\lambda)x} s_1(x, \lambda), \quad y_2(x, \lambda) = e^{-\beta(\lambda)x} s_2(x, \lambda),$$

where $\beta(\lambda)$ is a positive real number depending on λ and $s_1(x, \lambda)$ and $s_2(x, \lambda)$ are semi-periodic functions with semi-period a.

Again, due to Theorem 2.1, the zeros of $s_1(x, \lambda)$ are separated from the zeros of $s_2(x, \lambda)$. If there is a nontrivial solution $y(x, \Lambda)$ in this case, simple mathematics from (4.5) and (4.6) gives that we must have either

$$s_1(\tau, \Lambda) = 0 \text{ and } c_2 = 0, \tag{E.1}$$

or

$$s_2(\tau, \Lambda) = 0 \text{ and } c_1 = 0. \tag{E.2}$$

$(E.1)$ and $(E.2)$ indicate that if a solution of (4.3) and (4.4) exists inside a band gap at $k = \frac{\pi}{a}$, the corresponding wave function $y(x, \Lambda)$ of the confined electronic state must be a product of an exponential function and a semi-periodic function. The semi-periodic function must have a zero at the crystal boundary τ (and also $\tau + L$). Note that $(E.1)$ and $(E.2)$ cannot be true simultaneously.

From the discussions of Cases D and E, we can see that the existence of a non-trivial solution of (4.6) for a band gap $[\varepsilon_{2m}(\frac{\pi}{a}), \varepsilon_{2m+1}(\frac{\pi}{a})]$ requires that either one of $(D.1)$, $(E.1)$, or $(E.2)$ must be true. Since all functions $s_i(x, \lambda)$ in $(D.1)$, $(E.1)$, and $(E.2)$ are semi-periodic functions with a semi-period a, we always have $y(\tau + a, \Lambda) = 0$ if we have $y(\tau, \Lambda) = 0$. We are led to the same equation (4.8) as a necessary and sufficient condition for having a solution of (4.6) for a band gap at $k = \frac{\pi}{a}$. Therefore, (4.8) is a necessary and sufficient condition for having a solution of (4.6) corresponding to a band gap.

Theorem 2.8 indicates that for an arbitrary real number τ, there is always one and only one Λ for which (4.8) is true for each band gap $[\varepsilon_{2m}(\frac{\pi}{a}), \varepsilon_{2m+1}(\frac{\pi}{a})]$ or $[\varepsilon_{2m+1}(0), \varepsilon_{2m+2}(0)]$. Since no two linearly independent solutions of (4.1) with one Λ may have a same zero (Theorem 2.1), there is one and only one solution $\psi_n(x; \tau)$ of (4.3) and (4.4) in each gap.

Equation (4.8) does not contain the crystal length L; thus, an eigenvalue Λ of (4.3) and (4.4) in a band gap is only dependent on τ, but not on L.

Therefore, *for any real number τ, there is always one and only one $\Lambda_n(\tau)$ and one $\psi_n(x; \tau)$ as a solution of* (4.3) *and* (4.4) *in each band gap* $[\varepsilon_{2m}(\frac{\pi}{a}), \varepsilon_{2m+1}(\frac{\pi}{a})]$ *or* $[\varepsilon_{2m+1}(0), \varepsilon_{2m+2}(0)]$. *Such a $\Lambda_n(\tau)$ is dependent on τ but not on L*. We call these solutions τ-dependent states for simplicity; although only the eigenvalue $\Lambda_n(\tau)$ of such a state depends only on τ, the wave function $\psi_n(x; \tau)$ of such a state depends on both τ and L.

The eigenvalues $\Lambda_n(\tau)$ of these τ-dependent states $\psi_n(x; \tau)$ are exactly the $\Lambda_{\tau,2m}$ or $\Lambda_{\tau,2m+1}$ defined by (2.94). According to Theorem 2.8, $\Lambda_{\tau,2m}$ is in $[\varepsilon_{2m}(\frac{\pi}{a}), \varepsilon_{2m+1}(\frac{\pi}{a})]$ and $\Lambda_{\tau,2m+1}$ is in $[\varepsilon_{2m+1}(0), \varepsilon_{2m+2}(0)]$. That is, $\Lambda_{\tau,n}$ is in the nth band gap above the nth energy band.

Such a τ-dependent state has the form $\psi_n(x; \tau) = e^{\beta(\Lambda)x} f_n(x, \Lambda)$ in $[\tau, \tau + L]$, where β is either positive, negative, or zero, and $f_n(x, \Lambda)$ is a periodic function or a semi-periodic function depending on the band gap location. This state is either localized near one of the two ends of the finite crystal or a confined band edge state.

As an example, in Fig. 4.1, a comparison between the energy bands of (4.1) and the energies of the electronic states in a one-dimensional crystal (solutions of (4.3) and (4.4)) for a crystal length $L = 8a$ are shown.

Since each energy band n and each band gap exists alternatively where $n = 0, 1, 2, \ldots$ if (4.2) is true, the major results obtained in this section can be summarized as the following theorem:

Theorem 4.1 *For each energy band n of (4.1), there are two types of solutions of (4.3) and (4.4) if (4.2) is true. There are $N - 1$ stationary Bloch state solutions $\psi_{n,j}(x; \tau)$ whose energies $\Lambda_{n,j}$ are given by (4.7) and thus depend on the crystal length L but not on the crystal boundary τ. There is always one and only one solution $\psi_n(x; \tau)$ whose energy $\Lambda_n(\tau)$ is given by (4.8) and thus depend on the crystal boundary τ but not on the crystal length L and is in the nth band gap above the nth energy band.*

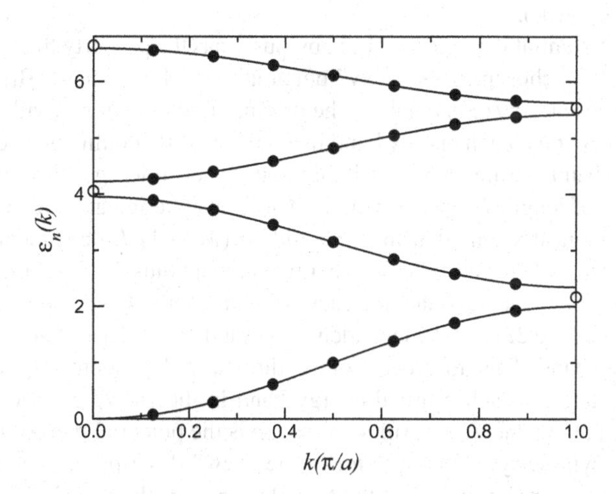

Fig. 4.1 A comparison between the energy bands $\varepsilon_n(k)$ of (4.1) (*solid lines*) and the energies Λ of the electronic states in a crystal of length $L = 8a$ (*solid circles*, L-dependent; *open circles*, τ-dependent). Note that the L-dependent energies map the energy bands exactly and satisfy (4.7); The τ-dependent energies satisfying (4.8) are in a band gap or at a band edge of (4.1). Reprinted with permission from S. Y. Ren: Ann. Phys.(NY) **301**, 22 (2002). Copyright by Elsevier

This τ-dependent state solution is either a state localized near one of the two ends of the finite crystal or a confined band-edge state.

There could also exist the cases of zero band gap in which $\varepsilon_{2m}(\frac{\pi}{a}) = \varepsilon_{2m+1}(\frac{\pi}{a})$ or $\varepsilon_{2m+1}(0) = \varepsilon_{2m+2}(0)$, where (4.2) is not generally true. Suppose in a specific case that $\varepsilon_{2m}(\frac{\pi}{a}) = \varepsilon_{2m+1}(\frac{\pi}{a})$; then two linearly independent solutions of (4.1) can be chosen according to (2.80) as

$$y_1\left[x, \varepsilon_{2m}\left(\frac{\pi}{a}\right)\right] = s_1\left[x, \varepsilon_{2m}\left(\frac{\pi}{a}\right)\right], \quad y_2\left[x, \varepsilon_{2m}\left(\frac{\pi}{a}\right)\right] = s_2\left[x, \varepsilon_{2m}\left(\frac{\pi}{a}\right)\right],$$

and $s_1[x, \varepsilon_{2m}(\frac{\pi}{a})]$ and $s_2[x, \varepsilon_{2m}(\frac{\pi}{a})]$ are semi-periodic functions with semi-period a. It is easy to see that the function

$$\begin{aligned} y_{2m}\left[x, \varepsilon_{2m}\right] = {} & s_2\left[\tau, \varepsilon_{2m}\left(\frac{\pi}{a}\right)\right] s_1\left[x, \varepsilon_{2m}\left(\frac{\pi}{a}\right)\right] \\ & - s_1\left[\tau, \varepsilon_{2m}\left(\frac{\pi}{a}\right)\right] s_2\left[x, \varepsilon_{2m}\left(\frac{\pi}{a}\right)\right] \end{aligned} \tag{4.9}$$

is a solution of (4.6). Since $s_1[\tau, \varepsilon_{2m}(\frac{\pi}{a})]$ and $s_2[\tau, \varepsilon_{2m}(\frac{\pi}{a})]$ are not both zero by Theorem 2.1, the function defined in (4.9) is a nontrivial solution of (4.3) and (4.4). This solution is a semi-periodic function whose energy $\Lambda = \varepsilon_{2m}(\frac{\pi}{a})$ depends on neither L nor τ. The cases where $\varepsilon_{2m+1}(0) = \varepsilon_{2m+2}(0)$ can be similarly discussed.

Therefore, in these cases, there is always a solution Λ of (4.3) and (4.4) that depends on neither L nor τ: $\Lambda = \varepsilon_{2m}(\frac{\pi}{a})$ or $\Lambda = \varepsilon_{2m+1}(0)$. $y(x, \Lambda)$ will be either a semi-periodic function (when $\varepsilon_{2m}(\frac{\pi}{a}) = \varepsilon_{2m+1}(\frac{\pi}{a})$) or a periodic function (when $\varepsilon_{2m+1}(0) = \varepsilon_{2m+2}(0)$).

A periodic potential $v(x+a) = v(x)$ obviously has the property that $v(x+2a) = v(x)$. If $\ell = 2a$ is chosen as the "new" potential period, the "new" Brillouin zone (with boundaries at $\pm\pi/\ell$) is a half of the original Brillouin zone (with boundaries at $\pm\frac{\pi}{a}$). Consequently, each energy band in the original Brillouin zone becomes two "new" energy bands in the "new" Brillouin zone (band-folding). Now, we consider a finite crystal of length $L = M\ell$, where M is a positive integer. According to the "new" description, it seems that there should be $(M-1)$ L-dependent states and one τ-dependent state for each "new" energy band and thus $2(M-1)$ L-dependent states and two τ-dependent states for each original energy band. From the original description, there are $2M-1$ L-dependent states and one τ-dependent state for each original energy band. This difference (one additional τ-dependent state and one less L-dependent state for each original energy band in the "new" description) comes from the fact that the "new" description ($\ell = 2a$ is the potential period) is not based on the whole symmetry of the system. In the "new" description, we always have $\varepsilon_{2m}(\pi/\ell) = \varepsilon_{2m+1}(\pi/\ell)$; that is, at the boundary π/ℓ of the "new" Brillouin zone, every band gap is a zero band gap. Therefore, in a finite crystal of length $L = M\ell$, there is always a state whose energy $\Lambda = \varepsilon_{2m}(\pi/\ell)$ depends on neither τ nor L. The "extra" τ-dependent state in the "new" description, in fact, is an L-dependent state with $j = M$ in the finite crystal of length $L = 2Ma$ in the original description. Its energy does not depend on τ since it is an L-dependent state. Its energy does not depend on L either, since, for each band, the state $j = M$ always exists in a finite crystal of length $2Ma$. We mention this point here since we will meet some relevant situations in Part III.

4.3 τ-Dependent States

It is well known that when one-dimensional plane waves are completely confined, all permitted states are stationary wave states. Thus, the very existence of the τ-dependent states in ideal one-dimensional finite crystals is a fundamental distinction of the quantum confinement of one-dimensional Bloch waves. In a one-dimensional finite crystal, such a τ-dependent state may have three different forms: a surface state localized near the left end τ of the crystal, a surface state localized near the right end $\tau + L$ of the crystal, or a band-edge state periodically distributed in the finite crystal—depending on the location of boundary τ.

We again take a band gap $[\varepsilon_{2m+1}(0), \varepsilon_{2m+2}(0)]$ at $k = 0$ as an example. In Sect. 2.6, we have seen how a τ-dependent eigenvalue $\Lambda_{\tau,n}$ changes as τ changes. As τ goes to the right continuously from a (any) zero $x_{1,2m+1}$ of $\phi_{2m+1}(0, x)$ to $x_{1,2m+2}$, the zero of $\phi_{2m+2}(0, x)$ next to $x_{1,2m+1}$ and then to $x_{2,2m+1}$, the next zero of $\phi_{2m+1}(0, x)$,[3] the corresponding $\Lambda_{\tau,2m+1}$ will also go continuously from $\varepsilon_{2m+1}(0)$ up to $\varepsilon_{2m+2}(0)$ and then back to $\varepsilon_{2m+1}(0)$. We can consider such an up and down of $\Lambda_{\tau,2m+1}$ as a basic undulation. Corresponding to a basic undulation, in $[x_{1,2m+1}, x_{2,2m+1})$ the function $y_{2m+1}(x, \Lambda)$ has different forms. Since for any solution of (4.3) and (4.4) in the band gap $[\varepsilon_{2m+1}(0), \varepsilon_{2m+2}(0)]$, one of $(B.1)$, $(C.1)$, or $(C.2)$ in Sect. 4.2 must be true, we have three different cases:

1. When $\tau = x_{1,2m+1}$, $(B.1)$ is true and $\Lambda_{\tau,2m+1} = \varepsilon_{2m+1}(0)$, the corresponding solution $y(x, \Lambda)$ in (4.8) has the form

$$y[x, \varepsilon_{2m+1}(0)] = \phi_{2m+1}(0, x)$$

and is a lower band-edge wave function of the band gap. Similarly, when $\tau = x_{1,2m+2}$, $(B.1)$ is true and $\Lambda_{\tau,2m+1} = \varepsilon_{2m+2}(0)$, the corresponding solution $y(x, \Lambda)$ in (4.8) has the form

$$y[x, \varepsilon_{2m+1}(0)] = \phi_{2m+2}(0, x)$$

and is an upper band-edge wave function of the band gap. Either one of these two subcases corresponds to a case that there is an electronic state in the finite crystal whose energy is a band-edge energy. The energy of such a state does not depend on the crystal length L, and its wave function inside the crystal is

---

[3]Remember that the zeros of $\phi_{2m+1}(0, x)$ and $\phi_{2m+2}(0, x)$ are distributed alternatively.

the band-edge wave function: The τ-dependent state $\psi_{2m+1}(x, \tau)$ is a confined band-edge state in the finite crystal.

2. When τ is in the interval $(x_{1,2m+1}, x_{1,2m+2})$, $\frac{\partial}{\partial \tau}\Lambda_{\tau,2m+1} > 0$; thus, according to our discussion in Chap. 3, a surface state can exist in a right semi-infinite crystal with a left boundary at τ. Correspondingly, $y(x, \Lambda_{\tau,2m+1})$ has the form $c_2 e^{-\beta(\Lambda_{\tau,2m+1})x}\ p_2(x, \Lambda_{\tau,2m+1})$, with $p_2(\tau, \Lambda_{\tau,2m+1}) = 0$ ($C.2$ is true). A function with the form of $c_2 e^{-\beta(\Lambda_{\tau,2m+1})x}\ p_2(x, \Lambda_{\tau,2m+1})$, in which $\beta(\Lambda_{\tau,2m+1}) > 0$ is mainly distributed near the left end τ of the finite crystal, due to the exponential factor. Thus, the τ-dependent state, in this case, is a surface state near the left end in the finite crystal introduced by the termination of the periodic potential.
3. When τ is in the interval $(x_{1,2m+2}, x_{2,2m+1})$, $\frac{\partial}{\partial \tau}\Lambda_{\tau,2m+1} < 0$; thus, according to our discussion in Chap. 3, a surface state can exist in a left semi-infinite crystal with a right boundary at τ. Correspondingly, $y(x, \Lambda_{\tau,2m+1})$ has the form $c_1 e^{\beta(\Lambda_{\tau,2m+1})x}\ p_1(x, \Lambda_{\tau,2m+1})$, with $p_1(\tau, \Lambda_{\tau,2m+1}) = 0$ ($C.1$ is true). A function with the form of $c_1 e^{\beta(\Lambda_{\tau,2m+1})x}\ p_1(x, \Lambda_{\tau,2m+1})$, in which $\beta(\Lambda_{\tau,2m+1}) > 0$, is mainly distributed near the right end $\tau + L$ of the finite crystal, due to the exponential factor. Thus, the τ-dependent state, in this case, is a surface state near the right end in the finite crystal introduced by the termination of the periodic potential.

Therefore, these three cases correspond to a wave function inside the crystal with a form $p(x, \Lambda)$, $e^{-\beta x} p(x, \Lambda)$ or $e^{\beta x} p(x, \Lambda)$, in which $\beta > 0$. The latter two correspond to a surface state located near either the left or the right end of the finite crystal. *Such a surface state is introduced into the band gap when the boundary τ is not a zero of either band-edge wave function of the Bloch waves.* As an example, Fig. 4.2 is shown $\Lambda_{\tau,1}$ as a function of τ in the interval $[x_{1,1}, x_{1,1} + a]$, where $x_{1,1}$ is a (any) zero of $\phi_1(0, x)$. In the figure, the two segments of a basic undulation are shown as a dotted line ($C.2$ is true) or a dashed line ($C.1$ is true), indicating two different locations of the surface state.

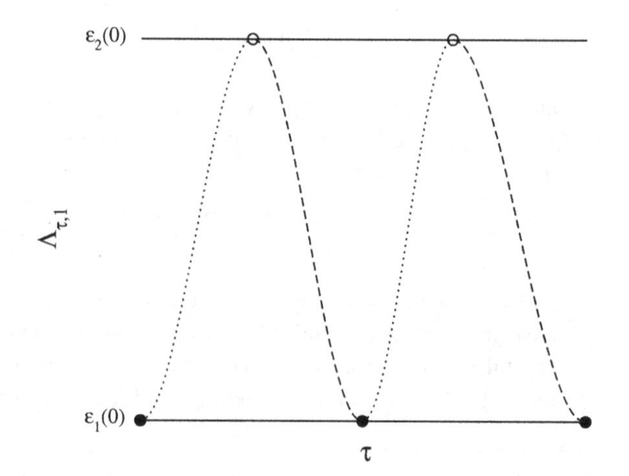

Fig. 4.2 $\Lambda_{\tau,1}$ as a function of τ in the interval $[x_{1,1}, x_{1,1} + a]$. The zeros of $\phi_1(0, x)$ are shown as *solid circles*, and the zeros of $\phi_2(0, x)$ are shown as *open circles*. Note that $\Lambda_{\tau,1}$ completes two basic undulations in $[x_{1,1}, x_{1,1} + a)$. The *dotted lines* and *dashed lines* indicate that a surface state is located near either the *left* or the *right* end of the finite crystal

Since according to Theorem 2.7, in the interval [0, a), both $\phi_{2m+1}(0, x)$ and $\phi_{2m+2}(0, x)$ have exactly $2m + 2$ zeros, then, in general, $\Lambda_{\tau,2m+1}$ as a function of τ always completes $2m + 2$ basic undulations in an interval of length a.

Correspondingly, in a band gap at $k = \frac{\pi}{a}$ ($\varepsilon_{2m}(\frac{\pi}{a}) < \Lambda_{\tau,2m} < \varepsilon_{2m+1}(\frac{\pi}{a})$), a surface state has the form of either $c_1 e^{\beta(\Lambda_{\tau,2m})x}$ $s_1(x, \Lambda_{\tau,2m})$ or $c_2 e^{-\beta(\Lambda_{\tau,2m})x}$ $s_2(x, \Lambda_{\tau,2m})$ inside the crystal, where $s_1(x, \Lambda_{\tau,2m})$ and $s_2(x, \Lambda_{\tau,2m})$ are semi-periodic functions. $\Lambda_{\tau,2m}$, as a function of τ, will always complete $2m+1$ basic undulations in an interval of length a.

These discussions are closely related to the discussions we had in Sect. 3.3. The three possibilities of a τ-dependent state—a surface state located near the left end τ of the finite crystal, a surface state located near the right end $\tau + L$ of the finite crystal, or a confined band-edge state—are determined by which one of the three sets $L(n)$, $R(n)$, or $M(n)$, τ is. Naturally, a band-edge state can also be considered as a special surface state with its energy equal to a band-edge energy and its decay factor $\beta = 0$.

Many years ago, Tamm [7] showed that the termination of the periodic potential at the surface of a semi-infinite Kronig–Penney crystal may cause a surface state to exist in each band gap below the barrier height outside the crystal. More than 60 years later, Zhang and Zunger [4], Zhang et al. [5], and Franceschetti and Zunger [10] observed the existence of a "zero confinement state" in their numerical calculations. Now, we understand that in one-dimensional cases, a "zero confinement state"—a confined band-edge state whose energy does not change as the crystal size changes—is simply a surface-like state with a decay factor $\beta = 0$. A surface state in the gap or a confined band-edge state is different consequences of the termination of the periodic potential at the crystal boundary. In the simplest "ideal" cases, it depends on whether the boundary location τ is a zero of a band-edge wave function.

Equation (4.8) has the similar form as Eq. (3.20). If Eq. (3.20) gives a solution of a surface state localized near the left boundary τ in a right semi-infinite one-dimensional crystal in the range of $[\tau, +\infty)$, then Eq. (4.8) gives a surface state localized near the left boundary τ of the finite crystal in the range of $[\tau, \tau + L]$. If Eq. (3.20) gives a solution of a surface state localized near the right boundary τ in a left semi-infinite one-dimensional crystal in the range of $(-\infty, \tau]$, then Eq. (4.8) gives a surface state localized near the right boundary $\tau + L$ of the finite crystal in the range of $[\tau, \tau + L]$. If Eq. (3.20) gives a solution of band-edge state in a semi-infinite one-dimensional crystal investigated there, Eq. (4.8) gives a band-edge state in the finite crystal in the range of $[\tau, \tau + L]$.

A slight change of the boundary location τ can change the properties of a τ-dependent state dramatically: It can change the τ-dependent state from a surface state located near one end of the crystal to a confined band-edge state, or to a surface state located near the other end. It can also change the energy of the surface state. This point can be clearly seen in Fig. 4.2: If τ is in the region corresponding to a dotted line, the τ-dependent state is a surface state near the left end of the crystal. If τ is in the region corresponding to a dashed line, it is a surface state near the right end of the crystal. If τ is a zero of a band-edge wave function (either a solid circle or an open circle in the figure), then $\tau + L$ is also a zero of the same band-edge wave function and, consequently, the τ-dependent state is a confined band-edge state. We can call these τ-dependent states surface-like states, in differentiation with the bulk-like states—the stationary Bloch states. The concept of surface-like states is an extended concept of the well-known surface states.

4.4 Stationary Bloch States

Although the stationary plane waves are well understood in physics, the corresponding understandings of the stationary Bloch states are relatively less. Based on the mathematical theory of periodic differential equations, we can have some primary general understandings of the one-dimensional stationary Bloch states.

The one-dimensional stationary Bloch states confined in the interval $[\tau, \tau + Na]$ generally can be written in the form

$$\psi_{n,j}(x) = C[\phi_n(k_j, x)\phi_n(-k_j, \tau) - \phi_n(-k_j, x)\phi_n(k_j, \tau)], \quad \tau \leq x \leq \tau + Na, \tag{4.10}$$

where $\phi_n(k, x)$ is a Bloch function in the nth band with a wave-vector k, and

$$k_j = j\frac{\pi}{Na}, \quad j = 1, 2, \ldots, N - 1. \tag{4.11}$$

j is the stationary state index and C is a normalization coefficient.

Since $\psi_{n,j}(x)$ in Eq. (4.10) has a zero at $x_0 = \tau$, $\psi^*_{n,j}(x)$ also has a zero at $x_0 = \tau$. By Theorem 2.1, $\psi_{n,j}(x)$ and $\psi^*_{n,j}(x)$ are linearly dependent to each other, $\psi_{n,j}(x)$ *can be chosen as a real function.* Therefore, in a one-dimensional case, the flux density of a stationary Bloch wave is always zero,

$$\psi^*_{n,j}(x)\frac{\mathrm{d}}{\mathrm{d}x}\psi_{n,j}(x) - \psi_{n,j}(x)\frac{\mathrm{d}}{\mathrm{d}x}\psi^*_{n,j}(x) = 0.$$

Furthermore, by the oscillation theorem [8, 9], $\psi_{n,j}(x)$ has exactly $[nN - 1 + j]$ zeros (if n is even) or $[(n + 1)N - 1 - j]$ zeros (if n is odd) in $(\tau, \tau + Na)$, in comparison to that a τ-dependent confined state $\psi_n(x; \tau)$ always has $[(n+1)N - 1]$ zeros in $(\tau, \tau + Na)$.

4.5 Electronic States in One-Dimensional Finite Symmetric Crystals

When a one-dimensional finite crystal is symmetric, the τ-dependent state for each gap is always a band-edge state, and the energies of all electronic states in the crystal can be obtained from the bulk band structure $\varepsilon_n(k)$.

A symmetric one-dimensional finite crystal means that (1) the crystal potential $v(x)$ has an inversion symmetry center and thus has an infinite number of inversion centers[4]; (2) the two ends of the crystal are also symmetric to one of these inversion centers. Therefore, this inversion center can be chosen to be the origin: $v(-x) = v(x)$. Now, the two ends of the crystal are equivalent: one end $\tau = -L/2$ and the other end $\tau + L = L/2$. If this is the case, the energies of confined electronic states have an especially simple form.

Because $v(-x) = v(x)$, a noteworthy point is that $x = a/2$ is also an inversion center of the crystal potential $v(x)$ because $v(-x - a/2) = v(x + a/2) = v(x - a/2)$. Since the crystal length $L = Na$ and N is a positive integer, the two ends of the finite crystal $x = \tau = -L/2$ and $x = \tau + L = L/2$ must also be an inversion center of $v(x)$. Correspondingly, a band-edge wave function will have a specific parity for an inversion about $x = L/2$ (and also to $x = -L/2$), either an even parity or an odd parity. Furthermore, according to Theorem 2.7, the two band-edge wave functions corresponding to a specific band gap have the same number of zeros in $[0, a)$ and thus must have two different parities for an inversion about $x = L/2$: One is even, and the other is odd. The band-edge wave function that has an odd parity for an inversion about $x = L/2$ (and also to $x = -L/2$) has a zero at the two ends of the finite crystal $x = L/2$ and $x = -L/2$; therefore, $\tau = -L/2$ is a zero of such a band-edge wave function. This situation corresponds to the case in Sect. 4.2 in which $(B.1)$ or $(D.1)$ is true. The band-edge wave function satisfies both (4.3) and (4.6). The energy of this band-edge state is irrelevant to the finite crystal length L. For a specific band gap, whether this is the upper band-edge or the lower band-edge depends on the crystal potential $v(x)$. For each band gap, there is always one band-edge state whose energy does not change as the crystal length L changes. In Figs. 4.3 and 4.4, the energies of two confined states are shown near the two lowest band gaps $[\varepsilon_0(\frac{\pi}{a}), \varepsilon_1(\frac{\pi}{a})]$ and $[\varepsilon_1(0), \varepsilon_2(0)]$ as functions of the crystal length L separately, obtained in [2].

Therefore, we can see that the existence of band-edge states whose energies do not change as the crystal length changes and were observed in numerical calculations in [4, 5, 10] actually can quite often occur in one-dimensional finite symmetric crystals. Although such a state was called as "zero-confinement state" in [4, 5, 10], these states are really confined states. Nevertheless, the energy of these states does not change as L changes. We prefer to call these states confined band-edge states. The fundamental reason for the existence of these confined band-edge states in a one-dimensional symmetric finite crystal is that due to the symmetry of the periodic potential, for each band gap there is always a band-edge state that naturally has a zero at both ends of the finite crystal. Whether this is the upper band-edge state or the lower band-edge state depends on the specific form of $v(x)$ and the location of the band gap.

Since the energy of each τ-dependent state in a one-dimensional symmetric finite crystal is always a band-edge energy, the energies of all electronic states in such

[4]Any point ℓa (ℓ: an integer) away from an inversion symmetry center of a periodic potential is also an inversion symmetry center of the periodic potential.

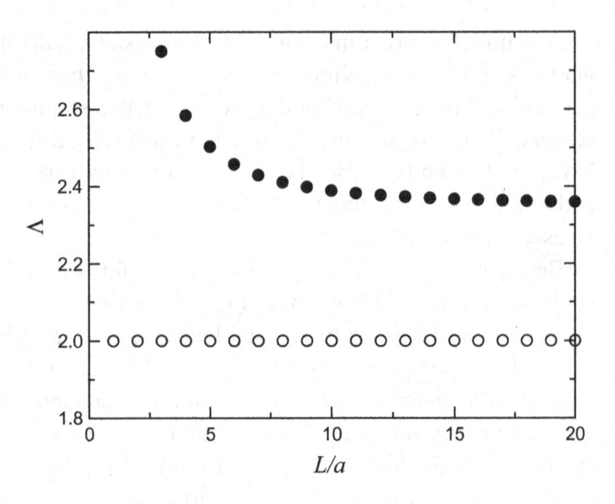

Fig. 4.3 The energies of two confined states near the lowest band gap as functions of the confinement length L. Note that the energy of the lower confined state is the band-edge energy and is constant; only the energy of the higher confined state changes as L changes

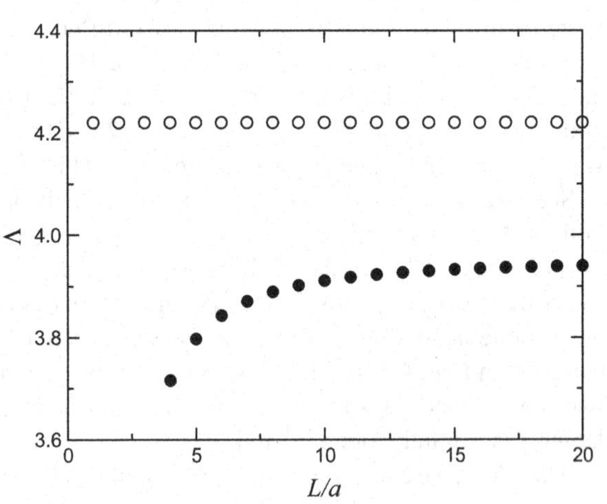

Fig. 4.4 The energies of two confined states near the second lowest band gap as functions of the crystal length L. Note that the energy of the higher confined state is the band-edge energy and is constant; only the energy of the lower confined state changes as L changes

a finite crystal can be obtained from the band structure $\varepsilon_n(k)$ of the corresponding infinite crystal. Figure 4.5 shows a comparison between the energy bands $\varepsilon_n(k)$ as the solutions of (4.1) and the energy spectrum of $\Lambda_{n,j}$ and $\Lambda_{\tau,n}$ as solutions of (4.3) and (4.4) for a symmetric crystal of length $L = 8a$, obtained in [2].

4.6 Comments on the Effective Mass Approximation

As we mentioned in Chap. 1, the effective mass approximation (EMA) has been widely used to investigate the quantum confinement of Bloch electrons. On the basis of a clearer understanding of the quantum confinement of one-dimensional Bloch waves, we can now make some comments on the use of EMA in the one-dimensional case.

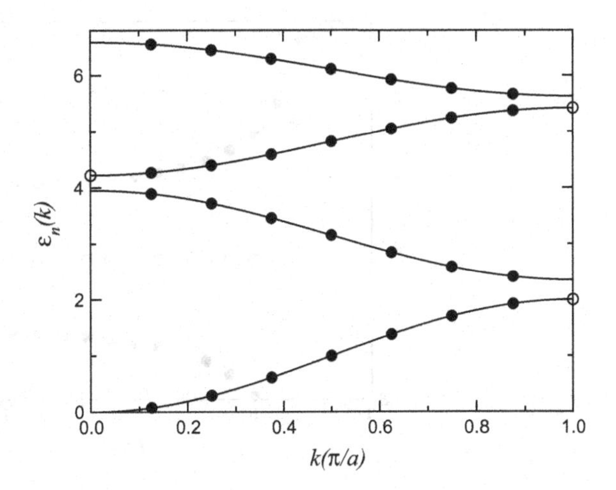

Fig. 4.5 A comparison between the energy spectrum of $\Lambda_{n,j}$ (*solid circles*) and $\Lambda_{\tau,n}$ (*open circles*) in a finite symmetric crystal of length $L = 8a$ and the energy bands $\varepsilon_n(k)$ (*solid lines*) for the lowest four bands. Note that (1) $\Lambda_{n,j}$ maps the energy bands exactly and (2) the existence of a confined band-edge state in each band gap

(1) We have understood that the complete quantum confinement of one-dimensional Bloch waves produces two different types of electronic states. The very existence of the τ-dependent states is a fundamental distinction of the quantum confinement of Bloch waves. *EMA completely misses the very existence of the τ-dependent states and thus misses a fundamental distinction of the quantum confinement of Bloch waves.*

(2) *EMA can be a good approximation for the L-dependent states.*

From (4.7), we know that the energies of L-dependent electronic states can be written as

$$\Lambda_{n,j} = \varepsilon_n \left(\frac{j\pi}{L} \right), \quad j = 1, 2, \ldots, N-1.$$

Near a band edge, $\varepsilon_n(k)$ can be approximated. For example, near a band edge at $k = 0$, we may approximate $\varepsilon_n(k)$ as

$$\varepsilon_n(k) \approx \varepsilon_n(0) + \frac{1}{2} \left. \frac{\mathrm{d}^2 \varepsilon_n(k)}{\mathrm{d}k^2} \right|_{k=0} k^2. \tag{4.12}$$

Thus, for the L-dependent states near the band edge, we have

$$\Lambda_{n,j} \approx \varepsilon_n(0) + \frac{1}{2} \left. \frac{\mathrm{d}^2 \varepsilon_n(k)}{\mathrm{d}k^2} \right|_{k=0} \frac{j^2 \pi^2}{L^2}. \tag{4.13}$$

This expression is the EMA result for the complete quantum confinement of one-dimensional Bloch waves. Thus, for L-dependent states near a band edge at $k = 0$, as long as (4.12) is a good approximation, the exact result (4.7) approximately gives the results (4.13), the same as EMA. A corresponding expression of EMA can be easily obtained for the confined electronic states near a band gap at $k = \frac{\pi}{a}$.

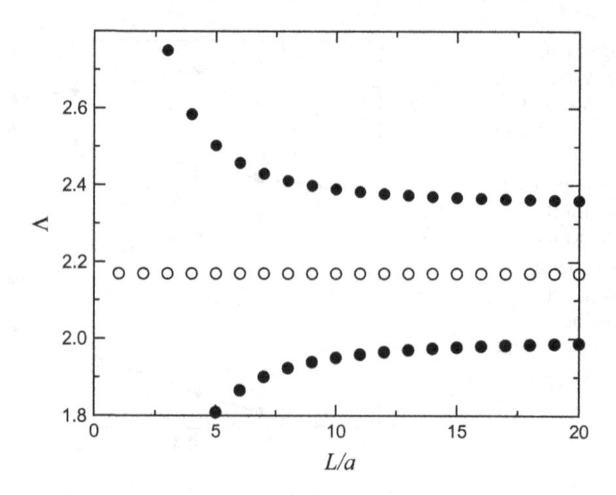

Fig. 4.6 The energies of three confined electronic states near or in the lowest band gap of (4.1) as functions of the crystal length L while τ is fixed. Note that the energy of the τ-dependent electronic state (*open circles*) in the gap is independent of L, and the energy of each L-dependent electronic state (*solid circles*) changes as L changes. EMA completely neglects the existence of the τ-dependent state; however, it may give a good description of the behaviors of the two L-dependent states (see the text). Reprinted with permission from S. Y. Ren: Ann. Phys. (NY) **301**, 22 (2002). Copyright by Elsevier

Originally, the EMA was developed for treating the electronic states near band edges in the presence of slowly varying and weak perturbations, such as an external electric and/or magnetic field as well as the potential of shallow impurities [11]. Nevertheless, we have seen here that in using the EMA to treat the *L-dependent states* in one-dimensional finite crystals, the *only* requirement is that the energy band $\varepsilon_n(k)$ near the band edge can be approximated by a parabolic energy band such as in (4.12). Even though in quantum confinement problems, the perturbation is neither weak nor slowly varying at the confinement boundaries, the original conditions for justifying the use of EMA are thus completely violated.[5] Figure 4.6 shows the energies of three electronic states in crystals of different lengths near or in the lowest band gap of the Bloch waves as functions of the crystal length L, obtained in [1]. The two points we commented on here can be clearly seen.

4.7 Comments on the Surface States

The surface states in one-dimensional crystals are the simplest surface states. Some properties of surface states in one-dimensional crystals are usually easy to analyze. Based on a clearer understanding of the electronic states in one-dimensional finite

[5]This is true only in the cases where the interested band edge is located either at the center or at the boundary of the Brillouin zone. It may not be true in the low-dimensional systems or finite crystals investigated in Part III.

crystals, we can also make comments on several interesting problems on the surface states in one-dimensional crystals.

Although investigations of the surface states have made significant progress since Tamm's classical work [7] many years ago, it seems that some fundamental problems on surface states have not yet been very well understood. One of such problems is how many surface states are there in a simplest one-dimensional finite crystal with two ends. Fowler [12] was the first to think that there are two surface states in each band gap in a one-dimensional finite crystal due to its two ends. This idea seemed natural and was soon accepted [13]. A long-standing belief of many people in the solid-state physics community is that there are two surface states in each gap of a one-dimensional finite crystal because it has two ends [14].

To the contrary, the general results on the electronic states in ideal one-dimensional finite crystals obtained here indicate that there is always one and only one electronic state whose energy depends on the crystal boundary but not the crystal length for each band gap. A surface state is one of the two possibilities of such a boundary-dependent state. Therefore, there is *at most* one surface state in each band gap in an ideal one-dimensional finite crystal. This result is, of course, different from the long-standing belief. The analytical results published in [1] were confirmed by numerical calculations [15, 16].

Why can a finite one-dimensional crystal of *two ends* only have *at most one* surface state in a band gap? The answer is because, (1) as we analyzed in Chap. 3 (and further in Appendix A), a termination of the periodic potential $v(x)$ may or may not cause a surface state in a band gap; (2) the two ends of a finite one-dimensional crystal, in general, are not equivalent.

Only in a symmetric one-dimensional finite crystal such as treated in Sect. 4.5 are the two ends of the crystal equivalent. We have understood that then there is always a confined band-edge state rather than a surface state in each gap.

In most cases, the two ends of a finite one-dimensional crystal are not equivalent. The properties of the τ-dependent state for *each gap* will depend on the relation between the crystal boundary τ and the zeros of the two corresponding band-edge wave functions. For a specific band gap, if τ is next to a zero of the upper band-edge wave function *in the finite crystal*, then $\tau + L$ will be next to a zero of the lower band-edge wave function *in the finite crystal*, and vice versa. Thus, the two ends of the finite crystals are not equivalent; it is possible that a surface state exists near only one of the two ends. From what we analyzed in Chap. 3, we can see that it is the end next to a zero of the upper band-edge wave function $\phi_{2m+1}(\frac{\pi}{a}, x)$ or $\phi_{2m+2}(0, x)$ *in the finite crystal* that may have a surface state localized near it.

Figure 4.2 can also be considered as obtained by combining Fig. 3.1 with Fig. 3.2. If τ is in a dotted region of the Λ_1–τ curve (τ is next to an open circle *in the finite crystal*), there is a surface state located near the left end τ of the finite crystal. If τ is in a dashed region of the Λ_1–τ curve ($\tau + L$ will be next to an open circle *in the finite crystal*), there is a surface state located near the right end $\tau + L$ of the finite crystal. If τ is located at one of the circles, then there is a confined band-edge state. Therefore, an ideal one-dimensional finite crystal bounded at τ and $\tau + L$ can have *at most* one surface state in each band gap, even though it always has two ends.

More mathematically, this is because that for a finite one-dimensional crystal bounded at τ and $\tau + Na$, *both* the left end τ and the right end $\tau + Na$ always belong to the *same one* of the three sets $L(n)$, $M(n)$, and $R(n)$ in Sect. 3.3. Thus, consequently, only one of the three possibilities is possible: a surface state on the left of the finite crystal, a band-edge state, or a surface state on the right of the finite crystal.

Having understood clearly the analysis presented here, we can see that the belief that a finite one-dimensional crystal always has two surface states in each gap is a misconception.[6]

In a tight-binding formalism, the number of permitted energy bands is determined by the number of states per unit cell. By using a nearest neighbor tight-binding formalism, Hatsugai [17] proved that in a finite linear chain with q states per unit cell, there is a total of $q - 1$ edge states, one in each of the $q - 1$ gaps. The properties of those edge states in [17] are somewhat similar to the τ-dependent states in this chapter—they can either be located near either end of the chain or be a band-edge state.[7]

A surface electronic state is usually understood as an electronic state that is mainly distributed near a specific surface of the crystal. Now, we have a more extended concept of the surface-like states: The electronic states whose properties and energies are determined by the surface location, that is the τ-dependent states discussed in this chapter. A confined band-edge state is merely a particular case of a surface-like state with its decay factor β being zero—in an ideal one-dimensional finite crystal, it happens when the surface location is a zero of a band-edge wave function.

The spatial extension of a surface state is determined by its decay factor $\beta(\Lambda)$. $\beta(\lambda)$ as a function of λ can be obtained from the discriminant $D(\lambda)$ of (4.1). In particular, for a surface state in a band gap at $k = 0$, from (2.78) we obtain

$$\beta(\lambda)a = \ln\left[\frac{D(\lambda) + \sqrt{D^2(\lambda) - 4}}{2}\right]. \tag{4.14}$$

Therefore, the surface state with an energy Λ at which $D(\lambda)$ takes a maximum has the largest decay factor and thus the smallest spatial extension in such a band gap.

Similarly, for a surface state in a band gap at $k = \frac{\pi}{a}$, from (2.82) we have

$$\beta(\lambda)a = \ln\left[\frac{-D(\lambda) + \sqrt{D^2(\lambda) - 4}}{2}\right]. \tag{4.14a}$$

[6] All of these discussions here are for ideal one-dimensional finite crystals defined by (4.3) and (4.4). An investigation of the electronic states in one-dimensional symmetric finite crystals with relaxed boundary conditions $(\psi'/\psi)_{x=\tau} = -(\psi'/\psi)_{x=L+\tau} = \sigma$ for finite σ can be found in Appendix B.

[7] By the theory in this book, the q states per unit cell give q permitted bands thus q τ-dependent states. It is *the tight-binding approximation* used by Hatsugai that gives $q - 1$ gaps thus only $q - 1$ edge states. It is well known that [18] in a tight-binding formalism with a single state per unit cell, a linear finite chain does not have a surface state. The reason is quite simple—there is no band gap ($q = 1$) in the band structure in that formalism.

Thus, the surface state with an energy Λ at which $D(\lambda)$ takes a minimum has the largest decay factor and thus the smallest spatial extension in such a band gap.

From Fig. 2.1, we can see that a surface state with its energy roughly near the mid-gap has a larger decay factor β and thus a smaller spatial extension and a surface state with its energy near a band edge has a smaller decay factor β and thus a larger spatial extension. By considering the energy of a surface state as a function of the surface position τ such as shown in Fig. 4.2, we can obtain the following qualitative understandings: A surface position near a zero of the lower band-edge wave function $\phi_{2m+1}(0, x)$ corresponds to a surface state with an energy near the lower band-edge $\varepsilon_{2m+1}(0)$ and a smaller decay factor β thus a larger spatial extension. A surface position near a zero of the upper band-edge wave function $\phi_{2m+2}(0, x)$ corresponds to a surface state with an energy near the upper band-edge $\varepsilon_{2m+2}(0)$ and also a smaller decay factor β thus a larger spatial extension. A surface position near a midpoint between two consecutive zeros of two band-edge wave functions corresponds to a surface state with an energy near the mid-gap and a larger decay factor β thus a smaller spatial extension. A surface state in a band gap at $k = \frac{\pi}{a}$ can be similarly analyzed.

The value of the decay factor β of a surface state is determined by the discriminant $D(\lambda)$, according to (4.14) or (4.14a). Therefore, by a consideration from either a limit of wide permitted bands and narrow band gap or a limit of the contrary and by referring to Fig. 2.1, we can obtain such qualitative conclusions: For a specific band gap, the smaller the two relevant permitted band widths are and/or the larger the band gap is, the larger the largest decay factor β_{max} in the band gap can be.[8]

However, some conclusions obtained for surface states in one-dimensional crystals are not valid for surface states in three-dimensional crystals. That a surface state is always in a band gap and decays most at the mid-gap is only a unique distinction of one-dimensional crystals.

4.8 Two Other Comments

4.8.1 A Comment on the Formation of the Energy Bands

The electron states in an infinite crystal with translational invariance have an energy band structure; in each permitted energy band, the energy spectrum is a continuum. On the other hand, the electron states in a finite system always have a discrete energy spectrum. One interesting question is, how are those energy bands formed as the number of atoms increases gradually? From (4.7), we can see that, in one-

[8]A near-zero band gap makes the largest possible numerator in (4.14) or (4.14a) small; therefore, the largest decay factor β_{max} for a surface state in the band gap is also small. On the contrary, a narrow permitted band width makes $|D'(\lambda)|$ at its band edge large and thus makes the largest possible numerator in (4.14) or (4.14a) large for the band gap, which leads to a large β_{max} in the band gap.

dimensional cases, the mapping of the energy bands by electronic states in finite crystals begins at $N = 2$ and linearly increases as N increases: A finite crystal of length $L = Na$ always has $N - 1$ stationary Bloch states in each energy band whose energies map the energy band exactly.

4.8.2 A Comment on the Boundary Locations

A consequence of the results obtained in this chapter is that the real boundary locations τ and $\tau + L$ of an ideal one-dimensional finite crystal discussed in this chapter are determined *only* by the τ-dependent electronic states. In our simplifying assumptions, the many-body effects between the electrons are neglected; the total energy of the system is simply the summation of the energies of all occupied single-electronic states, including the L-dependent states and τ-dependent states. Therefore, the real boundary locations τ and $\tau + L$ of a finite crystal with a fixed length L in our simplified model are determined by the condition that the summation of the energies of all occupied τ-dependent states takes the minimum.

4.9 Summary

In summary, based on the mathematical theory of ordinary differential equations with periodic coefficients, in particular, theorems on zeros of solutions of (4.1), we have obtained exact and general results on the properties of all electronic states in the simplest finite crystals—the ideal one-dimensional crystals of finite length. For a one-dimensional crystal bounded at τ and $\tau + L$ where $L = Na$, corresponding to each energy band of (4.1), there are two different types of electronic states: There are $N - 1$ Bloch stationary states in the energy band. Their energies Λ are given by (4.7) and thus depend on the crystal length L but not the crystal boundary location τ and map the energy band exactly. These stationary Bloch states can be considered as bulk-like states in the one-dimensional finite crystal; There is always one and only one electronic state in the band gap above the energy band, its energy Λ depends on the boundary location τ but not the crystal length L. Such a τ-dependent state is either a surface state in the band gap or a confined band-edge state, depending on whether τ is a zero of a band-edge wave function of (4.1). A slight change of the crystal boundary location τ could change the properties and the energy of this τ-dependent state dramatically. These τ-dependent states can be considered as surface-like states in the finite crystal. A confined band-edge state is a surface-like state with its decay factor $\beta = 0$.

The very existence of these τ-dependent surface-like states is a fundamental distinction of the quantum confinement of Bloch waves.

Some points disused here will be further illustrated by using a Kronig–Penney model in Appendix A.

The exact and general results obtained indicate that the primary difficulty or obstacle due to the lack of translational invariance in one-dimensional finite crystals, in fact, could be circumvented.

The general understandings obtained here provide a basis for further understanding of the quantum confinement of three-dimensional Bloch waves and the electronic states in low-dimensional systems and finite crystals in three-dimensional cases.

References

1. S.Y. Ren, Ann. Phys. (NY) **301**, 22 (2002)
2. S.Y. Ren, Phys. Rev. **B64**, 035322 (2001)
3. F.B. Pedersen, P.C. Hemmer, Phys. Rev. **B50**, 7724 (1994)
4. S.B. Zhang, A. Zunger, Appl. Phys. Lett. **63**, 1399 (1993)
5. S.B. Zhang, C.-Y. Yeh, A. Zunger, Phys. Rev. **B48**, 11204 (1993)
6. Z.V. Popovic, M. Cardona, E. Richter, D. Strauch, L. Tapfer, K. Ploog, Phys. Rev. **B40**, 1207 (1989); Z.V. Popovic, M. Cardona, E. Richter, D. Strauch, L. Tapfer, K. Ploog, Phys. Rev. **B40**, 3040 (1989); Z.V. Popovic, M. Cardona, E. Richter, D. Strauch, L. Tapfer, K. Ploog. Phys. Rev. **B41**, 5904 (1990)
7. I. Tamm, Phys. Z. Sowj. **1**, 733 (1932)
8. M.S.P. Eastham, *Theory of Ordinary Differential Equations* (Van Nostrand Reinhold, London, 1970)
9. L.D. Landau, E.M. Lifshitz, *Quantum Mechanics* (Pergamon Press, Paris, 1962)
10. A. Franceschetti, A. Zunger, Appl. Phys. Lett. **68**, 3455 (1996)
11. J.M. Luttinger, W. Kohn, Phys. Rev. **97**, 869 (1957); W. Kohn, *Solid State Physics*, vol. 5, ed. by F. Seitz, D. Turnbull (Academic Press, New York, 1955), pp. 257–320
12. R.H. Fowler, Proc. R. Soc. **141**, 56 (1933)
13. F. Seitz, *The Modern Theory of Solids* (McGraw-Hill, New York, 1940)
14. S.G. Davison, M. Stęślicka, *Basic Theory of Surface States* (Clarendon Press, Oxford, 1992)
15. Y. Zhang, private communications
16. Y.L. Xuan, P.W. Zhang, private communications
17. Y. Hatsugai, Phys. Rev. **B48**, 11851 (1993)
18. W.A. Harrison, Bull. Am. Phys. Soc. **47**, 367 (2002)

Part III
Low-Dimensional Systems and Finite Crystals

Chapter 5
Electronic States in Ideal Quantum Films

Starting from this chapter, we extend our investigations in Part II to three-dimensional crystals. The major difference between the problems treated in this part and in Part II is that the corresponding Schrödinger equation for the electronic states in a three-dimensional crystal is a *partial* differential equation; therefore, now the problem is a more difficult one. In comparison to the solutions of ordinary differential equations, the properties of solutions of partial differential equations are much less understood mathematically (e.g., [1]). The variety and complexity of the three-dimensional crystal structures and the shapes of three-dimensional finite crystals further make the cases more variational and more complicated.

Nevertheless, it is now understood that the band structure of a multidimensional crystal is fundamentally different from the band structure of a one-dimensional crystal [2–4].

A multidimensional crystal can be considered as the result of a multidimensional free space—a crystal with a zero potential (empty lattice)—reconstructed by a perturbation of the crystal potential $v(\mathbf{x})$, see, for example, [5]. The crystal with an empty lattice has a free-electron-like energy spectrum $\lambda(\mathbf{k}) = ck^2 = c(\mathbf{k} \cdot \mathbf{k})$ (c is a proportional constant) without any band gap in an extended Brillouin zone scheme. In the reduced Brillouin zone scheme, the energy spectrum has *significant band crossings and band overlaps* due to the band folding of $\lambda(\mathbf{k}) = ck^2$ at the multidimensional Brillouin zone boundary, since $\lambda(\mathbf{k}) = ck^2$ is not constant at the multidimensional Brillouin zone boundary. *The higher the dimensionality is and/or the higher* $\lambda(\mathbf{k})$ *is, the more significant are the band crossings and the band overlaps*. The Fourier components of the crystal potential $v(\mathbf{x})$ may open gaps (local in the $\mathbf{k}$ space) for specific Bloch waves, such as at the Brillouin zone center and other (often highly symmetrical) points at the Brillouin zone boundary, thus to change the energy spectrum formed by band folding of the free-electron-like $\lambda(\mathbf{k})$ to the band structure $\varepsilon_n(\mathbf{k})$ of the crystal. However, *The higher the dimensionality is and/or the higher* $\lambda(\mathbf{k})$ *is, the more difficult it is for these local gap openings to overcome the band crossings and the band overlaps* to dramatically change the energy spectrum. Therefore, in cases where either $\lambda(\mathbf{k})$ is higher enough (depending on the dimensionality) and/or the crystal potential $v(\mathbf{x})$ is not large enough, the local (in the $\mathbf{k}$ space) band gap

S.Y. Ren, *Electronic States in Crystals of Finite Size*, Springer Tracts in Modern Physics 270, DOI 10.1007/978-981-10-4718-3_5

openings would not be able to open *a true band gap in the whole Brillouin zone*. Such a case is significantly different from the one-dimensional case described in p. 43. It was conjectured by Bethe and Sommerfeld [6] first and then proved by Skriganov [7] (and also others [2–4]) that the number of band gaps in the energy spectrum of a multidimensional crystal is *always* finite. Furthermore, if the potential is small enough, there is no band gap at all in the energy spectrum. Consequently, unlike in a one-dimensional crystal where each permitted band and each band gap always exist alternatively as the energy increases, in a multidimensional crystal, the permitted bands are strongly overlapped, and the number of band gaps is finite.[1] The significant differences between the band structure of a multidimensional crystal and that of a one-dimensional crystal will naturally lead to the differences between the truncated multidimensional periodicity and the truncated one-dimensional periodicity, that is, the differences between the quantum confinements of two different dimensional Bloch waves.

The electronic states in a quantum film can be seen as the quantum confinement of three-dimensional Bloch waves in one specific direction. This case is the simplest case of the quantum confinement of three-dimensional Bloch waves. Our purpose in this chapter is to understand the similarities and the differences between the effects of the quantum confinement of three-dimensional Bloch waves in one specific direction and the quantum confinement of one-dimensional Bloch waves as treated in Chap. 4. Based on an extension of a mathematical theorem in [8], we demonstrate that in many simple but interesting cases, the properties of electronic states in an ideal quantum film can be understood, and the energies of many electronic states can be directly obtained from the energy band structure of the bulk. Again, the major obstacle due to the lack of translational invariance can be circumvented.[2]

This chapter is organized as follows: In Sect. 5.1, we present a basic theorem that corresponds to Theorem 2.8 in one-dimensional cases and is the basis of the theory in this chapter. In Sect. 5.2, we briefly discuss some consequences of the theorem. In Sects. 5.3–5.6, we obtain the electronic states in several ideal quantum films of different Bravais lattices by reasonings based on this theorem and physical intuition, by considering the quantum confinement of three-dimensional Bloch waves in one specific direction.[3] In Sect. 5.7, we compare the analytical results obtained here to previously published numerical results. In Sect. 5.8 are some further discussions.

[1] It was pointed out in [2] that many properties of the energy band structure of a one-dimensional crystal as summarized in p. 43 are closely related to the fact that a second-order ordinary differential equation cannot have more than two independent solutions. For a partial differential equation, these claims are generally incorrect.

[2] Part of the results of this chapter was published in [9].

[3] The terminology in Chaps. 5–7 of the first edition followed the one used in [8] where Eastham investigated a general N-dimensional case. Since we specifically treat three-dimensional cases in this book, we use more specific terms, such as parallelepiped, edge, vertex, etc., in this edition.

5.1 A Basic Theorem

The single-electron Schrödinger equation for a three-dimensional crystal can be written as

$$-\nabla^2 y(\mathbf{x}) + [v(\mathbf{x}) - \lambda]y(\mathbf{x}) = 0, \tag{5.1}$$

where $v(\mathbf{x})$ is a periodic potential:

$$v(\mathbf{x} + \mathbf{a}_1) = v(\mathbf{x} + \mathbf{a}_2) = v(\mathbf{x} + \mathbf{a}_3) = v(\mathbf{x}).$$

$\mathbf{a}_1$, $\mathbf{a}_2$, and $\mathbf{a}_3$ are three primitive lattice vectors of the crystal. The corresponding primitive lattice vectors in the $\mathbf{k}$ space are denoted as $\mathbf{b}_1$, $\mathbf{b}_2$, and $\mathbf{b}_3$ and $\mathbf{a}_i \cdot \mathbf{b}_j = \delta_{i,j}$; here, $\delta_{i,j}$ is the Kronecker symbol. The position vector $\mathbf{x}$ can be written as $\mathbf{x} = x_1\mathbf{a}_1 + x_2\mathbf{a}_2 + x_3\mathbf{a}_3$ and the $\mathbf{k}$ vector as $\mathbf{k} = k_1\mathbf{b}_1 + k_2\mathbf{b}_2 + k_3\mathbf{b}_3$.

The eigenfunctions of (5.1) satisfying the condition

$$\phi(\mathbf{k}, \mathbf{x} + \mathbf{a}_i) = \mathrm{e}^{\mathrm{i}k_i}\phi(\mathbf{k}, \mathbf{x}), \quad -\pi < k_i \leq \pi, \quad i = 1, 2, 3 \tag{5.2}$$

are three-dimensional Bloch functions. As solutions of (5.1), the three-dimensional Bloch functions and the energy bands in this book are denoted as $\phi_n(\mathbf{k}, \mathbf{x})$ and $\varepsilon_n(\mathbf{k})$: $\varepsilon_0(\mathbf{k}) \leq \varepsilon_1(\mathbf{k}) \leq \varepsilon_2(\mathbf{k}) \leq \cdots$. The energy band structure in the Cartesian system is denoted as $\varepsilon_n(k_x, k_y, k_z)$.

For the quantum films investigated in this chapter, we choose the primitive vectors $\mathbf{a}_1$ and $\mathbf{a}_2$ in the film plane and use $\hat{\mathbf{k}}$ to express a wave vector in the film plane: $\hat{\mathbf{k}} = k_1\hat{\mathbf{b}}_1 + k_2\hat{\mathbf{b}}_2$. $\hat{\mathbf{b}}_1$ and $\hat{\mathbf{b}}_2$ are in the film plane and $\mathbf{a}_i \cdot \hat{\mathbf{b}}_j = \delta_{i,j}$ for $i, j = 1, 2$.

The major mathematical basis for understanding the electronic states in one-dimensional finite crystals is Theorem 2.8. Correspondingly, the mathematical basis for understanding the quantum confinement of three-dimensional Bloch waves in a specific direction $\mathbf{a}_3$ is the following eigenvalue problem (5.3)—which corresponds to the problem defined by (2.94) in one-dimensional cases—and a relevant theorem.

Suppose A is a parallelepiped that has $\mathbf{a}_i$ forming the edges that meet at a vertex and has the bottom face defined by $x_3 = \tau_3$ and thus the top face defined by $x_3 = (\tau_3 + 1)$ (See Fig. 5.1.).[4] The function set $\hat{\phi}(\hat{\mathbf{k}}, \mathbf{x}; \tau_3)$ is defined by the conditions

$$\begin{cases} \hat{\phi}(\hat{\mathbf{k}}, \mathbf{x} + \mathbf{a}_i; \tau_3) = \mathrm{e}^{\mathrm{i}k_i}\hat{\phi}(\hat{\mathbf{k}}, \mathbf{x}; \tau_3), & -\pi < k_i \leq \pi, \quad i = 1, 2, \\ \hat{\phi}(\hat{\mathbf{k}}, \mathbf{x}; \tau_3) = 0, & \mathbf{x} \in \partial A_3, \end{cases} \tag{5.3}$$

---

[4]For a free-standing film with a boundary at $x_3 = \tau_3$, in general, we have neither a reason to require that τ_3 be a constant nor a reasonable way to assign $\tau_3(x_1, x_2)$ beforehand. However, since what we are interested in is mainly the quantum confinement effects, in this book it is assumed that the existence of the boundary τ_3 does not change the two-dimensional space group symmetry of the system, including but not limited to that $\tau_3 = \tau_3(x_1, x_2)$ must be a periodic function of x_1 and x_2: $\tau_3 = \tau_3(x_1, x_2) = \tau_3(x_1 + 1, x_2) = \tau_3(x_1, x_2 + 1)$.

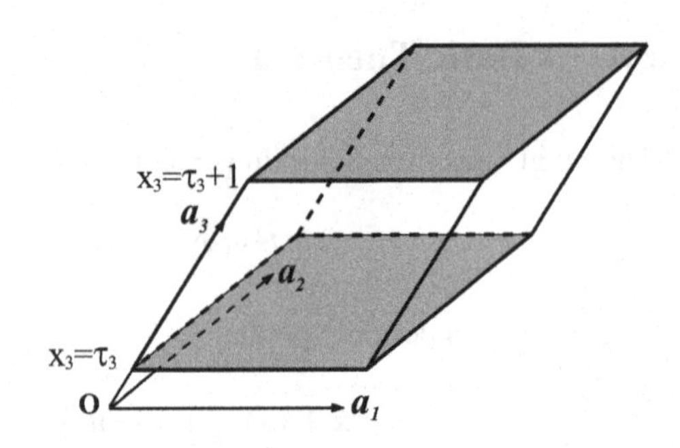

Fig. 5.1 The parallelepiped A for the eigenvalue problem of Eq. (5.1) under the boundary condition (5.3). The two shadowed faces of ∂A_3 are defined by $x_3 = \tau_3$ and $x_3 = (\tau_3 + 1)$ and are the two faces on which the function $\hat{\phi}(\hat{\mathbf{k}}, \mathbf{x}; \tau_3)$ is zero

where ∂A_i means two opposite faces of the boundary ∂A of A determined by the beginning and the end of $\mathbf{a}_i$. The eigenvalues and eigenfunctions of Eq. (5.1) under the boundary condition (5.3) are denoted by $\hat{\lambda}_n(\hat{\mathbf{k}}; \tau_3)$ and $\hat{\phi}_n(\hat{\mathbf{k}}, \mathbf{x}; \tau_3)$ respectively, where $\hat{\mathbf{k}}$ is a wave vector in the film plane and $n = 0, 1, 2, \ldots$.

For the eigenvalues of Eq. (5.1) in two different eigenvalue problems defined by Eqs. (5.2) and (5.3), the following theorem exists:

Theorem 5.1

$$\hat{\lambda}_n(\hat{\mathbf{k}}; \tau_3) \geq \varepsilon_n(\mathbf{k}), \qquad for\ (\mathbf{k} - \hat{\mathbf{k}}) \cdot \mathbf{a}_i = 0, \quad i = 1, 2. \tag{5.4}$$

In Eqs. (5.2)–(5.4), $\mathbf{k}$ is a three-dimensional wave vector defined in the whole Brillouin zone, and $\hat{\mathbf{k}}$ is in the film plane. In (5.4), $\mathbf{k}$ and $\hat{\mathbf{k}}$ have the same components in the film plane. In this book, we say that a three-dimensional wavevector $\mathbf{k}$ is related to the two-dimensional wavevector $\hat{\mathbf{k}}$ in the $\mathbf{a}_1, \mathbf{a}_2$ plane if $(\mathbf{k} - \hat{\mathbf{k}}) \cdot \mathbf{a}_i = 0, \ i = 1, 2$, or briefly, $\mathbf{k}$ is related to $\hat{\mathbf{k}}$.

Theorem 5.1 can be considered as an extension of Theorem 6.3.1 in [8], and the proof is similar, with some differences.

Proof We choose $\phi_n(\mathbf{k}, \mathbf{x})$ to be normalized over A: $\int_A \phi_n(\mathbf{k}; \mathbf{x})\phi_n^*(\mathbf{k}; \mathbf{x})\, d\mathbf{x} = 1$.

We denote $\mathscr{F}$ as the set of all complex-valued functions $f(\mathbf{x})$ that are continuous in A and have piecewise continuous first-order partial derivatives in A. The Dirichlet integral $J(f, g)$ in three dimensions is defined by

$$J(f, g) = \int_A \{\nabla f(\mathbf{x}) \cdot \nabla g^*(\mathbf{x}) + v(\mathbf{x}) f(\mathbf{x}) g^*(\mathbf{x})\}\, d\mathbf{x} \tag{5.5}$$

for $f(\mathbf{x})$ and $g(\mathbf{x})$ in $\mathscr{F}$. If in (5.5) $g(\mathbf{x})$ also has piecewise continuous second-order partial derivatives in A, from the Gauss's divergence theorem we have

$$J(f, g) = \int_A f(\mathbf{x})\{-\nabla^2 g^*(\mathbf{x}) + v(\mathbf{x}) g^*(\mathbf{x})\}\, d\mathbf{x} + \int_{\partial A} f(\mathbf{x}) \frac{\partial g^*(\mathbf{x})}{\partial n}\, dS, \tag{5.6}$$

where ∂A denotes the boundary of A, $\partial/\partial n$ denotes derivative along the outward normal to ∂A, and dS denotes an element of surface area of ∂A.

If $f(\mathbf{x})$ and $g(\mathbf{x})$ satisfy the conditions (5.2), the integral over ∂A in (5.6) is zero because the integrals over opposite faces of ∂A cancel out. Therefore, when $g(\mathbf{x}) = \phi_n(\mathbf{k}, \mathbf{x})$, (5.6) gives

$$J(f, g) = \varepsilon_n(\mathbf{k}) \int_A f(\mathbf{x})\phi_n^*(\mathbf{k}, \mathbf{x})\, d\mathbf{x}.$$

Therefore,

$$J[\phi_m(\mathbf{k}, \mathbf{x}), \phi_n(\mathbf{k}, \mathbf{x})] = \begin{cases} \varepsilon_n(\mathbf{k}), & m = n, \\ 0, & m \neq n. \end{cases}$$

Now, we consider the function set $\hat{\phi}(\hat{\mathbf{k}}, \mathbf{x}; \tau_3)$, which satisfy the boundary conditions (5.3). We also choose $\hat{\phi}(\hat{\mathbf{k}}, \mathbf{x}; \tau_3)$ to be normalized over A: $\int_A \hat{\phi}(\hat{\mathbf{k}}, \mathbf{x}; \tau_3)\hat{\phi}^*(\hat{\mathbf{k}}, \mathbf{x}; \tau_3)\, d\mathbf{x} = 1$.

Note that if $f(\mathbf{x}) = \hat{\phi}(\hat{\mathbf{k}}, \mathbf{x}; \tau_3)$ and $g(\mathbf{x}) = \phi_n(\mathbf{k}, \mathbf{x})$, the integral over ∂A in (5.6) is also zero because the integral over two opposite faces of ∂A_1 and ∂A_2 cancel out since $(\mathbf{k} - \hat{\mathbf{k}}) \cdot \mathbf{a}_i = 0$ for $i = 1, 2$ and the integral over each face of ∂A_3 is zero since $f(\mathbf{x}) = 0$ when $\mathbf{x} \in \partial A_3$.

Thus,

$$J[\hat{\phi}(\hat{\mathbf{k}}, \mathbf{x}; \tau_3), \phi_n(\mathbf{k}, \mathbf{x})] = \varepsilon_n(\mathbf{k}) f_n(\mathbf{k}),$$

where

$$f_n(\mathbf{k}) = \int_A \hat{\phi}(\hat{\mathbf{k}}, \mathbf{x}; \tau_3)\phi_n^*(\mathbf{k}, \mathbf{x})\, d\mathbf{x},$$

and

$$\sum_{n=0}^{\infty} |f_n(\mathbf{k})|^2 = 1$$

by the Parseval formula [8, 10, 11]. An important property of the function $\hat{\phi}(\hat{\mathbf{k}}, \mathbf{x}; \tau_3)$ defined by (5.3) is

$$J[\hat{\phi}(\hat{\mathbf{k}}, \mathbf{x}; \tau_3), \hat{\phi}(\hat{\mathbf{k}}, \mathbf{x}; \tau_3)] \geq \sum_{n=0}^{\infty} \varepsilon_n(\mathbf{k}) |f_n(\mathbf{k})|^2. \tag{5.7}$$

To prove (5.7), we assume $v(\mathbf{x}) \geq 0$ first. Then $J(f, f) \geq 0$ from (5.5) for any f in $\mathscr{F}$. Thus, for any positive integer N, we have

$$J[\hat{\phi}(\hat{\mathbf{k}}, \mathbf{x}; \tau_3) - \sum_{n=0}^{N} f_n(\mathbf{k})\phi_n(\mathbf{k}, \mathbf{x}),\ \hat{\phi}(\hat{\mathbf{k}}, \mathbf{x}; \tau_3) - \sum_{n=0}^{N} f_n(\mathbf{k})\phi_n(\mathbf{k}, \mathbf{x})] \geq 0;$$

that is,

$$J[\hat{\phi}(\hat{\mathbf{k}}, \mathbf{x}; \tau_3), \hat{\phi}(\hat{\mathbf{k}}, \mathbf{x}; \tau_3)] \geq \sum_{n=0}^{N} \varepsilon_n(\mathbf{k})|f_n(\mathbf{k})|^2.$$

N can be as large as needed, therefore,

$$J[\hat{\phi}(\hat{\mathbf{k}}, \mathbf{x}; \tau_3), \hat{\phi}(\hat{\mathbf{k}}, \mathbf{x}; \tau_3)] \geq \sum_{n=0}^{\infty} \varepsilon_n(\mathbf{k})|f_n(\mathbf{k})|^2, \quad \text{if } v(\mathbf{x}) \geq 0. \tag{5.8}$$

To prove (5.7) without the assumption that $v(\mathbf{x}) \geq 0$, let v_0 be a constant that is sufficiently large to make $v(\mathbf{x}) + v_0 \geq 0$ in A. Then (5.1) can be rewritten as

$$-\nabla^2 y(\mathbf{x}) + [V(\mathbf{x}) - \Lambda]y(\mathbf{x}) = 0, \tag{5.9}$$

where $V(\mathbf{x}) = v(\mathbf{x}) + v_0$ and $\Lambda = \lambda + v_0$. Since in (5.9), $V(\mathbf{x}) \geq 0$ in A, due to (5.8) we have

$$\int_A \{\nabla\hat{\phi}(\hat{\mathbf{k}}, \mathbf{x}; \tau_3) \cdot \nabla\hat{\phi}^*(\hat{\mathbf{k}}, \mathbf{x}; \tau_3) + [v(\mathbf{x}) + v_0]\hat{\phi}(\hat{\mathbf{k}}, \mathbf{x}; \tau_3)\hat{\phi}^*(\hat{\mathbf{k}}, \mathbf{x}; \tau_3)\}\, d\mathbf{x}$$
$$\geq \sum_{n=0}^{\infty} [\varepsilon_n(\mathbf{k}) + v_0]|f_n(\mathbf{k})|^2;$$

that is,

$$\int_A [\nabla\hat{\phi}(\hat{\mathbf{k}}, \mathbf{x}; \tau_3) \cdot \nabla\hat{\phi}^*(\hat{\mathbf{k}}, \mathbf{x}; \tau_3) + v(\mathbf{x})\hat{\phi}(\hat{\mathbf{k}}, \mathbf{x}; \tau_3)\hat{\phi}^*(\hat{\mathbf{k}}, \mathbf{x}; \tau_3)]\, d\mathbf{x}$$
$$\geq \sum_{n=0}^{\infty} \varepsilon_n(\mathbf{k})|f_n(\mathbf{k})|^2.$$

This is (5.7). Equation (5.4) can generally be proven on the basis of (5.7).

We consider

$$\hat{\phi}(\hat{\mathbf{k}}, \mathbf{x}; \tau_3) = c_0\hat{\phi}_0(\hat{\mathbf{k}}, \mathbf{x}; \tau_3) + c_1\hat{\phi}_1(\hat{\mathbf{k}}, \mathbf{x}; \tau_3) + \cdots + c_n\hat{\phi}_n(\hat{\mathbf{k}}, \mathbf{x}; \tau_3)$$

and choose $n + 1$ constants c_i to make

$$\sum_{i=0}^{n} |c_i|^2 = 1,$$

and

$$f_i(\mathbf{k}) = \int_A \hat{\phi}(\hat{\mathbf{k}}, \mathbf{x}; \tau_3)\phi_i^*(\mathbf{k}, \mathbf{x})\, d\mathbf{x} = 0, \quad i = 0, 1, \ldots, n-1. \tag{5.10}$$

Equation (5.10) corresponds to n homogeneous algebraic equations for $n+1$ constants $c_0, c_1, \ldots, c_n$. A choice of such c_i's is always possible. Therefore,

$$\hat{\lambda}_n(\hat{\mathbf{k}};\tau_3) \geq \sum_{i=0}^{n} |c_i|^2 \hat{\lambda}_i(\hat{\mathbf{k}};\tau_3) = J[\hat{\phi}(\hat{\mathbf{k}},\mathbf{x};\tau_3),\hat{\phi}(\hat{\mathbf{k}},\mathbf{x};\tau_3)]$$
$$\geq \sum_{i=0}^{\infty} |f_i(\mathbf{k})|^2 \varepsilon_i(\mathbf{k}) = \sum_{i=n}^{\infty} |f_i(\mathbf{k})|^2 \varepsilon_i(\mathbf{k}) \geq \varepsilon_n(\mathbf{k}) \sum_{i=n}^{\infty} |f_i(\mathbf{k})|^2 = \varepsilon_n(\mathbf{k}).$$

This is (5.4). □

5.2 Consequences of the Theorem

Theorem 5.1 indicates that for each bulk energy band n and each $\hat{\mathbf{k}}$, for any specific τ_3 there is always one and only one $\hat{\lambda}_n(\hat{\mathbf{k}};\tau_3)$ and thus one $\hat{\phi}_n(\hat{\mathbf{k}},\mathbf{x};\tau_3)$.

Theorem 5.1 gives a relationship between two sets of eigenvalues $\varepsilon_n(\mathbf{k})$ and $\hat{\lambda}_n(\hat{\mathbf{k}};\tau_3)$. On this basis, many consequences concerning the relationship between two sets of eigenfunctions $\phi_n(\mathbf{k},\mathbf{x})$ and $\hat{\phi}_n(\hat{\mathbf{k}},\mathbf{x};\tau_3)$ can be obtained.

Theorem 5.1 gives a lower limit of $\hat{\lambda}_n(\hat{\mathbf{k}};\tau_3)$. As a consequence of this, only the *one* Bloch function $\phi_n(\mathbf{k},\mathbf{x})$ corresponding to the energy maximum of $\varepsilon_n(\mathbf{k})$ with a $\mathbf{k}$ related to $\hat{\mathbf{k}}$ *may have the possibility of being* $\hat{\phi}_n(\hat{\mathbf{k}},\mathbf{x};\tau_3)$ to make the equality in (5.4) to be valid. If the equality in (5.4) is valid, such a Bloch function $\phi_n(\mathbf{k},\mathbf{x})$ has a nodal surface (the surface on which the Bloch function is zero) at $x_3=\tau_3$.

Theorem 5.1 does not give an upper limit of $\hat{\lambda}_n(\hat{\mathbf{k}};\tau_3)$. It means that the possibility that a Bloch function $\phi_{n'}(\mathbf{k},\mathbf{x})$ has a nodal surface at $x_3=\tau_3$ and thus is a $\hat{\phi}_n(\hat{\mathbf{k}},\mathbf{x};\tau_3)$ (in which $n<n'$) cannot be excluded. Such an $\varepsilon_{n'}(\mathbf{k})$ does not have to be the energy maximum with a $\mathbf{k}$ related to $\hat{\mathbf{k}}$.

In this chapter, we are mainly interested in the quantum films of crystals with a band structure $\varepsilon_n(\hat{\mathbf{k}}+k_3\mathbf{b}_3)=\varepsilon_n(\hat{\mathbf{k}}-k_3\mathbf{b}_3)$. In a case where $\hat{\phi}_n(\hat{\mathbf{k}},\mathbf{x};\tau_3)$ is a Bloch function $\phi_{n'}(\mathbf{k},\mathbf{x})$ in which $n\leq n'$, the corresponding wave vector $\mathbf{k}$ must be either $\mathbf{k}=\hat{\mathbf{k}}$ or $\mathbf{k}=\hat{\mathbf{k}}+\pi\mathbf{b}_3$.[5] Therefore, only a Bloch function $\phi_{n'}(\mathbf{k},\mathbf{x})$ with such a wave vector $\mathbf{k}$ might have a nodal surface at $x_3=\tau_3$.[6] In particular, in the cases where $\hat{\mathbf{k}}=0$, only a Bloch function $\phi_{n'}(\mathbf{k}=0,\mathbf{x})$ or $\phi_{n'}(\mathbf{k}=\pi\mathbf{b}_3,\mathbf{x})$ might have a nodal surface at $x_3=\tau_3$.

These points can find similarities in the one-dimensional case.

However, even for a wave vector $\mathbf{k}=\hat{\mathbf{k}}$ or $\mathbf{k}=\hat{\mathbf{k}}+\pi\mathbf{b}_3$, $\varepsilon_n(\mathbf{k})$ is the energy maximum for the $\hat{\mathbf{k}}$ related $\mathbf{k}$, the Bloch function $\phi_n(\mathbf{k},\mathbf{x})$ might or might not have

---

[5] In such a case, $\hat{\lambda}_n(\hat{\mathbf{k}};\tau_3)=\varepsilon_{n'}(\mathbf{k})=\varepsilon_{n'}(\hat{\mathbf{k}}+k_3\mathbf{b}_3)=\varepsilon_{n'}(\hat{\mathbf{k}}-k_3\mathbf{b}_3)$ is true. Only when either $k_3=0$ or $k_3=\pi$ is $\phi_{n'}(\hat{\mathbf{k}}\pm k_3\mathbf{b}_3,\mathbf{x})$ one single function and $\varepsilon_{n'}(\hat{\mathbf{k}}\pm k_3\mathbf{b}_3)$ one single eigenvalue.

[6] Note that we commented on the zeros of one-dimensional Bloch function $\phi_n(k,x)$ on p. 48, as a consequence of Theorem 2.8.

a nodal surface,[7] not to mention a specific nodal surface at $x_3 = \tau_3$. Only when the specific $\varepsilon_n(\mathbf{k})$ is the energy maximum for a $\mathbf{k}$ related to $\hat{\mathbf{k}}$, *and* the corresponding Bloch function $\phi_n(\mathbf{k}, \mathbf{x})$ does have a nodal surface at $x_3 = \tau_3$, can the equality in Theorem 5.1 be valid. If this is the case, the Bloch function $\phi_n(\mathbf{k}, \mathbf{x})$ is $\hat{\phi}_n(\hat{\mathbf{k}}, \mathbf{x}; \tau_3)$ for that specific τ_3.

Theorem 5.1 is not as strong as Theorem 2.8: No upper limit of $\hat{\lambda}_n(\hat{\mathbf{k}}; \tau_3)$ is given except $\hat{\lambda}_n(\hat{\mathbf{k}}; \tau_3) \leq \hat{\lambda}_{n+1}(\hat{\mathbf{k}}; \tau_3)$. It is this point that leads to a significant difference between a surface-like state in an ideal quantum film and a surface-like state in a one-dimensional finite crystal.[8]

More on these points and consequences of Theorem 5.1 will be seen later in this chapter. Essentially, it is the consequences of Theorem 5.1 that lead to the similarities and differences of the quantum confinement of three-dimensional Bloch waves in one specific direction in comparison with the results obtained in Chap. 4.

Since $v(\mathbf{x} + \mathbf{a}_3) = v(\mathbf{x})$, the function $\hat{\phi}_n(\hat{\mathbf{k}}, \mathbf{x}; \tau_3)$ has the form

$$\hat{\phi}_n(\hat{\mathbf{k}}, \mathbf{x} + \mathbf{a}_3; \tau_3) = e^{ik_3} \hat{\phi}_n(\hat{\mathbf{k}}, \mathbf{x}; \tau_3). \tag{5.11}$$

Depending on n, $\hat{\mathbf{k}}$, and τ_3, k_3 in (5.11) can be complex or real. If in (5.11) k_3 is real, then $\hat{\phi}_n(\hat{\mathbf{k}}, \mathbf{x}; \tau_3)$ is a Bloch function. There exist such cases, but probably in most cases, $\hat{\phi}_n(\hat{\mathbf{k}}, \mathbf{x}; \tau_3)$ is not a Bloch function. Even though a specific Bloch function $\phi_{n'}(\mathbf{k}, \mathbf{x})$ could have a nodal surface at a specific $x_3 = \tau_3$ and thus is a $\hat{\phi}_n(\hat{\mathbf{k}}, \mathbf{x}; \tau_3)$, the possibility that other Bloch functions with a different n' and/or $\hat{\mathbf{k}}$ will have a same nodal surface (see later in this chapter) is small. In the theorem τ_3 is treated as a general variable, k_3 in (5.11) can be real only in special cases where $\hat{\phi}_n(\hat{\mathbf{k}}, \mathbf{x}; \tau_3)$ is a Bloch function for a specific τ_3; in most cases, k_3 is complex with a nonzero imaginary part.

Depending on n, $\hat{\mathbf{k}}$, and τ_3, the imaginary part of k_3 in (5.11) can be either positive or negative, corresponding to whether $\hat{\phi}_n(\hat{\mathbf{k}}, \mathbf{x}; \tau_3)$ decays in either the positive or the negative direction of $\mathbf{a}_3$. Such states $\hat{\phi}_n(\hat{\mathbf{k}}, \mathbf{x}; \tau_3)$ with a nonzero imaginary part of k_3 cannot exist in a bulk crystal with three-dimensional translational invariance since they are divergent in either the negative or the positive direction of $\mathbf{a}_3$. However, they can play a significant role in a quantum film of finite thickness.

---

[7]In one-dimensional cases, it is Theorem 2.7 that warrants that a band-edge Bloch function $\phi_n(k, x)$ at $k = 0$ or $k = \frac{\pi}{a}$ ($\phi_0(0, x)$ excluded) always has zeros.

[8]Mathematically, it is the Floquet theory and Theorem 2.8 that limits the energy range of any surface-like state in a one-dimensional finite crystal always in a band gap. This is also related to that in a one-dimensional crystal, each permitted energy band and each band gap always exist alternatively as the energy increases. By [2], the origin is that a second-order ordinary differential equation cannot have more than two independent solutions.

5.3 Basic Considerations on the Electronic States in an Ideal Quantum Film

For the electronic states in an ideal low-dimensional system such as in an ideal quantum film, wire, dot, or finite crystal, we treat them based on two assumptions. (1) The potential $v(\mathbf{x})$ *inside* the low-dimensional system is the same as in (5.1): (2) The electronic states are completely confined in the system.

In an ideal quantum film with N_3 layers in the $\mathbf{a}_3$ direction, the electronic states $\hat{\psi}(\hat{\mathbf{k}}, \mathbf{x})$ are solutions of the following two equations:

$$\begin{cases} -\nabla^2 \hat{\psi}(\hat{\mathbf{k}}, \mathbf{x}) + [v(\mathbf{x}) - \hat{\Lambda}]\hat{\psi}(\hat{\mathbf{k}}, \mathbf{x}) = 0, & \tau_3 < x_3 < \tau_3 + N_3, \\ \hat{\psi}(\hat{\mathbf{k}}, \mathbf{x}) = 0, & x_3 \leq \tau_3 \text{ or } x_3 \geq \tau_3 + N_3, \end{cases} \tag{5.12}$$

where $x_3 = \tau_3$ defines the bottom of the film and N_3 is a positive integer indicating the film thickness.[9] These electronic states $\hat{\psi}(\hat{\mathbf{k}}, \mathbf{x})$ are two-dimensional Bloch waves in the film plane, with additional index(es) indicating the confinement in the $\mathbf{a}_3$ direction.

In the following, we try to find solutions of (5.12) in some simple cases by reasonings based on Theorem 5.1 and physical intuition. Our main purpose is to understand the underlying physics of the electronic states in low-dimensional systems in simple examples, rather than to explore more possibly treatable quantum films.

5.4 Stationary Bloch States

We can expect that in some interesting cases, one type of solutions of (5.12) should be linear combinations of three-dimensional Bloch functions: In the film, they are stationary Bloch states, formed due to multiple reflections of the Bloch waves $\phi_n(\mathbf{k}, \mathbf{x})$ at the two boundary surfaces of the film, whereas they are two-dimensional Bloch waves in the film plane.[10]

---

[9] In this book, such a film is usually called a film with N_3 layers in the $\mathbf{a}_3$ direction, despite the fact that the film may actually have more atomic layers.

[10] Unlike in a one-dimensional case there the stationary Bloch state solutions of Eqs. (4.3) and (4.4) such as (4.10) type always exist for *any* τ; the author does not have an in-depth and comprehensive understanding on the existence and properties of solutions of periodic partial differential equations mathematically to assert that for what τ_3 the stationary Bloch state solutions of (5.12) exist. However, he expects that the existence of such type of states is reasonable for a physical film with a physical boundary τ_3.

5.4.1 The Simplest Cases

The simplest cases are films of the crystals with a band structure having the following symmetry:

$$\varepsilon_n(k_1\mathbf{b}_1 + k_2\mathbf{b}_2 + k_3\mathbf{b}_3) = \varepsilon_n(k_1\mathbf{b}_1 + k_2\mathbf{b}_2 - k_3\mathbf{b}_3). \tag{5.13}$$

If (5.13) is true, we can expect that in some interesting cases, the stationary Bloch states in such a quantum film can be obtained from the linear combinations of $\phi_n(k_1\mathbf{b}_1 + k_2\mathbf{b}_2 + k_3\mathbf{b}_3, \mathbf{x})$ and $\phi_n(k_1\mathbf{b}_1 + k_2\mathbf{b}_2 - k_3\mathbf{b}_3, \mathbf{x})$, since, in general,

$$\begin{aligned} f_{n,k_1,k_2,k_3}(\mathbf{x}) = c_+\phi_n(k_1\mathbf{b}_1 + k_2\mathbf{b}_2 + k_3\mathbf{b}_3, \mathbf{x}) \\ + c_-\phi_n(k_1\mathbf{b}_1 + k_2\mathbf{b}_2 - k_3\mathbf{b}_3, \mathbf{x}), \quad 0 < k_3 < \pi \end{aligned}$$

—where $c_\pm$ are nonzero constant coefficients—is a nontrivial solution of (5.1) due to (5.13). To be a solution of (5.12), the function $f_{n,k_1,k_2,k_3}(\mathbf{x}; \tau_3)$ is required to be zero at the bottom of the film $x_3 = \tau_3$ and at the top of the film $x_3 = \tau_3 + N_3$. By writing $\mathbf{x} = \hat{\mathbf{x}} + x_3\mathbf{a}_3$, where $\hat{\mathbf{x}} = x_1\mathbf{a}_1 + x_2\mathbf{a}_2$, we should have

$$\begin{aligned} c_+\phi_n(k_1\mathbf{b}_1 + k_2\mathbf{b}_2 + k_3\mathbf{b}_3, \hat{\mathbf{x}} + \tau_3\mathbf{a}_3) \\ + c_-\phi_n(k_1\mathbf{b}_1 + k_2\mathbf{b}_2 - k_3\mathbf{b}_3, \hat{\mathbf{x}} + \tau_3\mathbf{a}_3) = 0, \\ c_+\phi_n[k_1\mathbf{b}_1 + k_2\mathbf{b}_2 + k_3\mathbf{b}_3, \hat{\mathbf{x}} + (\tau_3 + N_3)\mathbf{a}_3] \\ + c_-\phi_n[k_1\mathbf{b}_1 + k_2\mathbf{b}_2 - k_3\mathbf{b}_3, \hat{\mathbf{x}} + (\tau_3 + N_3)\mathbf{a}_3] = 0. \end{aligned} \tag{5.14}$$

However, we have

$$\begin{aligned} \phi_n[k_1\mathbf{b}_1 + k_2\mathbf{b}_2 + k_3\mathbf{b}_3, \hat{\mathbf{x}} + (\tau_3 + N_3)\mathbf{a}_3] \\ = \mathrm{e}^{\mathrm{i}k_3N_3}\phi_n(k_1\mathbf{b}_1 + k_2\mathbf{b}_2 + k_3\mathbf{b}_3, \hat{\mathbf{x}} + \tau_3\mathbf{a}_3), \end{aligned}$$

and

$$\begin{aligned} \phi_n[k_1\mathbf{b}_1 + k_2\mathbf{b}_2 - k_3\mathbf{b}_3, \hat{\mathbf{x}} + (\tau_3 + N_3)\mathbf{a}_3] \\ = \mathrm{e}^{-\mathrm{i}k_3N_3}\phi_n(k_1\mathbf{b}_1 + k_2\mathbf{b}_2 - k_3\mathbf{b}_3, \hat{\mathbf{x}} + \tau_3\mathbf{a}_3). \end{aligned}$$

Therefore, for $c_\pm$ in (5.14) not both zero, $\mathrm{e}^{\mathrm{i}k_3N_3} - \mathrm{e}^{-\mathrm{i}k_3N_3} = 0$ has to be true for these stationary Bloch states.

Each stationary Bloch state solution of (5.12) has the form

$$\hat{\psi}_{n,j_3}(\hat{\mathbf{k}}, \mathbf{x}; \tau_3) = \begin{cases} f_{n,k_1,k_2,\kappa_3}(\mathbf{x}; \tau_3), & \tau_3 < x_3 < \tau_3 + N_3, \\ 0, & x_3 \leq \tau_3 \text{ or } x_3 \geq \tau_3 + N_3, \end{cases} \tag{5.15}$$

where

$$f_{n,k_1,k_2,k_3}(\mathbf{x};\, \tau_3) = c_{n,k_1,k_2,k_3;\tau_3}\phi_n(k_1\mathbf{b}_1 + k_2\mathbf{b}_2 + k_3\mathbf{b}_3, \mathbf{x}) \\ + c_{n,k_1,k_2,-k_3;\tau_3}\phi_n(k_1\mathbf{b}_1 + k_2\mathbf{b}_2 - k_3\mathbf{b}_3, \mathbf{x}), \quad (5.16)$$

$c_{n,k_1,k_2,\pm k_3;\tau_3}$ depend on τ_3, $\hat{\mathbf{k}} = k_1\hat{\mathbf{b}}_1 + k_2\hat{\mathbf{b}}_2$, and

$$\kappa_3 = j_3\pi/N_3, \quad j_3 = 1, 2, \ldots, N_3 - 1. \quad (5.17)$$

Here, j_3 is the stationary state index. It is easy to see that $f_{n,k_1,k_2,k_3}(\mathbf{x};\, \tau_3)$ defined in (5.16) is a two-dimensional Bloch wave with a wave vector $\hat{\mathbf{k}} = k_1\hat{\mathbf{b}}_1 + k_2\hat{\mathbf{b}}_2$ in the film plane:

$$f_{n,k_1,k_2,k_3}(\mathbf{x} + \mathbf{a}_i;\, \tau_3) = \mathrm{e}^{\mathrm{i}k_i} f_{n,k_1,k_2,k_3}(\mathbf{x};\, \tau_3), \quad -\pi < k_i \leq \pi,\ i = 1, 2. \quad (5.18)$$

The stationary Bloch state $\hat{\psi}_{n,j_3}(\hat{\mathbf{k}}, \mathbf{x};\, \tau_3)$ has the energy

$$\hat{\Lambda}_{n,j_3}(\hat{\mathbf{k}}) = \varepsilon_n(k_1\mathbf{b}_1 + k_2\mathbf{b}_2 + \kappa_3\mathbf{b}_3). \quad (5.19)$$

The (001) quantum films of crystals with a simple cubic (sc), a tetragonal (tetr), or an orthorhombic (ortho) Bravais lattice have $\hat{\mathbf{b}}_1 = \mathbf{b}_1$ and $\hat{\mathbf{b}}_2 = \mathbf{b}_2$. Such quantum films with a bulk band structure having the symmetry (5.13) are the simplest cases to which the theory in this section can be applied. There are $N_3 - 1$ stationary Bloch state solutions $\hat{\psi}_{n,j_3}(\hat{\mathbf{k}}, \mathbf{x};\, \tau_3)$ for each n and each $\hat{\mathbf{k}}$ in such a film of N_3 layers. They are two-dimensional Bloch waves with a wave vector $\hat{\mathbf{k}}$ in the film plane. Their energies $\hat{\Lambda}_{n,j_3}(\hat{\mathbf{k}})$ depend on the film thickness N_3 and map the bulk energy band structure $\varepsilon_n(\mathbf{k})$ exactly. These states can be seen as the bulk-like states in the quantum film.

5.4.2 *More General Cases*

In general, (5.13) is not true for many crystals. The arguments in Sect. 5.4.1 are not valid for general films. Nevertheless, it can be shown that if the band structure of a bulk crystal has the symmetry

$$\varepsilon_n(k_1\hat{\mathbf{b}}_1 + k_2\hat{\mathbf{b}}_2 + k_3\mathbf{b}_3) = \varepsilon_n(k_1\hat{\mathbf{b}}_1 + k_2\hat{\mathbf{b}}_2 - k_3\mathbf{b}_3), \quad (5.20)$$

in a film of N_3 layers, in some interesting cases, there are $N_3 - 1$ stationary Bloch states for each n and each $\hat{\mathbf{k}} = k_1\hat{\mathbf{b}}_1 + k_2\hat{\mathbf{b}}_2$, similar to the cases in Sect. 5.4.1. The energies of these stationary Bloch states can be similarly obtained.

If (5.20) is true, we can expect that in some interesting cases, the stationary Bloch states in such a quantum film can be obtained from the linear combinations of $\phi_n(k_1\hat{\mathbf{b}}_1 + k_2\hat{\mathbf{b}}_2 + k_3\mathbf{b}_3, \mathbf{x})$ and $\phi_n(k_1\hat{\mathbf{b}}_1 + k_2\hat{\mathbf{b}}_2 - k_3\mathbf{b}_3, \mathbf{x})$, since, in general,

$$f_{n,k_1,k_2,k_3}(\mathbf{x}) = c_+\phi_n(k_1\hat{\mathbf{b}}_1 + k_2\hat{\mathbf{b}}_2 + k_3\mathbf{b}_3, \mathbf{x}) \\ + c_-\phi_n(k_1\hat{\mathbf{b}}_1 + k_2\hat{\mathbf{b}}_2 - k_3\mathbf{b}_3, \mathbf{x}), \quad 0 < k_3 < \pi, \tag{5.21}$$

—where $c_\pm$ are nonzero constant coefficients—is a nontrivial solution of (5.1) due to (5.20). It is easy to see that $f_{n,k_1,k_2,k_3}(\mathbf{x})$ defined in (5.21) is a two-dimensional Bloch wave with a wave vector $\hat{\mathbf{k}} = k_1\hat{\mathbf{b}}_1 + k_2\hat{\mathbf{b}}_2$ in the film plane:

$$f_{n,k_1,k_2,k_3}(\mathbf{x} + \mathbf{a}_i) = \mathrm{e}^{\mathrm{i}k_i} f_{n,k_1,k_2,k_3}(\mathbf{x}), \quad -\pi < k_i \leq \pi, \;\; i = 1, 2. \tag{5.22}$$

To be a solution of (5.12), the function $f_{n,k_1,k_2,k_3}(\mathbf{x}; \tau_3)$ is required to be zero at the bottom of the film $x_3 = \tau_3$ and at the top of the film $x_3 = \tau_3 + N_3$. By writing $\mathbf{x} = \hat{\mathbf{x}} + x_3\mathbf{a}_3$, where $\hat{\mathbf{x}} = x_1\mathbf{a}_1 + x_2\mathbf{a}_2$, we should have

$$\begin{aligned} &c_+\phi_n(k_1\hat{\mathbf{b}}_1 + k_2\hat{\mathbf{b}}_2 + k_3\mathbf{b}_3, \hat{\mathbf{x}} + \tau_3\mathbf{a}_3) \\ &\qquad + c_-\phi_n(k_1\hat{\mathbf{b}}_1 + k_2\hat{\mathbf{b}}_2 - k_3\mathbf{b}_3, \hat{\mathbf{x}} + \tau_3\mathbf{a}_3) = 0, \\ &c_+\phi_n[k_1\hat{\mathbf{b}}_1 + k_2\hat{\mathbf{b}}_2 + k_3\mathbf{b}_3, \hat{\mathbf{x}} + (\tau_3 + N_3)\mathbf{a}_3] \\ &\qquad + c_-\phi_n[k_1\hat{\mathbf{b}}_1 + k_2\hat{\mathbf{b}}_2 - k_3\mathbf{b}_3, \hat{\mathbf{x}} + (\tau_3 + N_3)\mathbf{a}_3] = 0. \end{aligned} \tag{5.23}$$

We can write $\hat{\mathbf{b}}_i = \mathbf{b}_i + \alpha_i\mathbf{b}_3$ $(i = 1, 2)$, i.e.

$$k_1\hat{\mathbf{b}}_1 + k_2\hat{\mathbf{b}}_2 + k_3\mathbf{b}_3 = k_1\mathbf{b}_1 + k_2\mathbf{b}_2 + (\alpha_1 k_1 + \alpha_2 k_2 + k_3)\mathbf{b}_3.$$

Then we have

$$\phi_n(k_1\hat{\mathbf{b}}_1 + k_2\hat{\mathbf{b}}_2 + k_3\mathbf{b}_3, \mathbf{x} + \mathbf{a}_3) = \mathrm{e}^{\mathrm{i}(\alpha_1 k_1 + \alpha_1 k_2 + k_3)}\phi_n(k_1\hat{\mathbf{b}}_1 + k_2\hat{\mathbf{b}}_2 + k_3\mathbf{b}_3, \mathbf{x}),$$

and

$$\phi_n(k_1\hat{\mathbf{b}}_1 + k_2\hat{\mathbf{b}}_2 - k_3\mathbf{b}_3, \mathbf{x} + \mathbf{a}_3) = \mathrm{e}^{\mathrm{i}(\alpha_1 k_1 + \alpha_1 k_2 - k_3)}\phi_n(k_1\mathbf{b}_1 + k_2\hat{\mathbf{b}}_2 - k_3\mathbf{b}_3, \mathbf{x}).$$

Therefore, for $c_\pm$ in (5.23) not both zero, $\mathrm{e}^{\mathrm{i}k_3N_3} - \mathrm{e}^{-\mathrm{i}k_3N_3} = 0$ has to be true for these stationary Bloch states.

Each stationary Bloch state solution $\hat{\psi}_{n,j_3}(\hat{\mathbf{k}}, \mathbf{x}; \tau_3)$ of Eq. (5.12) should have the form

$$\hat{\psi}_{n,j_3}(\hat{\mathbf{k}}, \mathbf{x}; \tau_3) = \begin{cases} f_{n,k_1,k_2,\kappa_3}(\mathbf{x}; \tau_3), & \tau_3 < x_3 < \tau_3 + N_3, \\ 0, & x_3 \leq \tau_3 \text{ or } x_3 \geq \tau_3 + N_3, \end{cases} \tag{5.24}$$

where

$$f_{n,k_1,k_2,k_3}(\mathbf{x}; \tau_3) = c_{n,k_1,k_2,k_3;\tau_3}\phi_n(k_1\hat{\mathbf{b}}_1 + k_2\hat{\mathbf{b}}_2 + k_3\mathbf{b}_3, \mathbf{x}) \\ + c_{n,k_1,k_2,-k_3;\tau_3}\phi_n(k_1\hat{\mathbf{b}}_1 + k_2\hat{\mathbf{b}}_2 - k_3\mathbf{b}_3, \mathbf{x}), \tag{5.25}$$

$c_{n,k_1,k_2,\pm k_3;\tau_3}$ depend on τ_3, and

$$\kappa_3 = j_3\pi/N_3, \quad j_3 = 1, 2, \ldots, N_3 - 1, \tag{5.26}$$

as in (5.17). j_3 is the stationary state index. Therefore, there are $N_3 - 1$ solutions of Eq. (5.12) in a film of N_3 layers for each bulk energy band n. Each solution $\hat{\psi}_{n,j_3}(\hat{\mathbf{k}}, \mathbf{x}; \tau_3)$ in (5.24)–(5.26) is a stationary Bloch state in the normal direction $\mathbf{b}_3$ of the film, whereas it is a two-dimensional Bloch wave with a wave vector $\hat{\mathbf{k}} = k_1\hat{\mathbf{b}}_1 + k_2\hat{\mathbf{b}}_2$ in the film plane:

$$\hat{\psi}_{n,j_3}(\hat{\mathbf{k}}, \mathbf{x} + \mathbf{a}_i; \tau_3) = e^{ik_i}\hat{\psi}_{n,j_3}(\hat{\mathbf{k}}, \mathbf{x}; \tau_3), \quad -\pi < k_i \leq \pi, \quad i = 1, 2, \tag{5.27}$$

due to (5.22). The corresponding energy for each such state is

$$\hat{\Lambda}_{n,j_3}(\hat{\mathbf{k}}) = \varepsilon_n(\hat{\mathbf{k}} + \kappa_3\mathbf{b}_3). \tag{5.28}$$

A (001) quantum film of a crystal with a sc, tetr, or ortho Bravais lattice and with a bulk band structure (5.13) can also be considered as a special and simple case of the more general cases discussed in this subsection. For such a film $\alpha_1 = \alpha_2 = 0$.

Equations (5.24)–(5.28) are similar to that there are $N - 1$ bulk-like states in each band in a one-dimensional finite crystal of length $L = Na$, as indicated by (4.7). The electronic states $\hat{\psi}_{n,j_3}(\hat{\mathbf{k}}, \mathbf{x}; \tau_3)$ in (5.24)–(5.26) can be seen as bulk-like electronic states in the quantum film and each $\hat{\Lambda}_{n,j_3}(\hat{\mathbf{k}})$ can be seen as a bulk-like subband: $\hat{\Lambda}_{n,j_3}(\hat{\mathbf{k}})$ maps the bulk energy band $\varepsilon_n(\mathbf{k})$ exactly by (5.28) and depends on the film thickness N_3.

5.5 τ_3-Dependent States

It is expected that for each n and each $\hat{\mathbf{k}}$, there are N_3 electronic states for an ideal quantum film of N_3 layers. For films in which (5.20) is true, $N_3 - 1$ states were obtained in (5.24)–(5.26); the other type of nontrivial solutions of Eq. (5.12) can be obtained from (5.11) by assigning

$$\hat{\psi}_n(\hat{\mathbf{k}}, \mathbf{x}; \tau_3) = \begin{cases} c_{N_3}\,\hat{\phi}_n(\hat{\mathbf{k}}, \mathbf{x}; \tau_3), & \tau_3 < x_3 < \tau_3 + N_3, \\ 0, & x_3 \leq \tau_3 \text{ or } x_3 \geq \tau_3 + N_3, \end{cases} \tag{5.29}$$

where c_{N_3} is a normalization constant. Correspondingly, the energy of such a state is given by

$$\hat{\Lambda}_n(\hat{\mathbf{k}}; \tau_3) = \hat{\lambda}_n(\hat{\mathbf{k}}; \tau_3). \tag{5.30}$$

There is one solution (5.29) of Eq. (5.12) for each energy band n and each $\hat{\mathbf{k}}$. Each $\hat{\psi}_n(\hat{\mathbf{k}}, \mathbf{x}; \tau_3)$ defined in (5.29) is an electronic state in the film whose energy $\hat{\Lambda}_n(\hat{\mathbf{k}}; \tau_3)$

(5.30) depends on the film boundary τ_3, but not on the film thickness N_3. By Theorem 5.1, $\hat{\Lambda}_n(\hat{\mathbf{k}}; \tau_3)$ is either above or, occasionally, at the energy maximum of $\varepsilon_n(\mathbf{k})$ for $\mathbf{k}$ related to $\hat{\mathbf{k}}$.

In the special cases where $\hat{\phi}_n(\hat{\mathbf{k}}, \mathbf{x}; \tau_3)$ in (5.29) is a Bloch function,

$$\hat{\phi}_n(\hat{\mathbf{k}}, \mathbf{x}; \tau_3) = \phi_{n'}(\mathbf{k}, \mathbf{x}), \quad n \leq n', \tag{5.31}$$

the corresponding Bloch function $\phi_{n'}(\mathbf{k}, \mathbf{x})$ has a nodal surface at $x_3 = \tau_3$ and thus has nodal surfaces at $x_3 = \tau_3 + \ell$, where $\ell = 1, 2, \ldots, N_3$. As we pointed out in Sect. 5.2, the wave vector has to be either $\mathbf{k} = \hat{\mathbf{k}}$ or $\mathbf{k} = \hat{\mathbf{k}} + \pi\mathbf{b}_3$. If $n = n'$, $\varepsilon_n(\mathbf{k})$ has to be the energy maximum for $\mathbf{k}$ related to $\hat{\mathbf{k}}$.

In most cases, $\hat{\phi}_n(\hat{\mathbf{k}}, \mathbf{x}; \tau_3)$ in (5.29) is not a Bloch function. Consequently, in these cases there is a nonzero imaginary part of k_3 in (5.11), indicating that $\hat{\psi}_n(\hat{\mathbf{k}}, \mathbf{x}; \tau_3)$ now is a surface state located near either the top or the bottom of the film. Correspondingly, the energy of such a state

$$\hat{\Lambda}_n(\hat{\mathbf{k}}; \tau_3) > \varepsilon_n(\mathbf{k}), \quad (\mathbf{k} - \hat{\mathbf{k}}) \cdot \mathbf{a}_i = 0, \quad i = 1, 2, \tag{5.32}$$

by Theorem 5.1. However, there is no reason to expect that $\hat{\Lambda}_n(\hat{\mathbf{k}}; \tau_3)$ has to be in a band gap like that Theorem 2.8 requires of $\Lambda_{\tau,n}$ in one-dimensional cases.

Each $\hat{\psi}_n(\hat{\mathbf{k}}, \mathbf{x}; \tau_3)$ can be seen as a surface-like state in the film, in differentiation with the bulk-like states $\hat{\psi}_{n,j_3}(\hat{\mathbf{k}}, \mathbf{x}; \tau_3)$. Therefore, for an ideal quantum film bounded at $x_3 = \tau_3$ and $x_3 = (\tau_3 + N_3)$ in which (5.20) is true, for each bulk energy band n, there are $N_3 - 1$ bulk-like subbands with energies given by (5.28) and one surface-like subband with energies given by (5.30) in the quantum film.

Because of (5.4), (5.28), and (5.30), in the ideal quantum film discussed here, for each bulk energy band n the corresponding surface-like subband $\hat{\Lambda}_n(\hat{\mathbf{k}}; \tau_3)$ is *always above* the bulk-like subbands $\hat{\Lambda}_{n,j_3}(\hat{\mathbf{k}})$[11]:

$$\hat{\Lambda}_n(\hat{\mathbf{k}}; \tau_3) > \hat{\Lambda}_{n,j_3}(\hat{\mathbf{k}}). \tag{5.33}$$

The electronic states in each subband are two-dimensional Bloch waves in the film plane.

These results might be valid for (001) films of crystals with a sc, a tetr, or an ortho Bravais lattice for which (5.13) is true. More generally, they might also be correct for films of crystals for which (5.20) is true. In particular, they might be correct for ideal (001) or (110) quantum films of crystals with a face-centered-cubic (fcc) or a body-centered-cubic (bcc) Bravais lattice for which (5.20) is true.

[11] $\hat{\Lambda}_{n,j_3}(\hat{\mathbf{k}})$ can never be equal to $\hat{\Lambda}_n(\hat{\mathbf{k}}; \tau_3)$: $\hat{\mathbf{k}} + \kappa_3\mathbf{b}_3$ is neither $\hat{\mathbf{k}}$ nor $\hat{\mathbf{k}} + \pi\mathbf{b}_3$. The equality in (5.4) can be excluded in (5.33).

5.6 Several Practically More Interesting Films

All cubic semiconductors and many metals have an fcc Bravais lattice. All alkali metals (Li, Na, K, Rb, Cs, Fr) and many other metals have a bcc Bravais lattice. Therefore, the quantum films of crystals with an fcc Bravais lattice or a bcc Bravais lattice often are practically more interesting.

5.6.1 (001) Films with an fcc Bravais Lattice

For an fcc (001) film, the primitive lattice vectors can be chosen as

$$\mathbf{a}_1 = a/2(1, -1, 0), \quad \mathbf{a}_2 = a/2(1, 1, 0), \quad \mathbf{a}_3 = a/2(1, 0, 1); \tag{5.34}$$

thus, $\mathbf{b}_1 = 1/a(1, -1, -1)$, $\mathbf{b}_2 = 1/a(1, 1, -1)$, and $\mathbf{b}_3 = 1/a(0, 0, 2)$. Correspondingly, $\hat{\mathbf{b}}_1 = 1/a(1, -1, 0)$ and $\hat{\mathbf{b}}_2 = 1/a(1, 1, 0)$. Here, a is the lattice constant. This corresponds to $\alpha_1 = \alpha_2 = 1/2$ in Sect. 5.4.2.

In general, the band structure of a cubic semiconductor or an fcc metal has the symmetry

$$\varepsilon_n(k_x, k_y, k_z) = \varepsilon_n(k_x, k_y, -k_z);$$

thus, for such a (001) film, (5.20) is true. Therefore, the results in Sects. 5.4.2 and 5.5 might be applied to these films: For a film of N_3 layers, there are $N_3 - 1$ bulk-like subbands and one surface-like subband in the film for each bulk energy band. Equation (5.28) for (001) films can be written as

$$\hat{\Lambda}_{n,j_3}(\hat{\mathbf{k}}) = \varepsilon_n \left[k_1\hat{\mathbf{b}}_1 + k_2\hat{\mathbf{b}}_2 + \frac{j_3\pi}{N_3 a}(0, 0, 2) \right] \tag{5.35}$$

for any $\hat{\mathbf{k}} = k_1\hat{\mathbf{b}}_1 + k_2\hat{\mathbf{b}}_2$, where $j_3 = 1, 2, \ldots, N_3 - 1$, given by (5.26).

Now, τ_3 can be written as τ_{001}. By (5.33), each surface-like subband $\hat{\Lambda}_n(\hat{\mathbf{k}}; \tau_{001})$ is always above each relevant bulk-like subband $\hat{\Lambda}_{n,j_3}(\hat{\mathbf{k}})$. If a $\hat{\psi}_n(\hat{\mathbf{k}} = 0, \mathbf{x}; \tau_{001})$ inside the film is a Bloch function $\phi_n(\mathbf{k}, \mathbf{x})$, either $\mathbf{k} = 0$ or $\mathbf{k} = \frac{2\pi}{a}(0, 0, 1)$ must be true, and the corresponding energy is the energy maximum of $\varepsilon_n(k\mathbf{b}_3)$, since any $\mathbf{k} = k\mathbf{b}_3$ is related to $\hat{\mathbf{k}} = 0$. It is assumed that the existence of boundary faces of the film does not change the two-dimensional space group symmetry of the system (Footnote 4 on page 93). For each surface-like subband $\hat{\Lambda}_n(\hat{\mathbf{k}}; \tau_{001})$, it is expected that $\hat{\Lambda}_n(k_1\hat{\mathbf{b}}_1 + k_2\hat{\mathbf{b}}_2; \tau_{001}) = \hat{\Lambda}_n(k_1\hat{\mathbf{b}}_1 - k_2\hat{\mathbf{b}}_2; \tau_{001})$ is true.

A "new" way of choosing the primitive lattice vectors is

$$\mathbf{a}_1 = a/2(1, -1, 0), \quad \mathbf{a}_2 = a/2(1, 1, 0), \quad \mathbf{a}_3 = a(0, 0, 1), \tag{5.34a}$$

and thus $\mathbf{b}_1 = \hat{\mathbf{b}}_1 = 1/a(1, -1, 0)$, $\mathbf{b}_2 = \hat{\mathbf{b}}_2 = 1/a(1, 1, 0)$, and $\mathbf{b}_3 = 1/a(0, 0, 1)$.

By this "new" way of choosing the primitive lattice vectors, the "new" Brillouin zone (with two boundaries at $(0, 0, \pm 1)\frac{\pi}{a}$) is a half of the original Brillouin zone (with two boundaries at $(0, 0, \pm 2)\frac{\pi}{a}$). Therefore, each original bulk energy band now becomes two "new" energy bands in the "new" Brillouin zone (band folding). Accordingly, for a (001) film of thickness $N_{001}a$, where N_{001} is a positive integer, by the "new" description, there are $(N_{001} - 1)$ bulk-like subbands and one surface-like subband for each "new" energy band and thus $2(N_{001} - 1)$ bulk-like subbands and two surface-like subbands for each original energy band. From the original description, since $N_3 = 2N_{001}$, there are $2N_{001} - 1$ bulk-like subbands and one surface-like subband for each original energy band. This difference (one extra surface-like subband and one less bulk-like subband for each original bulk energy band in the "new" description) comes from the fact that the "new" description (5.34a) is only based on half of the whole symmetry of the film in the [001] direction. Thus, the "extra" surface-like subband in the "new" description, in fact, is a bulk-like subband in the original primitive lattice vector system (5.34) where the full symmetry of the film in the [001] direction is used. We mention this point here since we will meet some relevant situations later.

5.6.2 *(110) Films with an fcc Bravais Lattice*

For an fcc (110) film, the primitive lattice vectors can be chosen as

$$\mathbf{a}_1 = a/2(1, -1, 0), \quad \mathbf{a}_2 = a(0, 0, -1), \quad \mathbf{a}_3 = a/2(0, 1, 1); \tag{5.36}$$

thus, $\mathbf{b}_1 = 1/a(2, 0, 0)$, $\mathbf{b}_2 = 1/a(1, 1, -1)$, and $\mathbf{b}_3 = 1/a(2, 2, 0)$. Correspondingly, $\hat{\mathbf{b}}_1 = 1/a(1, -1, 0)$ and $\hat{\mathbf{b}}_2 = 1/a(0, 0, -1)$. This corresponds to $\alpha_1 = \alpha_2 = -1/2$ in Sect. 5.4.2.

In general, the band structure of a cubic semiconductor or an fcc metal has the symmetry

$$\varepsilon_n(k_x, k_y, k_z) = \varepsilon_n(k_y, k_x, k_z);$$

thus, for such an (110) film, (5.20) is true. Therefore, the results in Sects. 5.4.2 and 5.5 might be applied to these films: For a film of N_3 layers, there are $N_3 - 1$ bulk-like subbands and one surface-like subband in the film for each bulk energy band. Equation (5.28) can also be written as

$$\hat{\Lambda}_{n,j_3}(\hat{\mathbf{k}}) = \varepsilon_n \left[k_1\hat{\mathbf{b}}_1 + k_2\hat{\mathbf{b}}_2 + \frac{j_3\pi}{N_3 a}(2, 2, 0) \right] \tag{5.37}$$

for any $\hat{\mathbf{k}} = k_1\hat{\mathbf{b}}_1 + k_2\hat{\mathbf{b}}_2$, where $j_3 = 1, 2, \ldots, N_3 - 1$, given by (5.26).

Now, τ_3 can be written as τ_{110}. Because of (5.33), each surface-like subband $\hat{\Lambda}_n(\hat{\mathbf{k}}; \tau_{110})$ is always above each relevant bulk-like subband $\hat{\Lambda}_{n,j_3}(\hat{\mathbf{k}})$. If a $\hat{\psi}_n(\hat{\mathbf{k}} =$

$0, \mathbf{x}; \tau_{110})$ inside the film is a Bloch function $\phi_n(\mathbf{k}, \mathbf{x})$, either $\mathbf{k} = 0$ or $\mathbf{k} = \frac{\pi}{a}(2, 2, 0)$ must be true, and the corresponding energy is the energy maximum of $\varepsilon_n(k\mathbf{b}_3)$, since any $\mathbf{k} = k\mathbf{b}_3$ is related to $\hat{\mathbf{k}} = 0$. It is assumed that the existence of boundary faces of the film does not change the two-dimensional space group symmetry of the system (Footnote 4 on page 93). For each surface-like subband $\hat{\Lambda}_n(\hat{\mathbf{k}}; \tau_{110})$, it is expected that $\hat{\Lambda}_n(k_1\hat{\mathbf{b}}_1 + k_2\hat{\mathbf{b}}_2; \tau_{110}) = \hat{\Lambda}_n(k_1\hat{\mathbf{b}}_1 - k_2\hat{\mathbf{b}}_2; \tau_{110})$ is true.

Similar to Sect. 5.6.1, there is a "new" way of choosing the primitive lattice vectors as

$$\mathbf{a}_1 = a/2(1, -1, 0), \quad \mathbf{a}_2 = a(0, 0, -1), \quad \mathbf{a}_3 = a/2(1, 1, 0), \tag{5.36a}$$

and, thus, $\mathbf{b}_1 = \hat{\mathbf{b}}_1 = 1/a(1, -1, 0)$, $\mathbf{b}_2 = \hat{\mathbf{b}}_2 = 1/a(0, 0, -1)$, and $\mathbf{b}_3 = 1/a\ (1, 1, 0)$.

By this "new" way of choosing the primitive lattice vectors, the "new" Brillouin zone (with two boundaries at $\pm(1, 1, 0)\frac{\pi}{a}$) is a half of the original Brillouin zone (with two boundaries at $\pm(2, 2, 0)\frac{\pi}{a}$) and each original energy band now becomes two "new" energy bands in the "new" Brillouin zone (band folding). Accordingly, for an (110) film of thickness $N_{110}\sqrt{2}a/2$, where N_{110} is a positive integer, by the "new" description, there are $(N_{110} - 1)$ bulk-like subbands and one surface-like subband for each "new" energy band. That is, there are $2(N_{110} - 1)$ bulk-like subbands and two surface-like subbands for each original bulk energy band. From the original description, since $N_3 = 2N_{110}$ there are $2N_{110} - 1$ bulk-like subbands and one surface-like subband for each original bulk energy band. This difference (one extra surface-like subband and one less bulk-like subband for each original bulk energy band in the "new" description) comes from the fact that the "new" description (5.36a) is only based on half of the whole symmetry of the film in the [110] direction. Thus, the "extra" surface-like subband in the "new" description, in fact, is a bulk-like subband in the original primitive lattice vector system (5.36), where the full symmetry of the film in the [110] direction is used. We mention this point here since we will meet some relevant situations later.

5.6.3 (001) Films with a bcc Bravais Lattice

For a bcc (001) film, the primitive lattice vectors can be chosen as

$$\mathbf{a}_1 = a(1, 0, 0), \quad \mathbf{a}_2 = a(0, 1, 0), \quad \mathbf{a}_3 = a/2(1, 1, 1);$$

thus, $\mathbf{b}_1 = 1/a(1, 0, -1)$, $\mathbf{b}_2 = 1/a(0, 1, -1)$, and $\mathbf{b}_3 = 1/a(0, 0, 2)$. Correspondingly, $\hat{\mathbf{b}}_1 = 1/a(1, 0, 0)$, $\hat{\mathbf{b}}_2 = 1/a(0, 1, 0)$. This corresponds to that $\alpha_1 = \alpha_2 = 1/2$ in Sect. 5.4.2. In general, the band structure of a bcc metal has the symmetry $\varepsilon_n(k_x, k_y, k_z) = \varepsilon_n(k_x, k_y, -k_z)$; thus, for such (001) films, (5.20) is true. Therefore, the results in Sects. 5.4.2 and 5.5 might be applied to these films: For a film of N_3 layers, there are $N_3 - 1$ bulk-like subbands and one surface-like subband in the film for each bulk energy band. In the Cartesian system, Eq. (5.28) for bcc (001) films can be written as

$$\hat{\Lambda}_{n,j_3}(k_x, k_y) = \varepsilon_n(k_x, k_y, 2\kappa_3/a) \tag{5.38}$$

for any k_x and k_y, where κ_3 is given by (5.26).

Because of (5.33), each surface-like subband $\hat{\Lambda}_n(\hat{\mathbf{k}}; \tau_3)$ is always above each relevant bulk-like subband $\hat{\Lambda}_{n,j_3}(\hat{\mathbf{k}})$.

5.6.4 (110) Films with a bcc Bravais Lattice

For a bcc (110) film, the primitive lattice vectors can be chosen as

$$\mathbf{a}_1 = a/2(1, -1, 1), \quad \mathbf{a}_2 = a/2(1, -1, -1), \quad \mathbf{a}_3 = a/2(1, 1, 1);$$

thus $\mathbf{b}_1 = 1/a(0, -1, 1)$, $\mathbf{b}_2 = 1/a(1, 0, -1)$, and $\mathbf{b}_3 = 1/a(1, 1, 0)$. Correspondingly, $\hat{\mathbf{b}}_1 = 1/a(1/2, -1/2, 1)$ and $\hat{\mathbf{b}}_2 = 1/a(1/2, -1/2, -1)$. This corresponds to that $\alpha_1 = -\alpha_2 = 1/2$ in Sect. 5.4.2. In general, the band structure of a bcc metal has the symmetry $\varepsilon_n(k_x, k_y, k_z) = \varepsilon_n(k_y, k_x, k_z)$; thus, for such (110) films, (5.20) is true. Therefore, the results in Sects. 5.4.2 and 5.5 might be applied to these films: For a film of N_3 layers, there are $N_3 - 1$ bulk-like subbands and one surface-like subband in the film for each bulk energy band. Equation (5.28) can also be written as

$$\hat{\Lambda}_{n,j_3}(k_1\hat{\mathbf{b}}_1 + k_2\hat{\mathbf{b}}_2) = \varepsilon_n[(\kappa_3 + k_1/2 + k_2/2)/a, (\kappa_3 - k_1/2 - k_2/2)/a, (k_1 - k_2)/a] \tag{5.39}$$

for any k_1 and k_2, where κ_3 is given by (5.26).

Because of (5.33), each surface-like subband $\hat{\Lambda}_n(\hat{\mathbf{k}}; \tau_3)$ is always above each relevant bulk-like subband $\hat{\Lambda}_{n,j_3}(\hat{\mathbf{k}})$.

5.7 Comparisons with Previous Numerical Results

There are some previously published numerical results [12–15] to which our results obtained in this chapter can be compared.

5.7.1 Si (001) Films

Equations (5.28) and (5.35) can be used for Si (001) films. Zhang and Zunger [12] and Zhang et al. [13] calculated the electronic structure of thin Si (001) films using

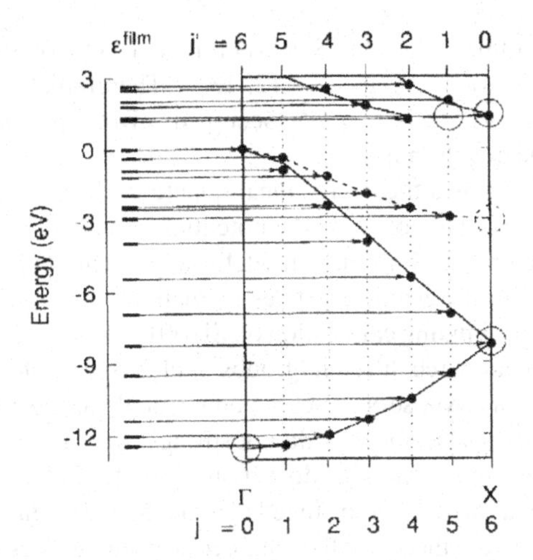

Fig. 5.2 Mapping of the directly calculated film energy levels in a Si (001) film of 12 monolayers at the center of the two-dimensional Brillouin zone by Zhang et al. (*solid dots*) onto the energy bands of the bulk Si. A solid (*dashed*) *open circle* indicates that one (two) state(s) does not exist in the film. Reprinted with permission from S.B. Zhang, C.-Y. Yeh, and A. Zunger: Phys. Rev. **B48**, 11204 (1993). Copyright by the American Physical Society

a pseudopotential method. Their results for even numbers N_f of monolayers can be directly compared with (5.35): The N_3 in (5.35) is equal to their $N_f/2$. Their "central observation" in [12] is that the energy spectrum of electronic states in a Si (001) quantum film ($N_f = 12$) maps the energy band structure of Si approximately, as shown in Figs. 1.4 and 5.2. Equation (9) in [12], which Zhang and Zunger obtained from their numerical results, is a special case of (5.35) with $k_1 = k_2 = 0$. Therefore, (5.35) is a more general prediction.

One of the triply degenerated VBM states and one of the doubly degenerated X_{1v} states in the valence bands may have a (001) nodal surface; thus, for these Bloch states, (5.31) can be true. The numerical results in [12, 13] for Si (001) films such as in Figs. 1.3, 1.4, and 5.2 are consistent with this understanding. Note that these Bloch states (X_{1v} or VBM) have the highest energy for that energy band ($n = 0$ or $n = 1$) and that $\hat{\mathbf{k}} = 0$, corresponding to that, (5.31) is true for these two cases: $n = n' = 0$ for one of the doubly degenerated states X_{1v}, and $n = n' = 1$ for one of the triply degenerated states VBM.

Although the VBM in Si (Γ'_{25}) is triply degenerated, only one of the VBM states may have a (001) nodal surface to make (5.31) true; thus, according to the theory in this chapter, there should be *only one* VBM band-edge state in (001) films whose energy does not depend on the film thickness in those figures in [12, 13].

Each one of the other two VBM states has the highest energy for that energy band ($n = 2$ or $n = 3$) and that $\hat{\mathbf{k}} = 0$ and each has a nodal surface, but not in the (001) plane. Consequently, $\Lambda_{n=2,3}(\hat{\mathbf{k}} = 0; \tau_3) > \varepsilon_n(k_3\mathbf{b}_3)$ ((5.32) for $\hat{\mathbf{k}} = 0$) is true for any k_3; thus, there must be two occupied surface-like states in the (001) films whose energies $\Lambda_{n=2,3}(\hat{\mathbf{k}} = 0; \tau_3)$ are *above* the VBM and do not depend on the film thickness. This is the reason that two occupied surface bands were observed in a Si quantum (001) film, such as the Si (001) film of 12 monolayers (corresponding $N_3 = 6$ in our notations) investigated in [12, 13].[12] Therefore, the VBM state shown in Figs. 1.3, 1.4, or Fig. 5.2 is a *single* occupied VBM state but *not* the highest occupied state in the quantum films.

Although our theory is a theory for ideal quantum films and the numerical calculations in [12, 13] used a more realistic potential outside the film, the rather good agreements indicate that the simplified model used here might have correctly given the essential physics of the electronic states in quantum films.

An interesting tight-binding calculation on Si (001) films by Gavrilenko and Koch [14] found that there are three different groups of electronic states in the films. They are: (i) bulk-related states whose energies depend strongly on the thickness of the film; (ii) surface-like states whose energies do not strongly depend on the film thickness; (iii) electronic states whose energies do not strongly depend on the film thickness and whose wave functions are not localized near the boundary faces of the film. This is also consistent with the theory in this chapter: The energies $\hat{\Lambda}_{n,j_3}(\hat{\mathbf{k}})$ of bulk-like states depend on the film thickness N_3, whereas the energies $\hat{\Lambda}_n(\hat{\mathbf{k}}; \tau_3)$ of surface-like states do not depend on the film thickness N_3. The corresponding wave functions $\hat{\psi}_n(\hat{\mathbf{k}}, \mathbf{x}; \tau_3)$ of the surface-like states may be either localized near one boundary surface of the film (the imaginary part of k_3 in (5.11) is not zero) or delocalized (the imaginary part of k_3 in (5.11) is zero).

5.7.2 *Si (110) Films and GaAs (110) Films*

In cases where $k_1 = k_2 = 0$, (5.37) gives

$$\hat{\Lambda}_{n,j_3}(0) = \varepsilon_n \left(\frac{2j_3\pi}{N_3 a}, \frac{2j_3\pi}{N_3 a}, 0 \right).$$

This is what was observed in the numerical calculations on a six-layer Si (110) film and a six-layer GaAs (110) film in [13], as shown in Fig. 5.3. Again, Eq. (5.37) is a more general prediction for fcc (110) films.

One of the triply degenerated VBM states (Γ'_{25} in Si or Γ_{15} in GaAs) may have a nodal surface in the (110) plane; therefore, there may be one VBM state in free-standing Si (110) or GaAs (110) films whose energy does not depend on the film thickness in [13, 15].

Each one of the other two VBM states does have a nodal surface, but not in the (110) plane. Correspondingly, $\hat{\Lambda}_{n=2,3}(\hat{\mathbf{k}} = 0; \tau_3) > \varepsilon_n(k_3\mathbf{b}_3)$ ((5.32) for $\hat{\mathbf{k}} = 0$) is true for any k_3; thus, there are also two occupied surface-like states in the (110) films whose energies $\hat{\Lambda}_n(\hat{\mathbf{k}} = 0; \tau_3)$ are above the VBM and do not depend on the film thickness. Therefore, the VBM states shown in Fig. 5.3 (and also in Fig. 5.4) are

[12] Therefore, the existence of two occupied surface bands above the VBM is because that (5.32) is true for two valence bands, rather than because that the film has two surfaces.

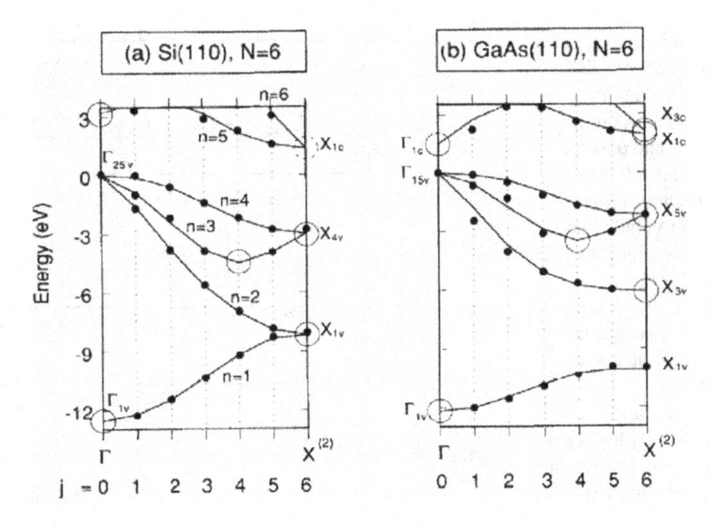

Fig. 5.3 Mapping of the directly calculated film energy levels in [13] (*solid dots*) at the center of the two-dimensional Brillouin zone on to the energy bands of the bulk for **a** a six-layer Si (110) film and **b** a six-layer GaAs film. The legends are the same as in Fig. 5.2. Reprinted with permission from S.B. Zhang, C.-Y. Yeh, and A. Zunger: Phys. Rev. **B48**, 11204 (1993). Copyright by the American Physical Society

a single VBM state, they are not the highest occupied state in these quantum films either.

Franceschetti and Zunger [15] investigated the band-edge energies in free-standing and AlAs-embedded GaAs films, quantum wires, and quantum dots, as shown in Fig. 5.4.

Note that in the numerical results on the electronic states in free-standing GaAs (110) films and AlAs-embedded GaAs (110) films shown in Fig. 5.4a, the valence band-edge energy of the free-standing GaAs (110) films stays almost unchanged, but the valence band-edge energy of AlAs-embedded GaAs (110) films is always below the valence band-edge energy of free-standing GaAs (110) films of the same thickness as the film thickness decreases. If one thinks in an EMA way, the figure shows a stronger quantum confinement effect in AlAs-embedded GaAs (110) films than in free-standing GaAs (110) films since the effective mass at the VBM is negative. However, the free-standing GaAs (110) films have a much stronger confinement potential. This comparison leads to a real puzzle for EMA: A weaker confinement potential leads to a stronger quantum confinement effect. This is another example indicating that EMA might be qualitatively incorrect in some cases. However, this is a natural consequence of our theory: The stronger confinement potential in free-standing GaAs (110) films makes the corresponding energy in these films higher.

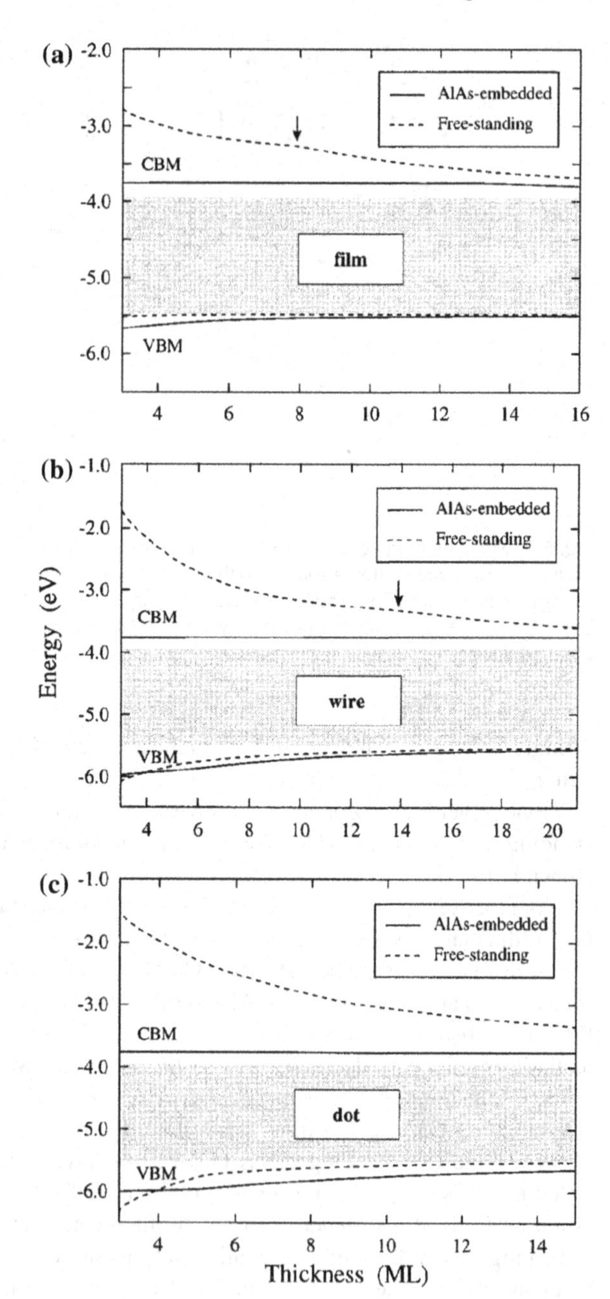

Fig. 5.4 Band-edge energies of AlAs-embedded (*solid lines*) and free-standing (*dashed lines*) GaAs films (**a**), wires (**b**), and dots (**c**) from numerical calculations in [15]. The *shaded areas* denote the GaAs bulk band gap. The *arrows* indicate the critical size for the direct/indirect transition in free-standing quantum films and wires. Reprinted with permission from A. Franceschetti and A. Zunger: Appl. Phys. Lett. **68**, 3455 (1996). Copyright by American Institute of Physics

5.8 Further Discussions

We have seen that in ideal (001) or (110) quantum films of cubic semiconductors investigated in [12, 13, 15], the existence of a VBM state whose energy does not change as the film thickness changes, as shown in Figs. 1.3, 1.4, 5.2, 5.3, and 5.4, simply is because there is always one VBM state that may have a nodal surface in the (001) plane or in the (110) plane. However, for the quantum confinement of the three-dimensional Bloch waves in quantum films, an even more interesting point is the existence of the other two higher occupied surface-like bands—simply because the other two VBM states may not have the same nodal surface in the (001) or in the (110) plane. In contrast with an EMA picture such as shown in Fig. 1.2, the highest occupied electronic states in an ideal (001) or (110) quantum film of a cubic semiconductor—the maximum of the surface-like subbands $\hat{\Lambda}_n(\hat{\mathbf{k}}; \tau_3)$ for $n = 2, 3$—are *above* the VBM, the highest occupied states in the bulk without quantum confinement.

It is the energies of the highest occupied bulk-like states that are always below the VBM and decrease as the film thickness decreases. The widely believed result in the solid physics community that the band gap increases as the film thickness decreases, in fact, is true only for the bulk-like electronic states in semiconductor quantum films. If the surface-like occupied subbands $\hat{\Lambda}_n(\hat{\mathbf{k}}; \tau_3)$ for $n = 2, 3$ are also considered, the real band gap in a semiconductor quantum film may, in fact, be *smaller* than the band gap of the bulk. It may even be possible that for some $\hat{\mathbf{k}}$ and $\hat{\mathbf{k}}'$,

$$\hat{\Lambda}_{n=2,3}(\hat{\mathbf{k}}; \tau_3) > \hat{\Lambda}_{4,j_3}(\hat{\mathbf{k}}'). \tag{5.40}$$

That is, some electronic states in the two surface-like subbands originating from two valence bands ($n = 2, 3$) may be energetically higher than the electronic states in the bulk-like subbands originating from the lowest conduction band ($n = 4$). If this happens, a film of a semiconductor crystal may, in fact, become a film with the electrical conductivity of a metal: The equality of the Fermi level in the film will force electrons to move from the surface-like subbands $\hat{\Lambda}_{n=2,3}(\hat{\mathbf{k}}; \tau_3)$ into the bulk-like subbands $\hat{\Lambda}_{4,j_3}(\hat{\mathbf{k}}')$, possibly to make the film an electrical conductor. These predictions are based on Theorem 5.1 and the properties of the VBM of cubic semiconductors. The thinner the film is, the more significant are the contributions from those surface-like states to the physical properties of the film. In the numerical calculations on a Si (001) quantum film in [13], the surface-like states were found at 1.2–1.6 eV above the VBM, indicating that the condition (5.40) is realistic.

Further theoretical and numerical investigations on this possibility and carefully designed experimental investigations to explore these possible physical phenomena will be very interesting.

Our predictions for ideal (001) and (110) films with an fcc Bravais lattice are general predictions. Although many of them are consistent with numerical calculations for Si (001) films, Si (110) films, and GaAs (110) films in [12, 13, 15], there are also some differences.

Since the VBM of a cubic semiconductor is triply degenerated and there is only one band (the lowest valence band $n = 0$ in our notation) below the three upper valence bands, as a consequence of Theorem 5.1, we have

$$\hat{\Lambda}_1(\hat{\mathbf{k}} = 0; \tau_3) \geq \text{VBM}$$

in these films; that is, only one $\hat{\Lambda}_{n=0}(\hat{\mathbf{k}} = 0; \tau_3)$ can exist below the VBM. In particular for the cases investigated in [12, 13], there could be only one $\hat{\Lambda}_{n=0}(\hat{\mathbf{k}} = 0; \tau_3)$ with an energy below the VBM at the boundary of the Brillouin zone in those films. This is what was observed in the numerical results on Si (001) films in [12, 13]: Only one of the doubly degenerated X_{1v} states exists in Si (001) films, as shown in Figs. 1.4 and 5.2. However, in the numerical results on a six-layer Si(110) film and a six-layer GaAs (110) film in [13], two such states were presented in both films. They are, X_{1v} and X_{4v} in the Si (110) film, X_{1v} and X_{5v} in the GaAs (110) film, as shown in Fig. 5.3a, b. In ideal films treated by the theory in this chapter, the very existence of two such states is contradictory to Theorem 5.1.[13]

Another noticeable difference is that, according to our theory, in an ideal quantum film of N_3 layers, for each energy band n and each $\hat{\mathbf{k}}$, there are $N_3 - 1$ bulk-like stationary Bloch states $\hat{\psi}_{n,j_3}(\hat{\mathbf{k}}, \mathbf{x}; \tau_3)$ where $j_3 = 1, 2, \ldots, N_3 - 1$. In the numerical calculations in [12, 13], the results are somewhat different: The $n = 5, j = 5$ (corresponding to our $n = 4$, $j_3 = 5$) state in Figs. 1.4 and 5.2 and the $n = 3, j = 4$ (corresponding to our $n = 2$, $j_3 = 4$)[14] state in Fig. 5.3a, b are missing.

A clear understanding of the origin of these differences will be helpful to improve our understandings on relevant physics.

We have understood that, in an ideal quantum film, each surface state with a specific n and a specific $\hat{\mathbf{k}}$ is located near either the top or the bottom of the film. Nevertheless, since there is not a clear understanding of the properties of solutions of the second-order partial differential equations with periodic coefficients, *there is no reason to expect that all surface states in one surface-like subband are located near the same surface of the film.* Depending on τ_3 and $\hat{\mathbf{k}}$, some of them may be located near the top surface of the film, and the others may be located near the bottom surface of the film, even though they all belong to the same surface-like subband in the film.[15] This is different from the surface-like states in one-dimensional finite crystals, where a surface state in a specific band gap can only be located near either one end of the finite crystal.

In the ideal quantum films discussed here, for each bulk energy band $\varepsilon_n(\mathbf{k})$ there are one surface-like subband $\hat{\Lambda}_n(\hat{\mathbf{k}}; \tau_3)$ and $N_3 - 1$ bulk-like subbands $\hat{\Lambda}_{n,j_3}(\hat{\mathbf{k}})$ in

[13]Neither any one of the doubly degenerated X_{1v} states in Si nor the X_{1v} state in GaAs could have a nodal (110) plane and, thus, the X_{1v} state in an ideal Si (110) film and the X_{1v} state in an ideal GaAs (110) film cannot exist. Therefore, we can only have $\hat{\Lambda}_0(\hat{\mathbf{k}} = 0; \tau_3) = X_{4v}$ for an ideal Si (110) film and $\hat{\Lambda}_0(\hat{\mathbf{k}} = 0; \tau_3) = X_{5v}$ for an ideal GaAs (110) film. These are two examples of the special cases mentioned on p. 97 and p. 104 in which $n = 0$ while $n' = 2$.

[14]Note that in our notations the lowest energy band index $n = 0$.

[15]As a consequence, one has no reason to expect that the surface states of a quantum film can be passivated from only one surface of it.

a film of N_3 layers. Therefore, a surface-like subband comes from a bulk permitted band at the truncation of periodicity; this fact indicates that the physical origin of a surface-like subband is related to a bulk permitted band rather than to a band gap. Such an understanding is not easy to obtain either in a one-dimensional analysis or an ordinary semi-infinite crystal analysis: Many previous theoretical investigations on surface electronic states, including a work by the author [16], considered that the surface-like states are related to the band gap(s). We have known that a significant difference from one-dimensional cases is that a surface-like state in a quantum film does not have to be in a band gap. Based on such an understanding, this significant difference is a consequence of the significant differences between a multidimensional band structure and a one-dimensional band structure. In a multidimensional crystal, the permitted bands are overlapped, there is not always a band gap above each permitted energy band. As a comparison, in a one-dimensional crystal, the nth band gap is always above the nth permitted band and below the $(n+1)$th permitted band. The significant differences between a multidimensional band structure and a one-dimensional band structure lead to that at the truncation of periodicity—which makes the boundary-dependent states higher— Theorem 5.1 does not give an upper limit for $\hat{\lambda}_n(\hat{\mathbf{k}};\, \tau_3)$ in three-dimensional cases, unlike Theorem 2.8 for $\Lambda_{\tau,n}$ in one-dimensional cases. Therefore, *only in a one-dimensional crystal must such a localized state be in a band gap.* A surface-like state in a film *generally* may have energy in the range of permitted energy bands of the bulk, and it still can decay in either the positive or the negative direction of $\mathbf{a}_3$. Consequently, *a surface state in a multidimensional crystal, in general, does not have to be in a band gap.*[16]

We mentioned in Sect. 4.6 that in a tight-binding formalism, a finite linear chain with a single state per unit cell does not have a surface state. According to the understandings obtained here, an ideal finite one-dimensional crystal always has a surface-like state for each bulk energy band. Furthermore, it is in the one-dimensional case that such a surface-like state must be in a band gap. It is *a consequence of the tight-binding approximation* that a crystal with a single state per unit cell has no band gap, thus, such a linear finite chain has no surface state.

The general relationship between the surface-like subband $\hat{\Lambda}_n(\hat{\mathbf{k}};\, \tau_3)$ and the bulk-like subbands $\hat{\Lambda}_{n,j_3}(\hat{\mathbf{k}})$ in a quantum film (5.33) will lead to some other interesting consequences.

Since, as a consequence of Theorem 5.1, there is one surface subband $\hat{\Lambda}_n(\hat{\mathbf{k}};\, \tau_3)$ for each bulk energy band $\varepsilon_n(\mathbf{k})$, an ideal (001) or (110) film of a cubic compound semiconductor can have *at most one* surface-like subband $\hat{\Lambda}_0(\hat{\mathbf{k}};\, \tau_3)$ in the minor band gap between the lowest valence band $n = 0$ and the upper valence bands, even though a film always has *two* surfaces.

Since the surface-like subbands $\hat{\Lambda}_n(\hat{\mathbf{k}};\, \tau_3)$ near the bulk band gap for an ideal semiconductor quantum film originate from the valence bands, in an everywhere neutral

[16] Surface states with energy in the range of a permitted energy band sometimes are referred as bound states in the continuum(BIC), in differentiation of surface states with energy in the gap which are referred as bound states outside the continuum(BOC). See, for example, [17, 18]. We have seen here that the existence of BIC might be quite general in multidimensional crystals, not only in some special systems.

semiconductor film these surface-like subbands should be occupied. Unoccupied states in these surface-like subbands cause the surfaces to be positively charged.

A similar effect probably might be more notable in (001) and (110) films of alkali metals since the conduction band is not fully occupied; the Fermi surface of an alkali metal is usually inside the conduction band. All alkali metals (Li, Na, K, Rb, Cs, Fr) have a bcc Bravais lattice. Corresponding to the conduction band in which the conducting electrons are occupied, there is a surface-like subband in a (001) or (110) quantum film. This surface-like subband in the film is, by (5.33), generally higher in energy than the corresponding bulk-like subbands. In an everywhere neutral alkali metal film, the surface-like subband should be equally occupied by electrons as the corresponding bulk-like subbands. However, the equality of the Fermi energy in the film must force some electrons to flow from the surface-like subband into the bulk-like subbands inside the film. Thus, the surfaces of the film could be positively charged.

This prediction seems to be supported by the positive surface-atom core-level shift in alkali metal (110) films, as reported by Riffe et al. (RWBC), e.g., [19]. Although the surface-atom core-level shifts in most other metals (transition metals and noble metals) were explained by a small charge flow between the surface and the inside of the metal [20], no reason for such a charge flow between the surface and the inside of alkali metals was published previously. The authors in [19] used the spill out of the conducting electrons to explain the fact that the surfaces of these alkali metals are positively charged. However, the theory in this chapter provided a possible reason for why the electrons in an alkali (110) film could flow from the surfaces into the film; such a flow can make the surfaces of the film positive charged. This reasoning gives an alternative explanation for the positive surface-atom core-level shift observed in alkali metal (110) films. If this explanation is correct, then the surface-atom core-level shift in metals can be understood on the same basis, as a charge flow between the surface and the inside of the metal.

In the RWBC model, the conductingis not justified electrons spill *outside* into the vacuum, whereas in the theory of this chapter, conducting electrons mainly flow *inside*. Therefore, the surface-atom top layer relaxation [21, 22] in an alkali metal is more likely to be an expansion if the RWBC model dominates. On the other hand, if conducting electrons flowing inside dominates, the surface-atom top layer relaxation in an alkali metal is more likely to be a contraction. Experimental investigations to explore this might be interesting.

There are also surface-like subbands corresponding to the bulk conduction bands in semiconductor films. Those surface-like subbands will be even higher in energy than the bulk conduction bands and, thus, will usually not be occupied. It seems unlikely that those surface-like subbands will have a significant effect on the properties of a semiconductor quantum film.

Similar to being stated in the comments in Sect. 4.5, the very existence of the boundary-dependent states in quantum films is neglected in the EMA. Furthermore, if the concerned energy extreme is not at the center of the Brillouin zone or the boundary of the Brillouin zone, such as the conduction band minima in Si or Ge,[17] the use of EMA, even for the bulk-like electronic states in ideal quantum films.

In summary, by considering the effects of quantum confinement of three-dimensional Bloch waves in one specific direction, the properties of the electronic states in some simple and interesting ideal quantum films can be understood. The effect of the lack of translational invariance in one specific direction is that there is always one and only one surface-like subband $\hat{\Lambda}_n(\hat{\mathbf{k}}; \tau_3)$ for each bulk energy band $\varepsilon_n(\mathbf{k})$.[18] Further, in some ideal interesting quantum films—such as in ideal (001) films of crystals with a sc, tetr, or ortho Bravais lattice for which (5.13) is true or in ideal (001) or (110) films of crystals with a fcc Bravais lattice or a bcc Bravais lattice for which (5.20) is true—exact and general results on the properties of all electronic states in such an ideal quantum film can be predicted: There are two types of electronic states in a film bounded at $x_3 = \tau_3$ and $x_3 = (\tau_3 + N_3)$ for each bulk energy band n and each wave vector $\hat{\mathbf{k}}$ in the film plane. There are $N_3 - 1$ bulk-like electronic states $\hat{\psi}_{n,j_3}(\hat{\mathbf{k}}, \mathbf{x}; \tau_3)$ by (5.24), whose energy $\hat{\Lambda}_{n,j_3}(\hat{\mathbf{k}})$ by (5.28) depends on the film thickness N_3 and can be directly obtained from the energy band structure $\varepsilon_n(\mathbf{k})$ of the bulk crystal. There is one surface-like electronic state $\hat{\psi}_n(\hat{\mathbf{k}}, \mathbf{x}; \tau_3)$ by (5.29) whose energy $\hat{\Lambda}_n(\hat{\mathbf{k}}; \tau_3)$ (5.30) is always above bulk-like states $\hat{\Lambda}_{n,j_3}(\hat{\mathbf{k}})$, by (5.33).[19] These are similar to the properties of the electronic states in a one-dimensional finite crystal. However, a surface-like state in a multidimensional crystal *generally* may have energy in the range of permitted energy bands of the bulk, and does not have to be in a band gap, due to the differences between the truncation of multidimensional periodicity and that of one-dimensional periodicity: Only in a one-dimensional crystal each band gap and each permitted band exist alternatively as the energy increases, due to that the second-order ordinary periodic differential equation for a one-dimensional crystal has only two linearly independent solutions.

The approach based on mathematical reasonings and physical intuitions used in this chapter can be naturally extended to investigate the effects of the further confinement of two-dimensional Bloch waves $\hat{\psi}_{n,j_3}(\hat{\mathbf{k}}, \mathbf{x}; \tau_3)$ and $\hat{\psi}_n(\hat{\mathbf{k}}, \mathbf{x}; \tau_3)$ in some simple and interesting ideal quantum wires.

---

[17]Conduction band minima in Ge are located at the boundary of the Brillouin zone in four equivalent [111] directions, the use of EMA is not justified for the (001) or (110) films investigated here.

[18]Even for films where neither (5.13) nor (5.20) is true, $\hat{\psi}_n(\hat{\mathbf{k}}, \mathbf{x}; \tau_3)$ in (5.29) and $\hat{\Lambda}_n(\hat{\mathbf{k}}; \tau_3)$ in (5.30) is a solution of Eq. (5.12).

[19]A detailed numerical verification of the corresponding theoretical results in a two-dimensional case was published in [23], where Ajoy and Karmalkar tried to verify the predictions of the analytical theory by numerically computing the subbands of zigzag ribbons and the corresponding bulk band structure in an artificial graphene. It was concluded that in many interesting cases, the analytical theory "predicts all the important subbands in these ribbons, and provides additional insight into the nature of their wavefunctions." Interesting investigations can also be seen in [24].

References

1. D. Gilbarg, N.S. Trudinger, *Elliptic Partial Differential Equations of Second Order* (Springer, Berlin, 1998)
2. P.A. Kuchment: Bull. (New Ser.) Am. Math. Soc. **53**, 343 (2016); P.A. Kuchment: *Floquet Theory For Partial Differential Equations*, Operator Theory: Advances and Applications, vol. 60, (Birkhauser Verlag, Basel, 1993)
3. Y.E. Karpeshina, *Perturbation Theory for the Schrödinger Operator with a Periodic Potential*. Lecture Notes in Mathematics, vol. 1663 (Springer, Berlin, 1997)
4. O. Veliev, *Multidimensional Periodic Schrödinger Operator: Perturbation Theory and Applications*. Springer Tracts in Modern Physics, vol. 263 (Springer, Berlin, 2015)
5. P.Y. Yu, M. Cardona, *Fundamentals of Semiconductors*, 2nd edn. (Springer, Berlin, 1999)
6. G. Bethe, A. Sommerfeld, *Elektronentheorie der Metalle* (Springer, Berlin, 1967)
7. M.M. Skriganov, Sov. Math. Dokl. **20**, 956 (1979); M.M. Skriganov, Invent. Math. **80**, 107 (1985)
8. M.S.P. Eastham, *The Spectral Theory of Periodic Differential Equations* (Scottish Academic Press, Edinburgh, 1973). (and references therein)
9. S.Y. Ren, Europhys. Lett. **64**, 783 (2003)
10. E.C. Titchmarsh, *Eigenfunction Expansions Associated with Second-Order Differential Equations, Part II* (Oxford University Press, Oxford, 1958)
11. F. Odeh, J.B. Keller, J. Math. Phys. **5**, 1499 (1964)
12. S.B. Zhang, A. Zunger, Appl. Phys. Lett. **63**, 1399 (1993)
13. S.B. Zhang, C.-Y. Yeh, A. Zunger, Phys. Rev. B **48**, 11204 (1993)
14. V.I. Gavrilenko, F. Koch, J. Appl. Phys. **77**, 3288 (1995)
15. A. Franceschetti, A. Zunger, Appl. Phys. Lett. **68**, 3455 (1996)
16. S.Y. Ren, Ann. Phys. (NY) **301**, 22 (2002)
17. S. Longhi, G. Della Valle, J. Phys. Condens. Matter **25**, 235601 (2013)
18. S. Longhi, G. Della Valle, Sci. Rep. **3**, 2219 (2013)
19. D.M. Riffe, G.K. Wertheim, D.N.E. Buchanan, P.H. Citrin, Phys. Rev. B **45**, 6216 (1992). (and references therein)
20. P.H. Citrin, G.K. Wertheim, Y. Baer, Phys. Rev. B **27**, 3160 (1983); P.H. Citrin, G.K. Wertheim. Phys. Rev. B **27**, 3176 (1983)
21. F. Bechstedt, *Principles of Surface Physics* (Springer, Berlin, 2003)
22. A. Groß, *Theoretical Surface Science: A Microscopic Perspective* (Springer, Berlin, 2003)
23. A. Ajoy, S. Karmalkar, J. Phys. Condens. Matter **22**, 435502 (2010)
24. A. Ajoy, Ph. D thesis, Indian Institute of Technology Madras (2013)

Chapter 6
Electronic States in Ideal Quantum Wires

In this chapter, we investigate the electronic states in ideal quantum wires. We are interested in the electronic states in rectangular quantum wires, which can be seen as the electronic states in a quantum film discussed in Chap. 5 further confined in one more direction. In particular, we are interested in those simple cases where the two primitive lattice vectors $\mathbf{a}_1$ and $\mathbf{a}_2$ in the film plane are perpendicular to each other. By using an approach similar to that used in Chap. 5, we try to understand the further quantum confinement effects of two-dimensional Bloch waves $\hat{\psi}_{n,j_3}(\hat{\mathbf{k}}, \mathbf{x}; \tau_3)$ in (5.24) and $\hat{\psi}_n(\hat{\mathbf{k}}, \mathbf{x}; \tau_3)$ in (5.29) in a quantum wire. It is found that each type of two-dimensional Bloch waves will produce two different types of one-dimensional Bloch waves in the quantum wire.

A rectangular quantum wire always has four boundary faces: two faces in the $(h_2k_2l_2)$ plane and two faces in the $(h_3k_3l_3)$ plane. The electronic states in such a quantum wire can be seen as that the electronic states in a quantum film with two boundary faces in the $(h_3k_3l_3)$ plane are further confined by two boundary faces in the $(h_2k_2l_2)$ plane. Or, equivalently, they can also be seen as that the electronic states in a quantum film with two boundary faces in the $(h_2k_2l_2)$ plane are further confined by two boundary faces in the $(h_3k_3l_3)$ plane. The results obtained in these two different confinement orders are equally valid and are complementary to each other. By combining the results obtained from the two different confinement orders, we can achieve a more comprehensive understanding of the electronic states in the quantum wire.

The simplest cases are the electronic states in an ideal rectangular quantum wire of crystals with a sc, tetr, or an ortho Bravais lattice. In these crystals, the three primitive lattice vectors $\mathbf{a}_1$, $\mathbf{a}_2$, and $\mathbf{a}_3$ are perpendicular to each other and are equivalent. Exact and general results on the electronic states in such a quantum wire in the direction of one specific primitive lattice vector $\mathbf{a}_1$ and with four faces in the (010) or in the (001) plane can be obtained by considering the electronic states in a quantum film with two boundary faces in the (001) plane further confined by the two boundary

S.Y. Ren, *Electronic States in Crystals of Finite Size*, Springer Tracts in Modern Physics 270, DOI 10.1007/978-981-10-4718-3_6

faces in the (010) plane, or, equivalently, as the electronic states in a quantum film with two boundary faces in (010) plane further confined by the two boundary faces in the (001) plane.

By the understandings of the further quantum confinement of two-dimensional Bloch waves $\hat{\psi}_{n,j_3}(\hat{\mathbf{k}}, \mathbf{x}; \tau_3)$ and $\hat{\psi}_n(\hat{\mathbf{k}}, \mathbf{x}; \tau_3)$ and considering two different confinement orders, we can also obtain predictions on the electronic states in some practically more interesting ideal rectangular quantum wires of crystals with an fcc or a bcc Bravais lattice. Electronic states in such a quantum wire can be seen as the two-dimensional Bloch waves in a quantum film discussed in Sect. 5.6 further confined in one more direction.

This chapter is organized as follows. After giving basic considerations of the problem in Sect. 6.1, in Sects. 6.2–6.3, we investigate the effects produced when the two types of two-dimensional Bloch waves obtained in Chap. 5 are further confined in one more specific direction. In Sects. 6.4–6.7, we obtain predictions on the electronic states in ideal quantum wires of crystals with several different Bravais lattices by applying the results obtained in Sects. 6.1–6.3 and by considering two different confinement orders. In Sect. 6.8 are a summary and some discussions on the results obtained.

6.1 Basic Considerations

In an ideal quantum film discussed in Chap. 5, there are two different types of electronic states: surface-like states $\hat{\psi}_n(\hat{\mathbf{k}}, \mathbf{x}; \tau_3)$ in (5.29) and bulk-like states $\hat{\psi}_{n,j_3}(\hat{\mathbf{k}}, \mathbf{x}; \tau_3)$ in (5.24). Both are two-dimensional Bloch waves in the film plane. Similar to the problem we treated in Chap. 5, in a quantum wire, each type of these two-dimensional Bloch waves will be further confined in one more direction.

We choose the primitive vector $\mathbf{a}_1$ in the wire direction. Such a rectangular quantum wire can be defined by a bottom face and a top face at $x_3 = \tau_3$ and $x_3 = \tau_3 + N_3$, a front face and a rear face perpendicularly intersecting the $\mathbf{a}_2$ axis at $x_2 = \tau_2$ and $x_2 = \tau_2 + N_2$. Here τ_2 and τ_3 define the boundary face locations of the wire, N_2 and N_3 are two positive integers indicating the wire size and shape. We use $\bar{\mathbf{k}}$ to express a wave vector in the wire direction: $\bar{\mathbf{k}} = k_1 \bar{\mathbf{b}}_1$, $\mathbf{a}_1 \cdot \bar{\mathbf{b}}_1 = 1$. Since in this chapter we are only interested in the cases where $\mathbf{a}_1 \cdot \mathbf{a}_2 = 0$, we have $\bar{\mathbf{b}}_1 = \hat{\mathbf{b}}_1$.

For the further confinement of two-dimensional Bloch waves $\hat{\psi}_n(\hat{\mathbf{k}}, \mathbf{x}; \tau_3)$ and $\hat{\psi}_{n,j_3}(\hat{\mathbf{k}}, \mathbf{x}; \tau_3)$, we look for the eigenvalues $\bar{\Lambda}$ and eigenfunctions $\bar{\psi}(\bar{\mathbf{k}}, \mathbf{x})$ of the following two equations:

$$\begin{cases} -\nabla^2 \bar{\psi}(\bar{\mathbf{k}}, \mathbf{x}) + [v(\mathbf{x}) - \bar{\Lambda}]\bar{\psi}(\bar{\mathbf{k}}, \mathbf{x}) = 0, & \mathbf{x} \in \text{the wire}, \\ \bar{\psi}(\bar{\mathbf{k}}, \mathbf{x}) = 0, & \mathbf{x} \notin \text{the wire}, \end{cases} \tag{6.1}$$

where $v(\mathbf{x})$ is a periodic potential:

$$v(\mathbf{x} + \mathbf{a}_1) = v(\mathbf{x} + \mathbf{a}_2) = v(\mathbf{x} + \mathbf{a}_3) = v(\mathbf{x}).$$

The solutions $\bar{\psi}(\bar{\mathbf{k}}, \mathbf{x})$ of (6.1) are one-dimensional Bloch waves with a wave vector $\bar{\mathbf{k}}$ in the wire direction $\mathbf{a}_1$.

The further quantum confinement of each type of two-dimensional Bloch waves $\hat{\psi}_n(\hat{\mathbf{k}}, \mathbf{x}; \tau_3)$ or $\hat{\psi}_{n,j_3}(\hat{\mathbf{k}}, \mathbf{x}; \tau_3)$ will have a new eigenvalue problem and a corresponding theorem, and will give two different types of electronic states in the quantum wire. Correspondingly, we will obtain four different sets of one-dimensional Bloch waves in the quantum wire.

6.2 Further Quantum Confinement of $\hat{\psi}_n(\hat{\mathbf{k}}, \mathbf{x}; \tau_3)$

For the quantum confinement of two-dimensional Bloch waves $\hat{\psi}_n(\hat{\mathbf{k}}, \mathbf{x}; \tau_3)$, we consider a rectangular parallelepiped B as shown in Fig. 6.1, with surfaces oriented in the $\mathbf{a}_1$, $\mathbf{a}_2$, or $\mathbf{b}_3$. It has a bottom face and a top face at $x_3 = \tau_3$ and $x_3 = \tau_3 + 1$, a front face and a rear face perpendicularly intersecting the $\mathbf{a}_2$ axis at $x_2 = \tau_2$[1] and $x_2 = \tau_2 + 1$, a left face and a right face perpendicular intersecting the $\mathbf{a}_1$ axis and separated by a_1. Since two-dimensional Bloch wave $\hat{\psi}_n(\hat{\mathbf{k}}, \mathbf{x}; \tau_3)$ are zero on ∂B_3 of the boundary B in the $\mathbf{b}_3$ direction, the function set $\bar{\phi}(\bar{\mathbf{k}}, \mathbf{x}; \tau_2, \tau_3)$ is defined by the conditions

$$\begin{cases} \bar{\phi}(\bar{\mathbf{k}}, \mathbf{x} + \mathbf{a}_1; \tau_2, \tau_3) = \mathrm{e}^{ik_1} \bar{\phi}(\bar{\mathbf{k}}, \mathbf{x}; \tau_2, \tau_3), & -\pi < k_1 \leq \pi, \\ \bar{\phi}(\bar{\mathbf{k}}, \mathbf{x}; \tau_2, \tau_3) = 0, & \mathbf{x} \in \partial B_2 \bigcup \partial B_3, \end{cases} \tag{6.2}$$

where ∂B_2 is the boundary ∂B of B in the $\mathbf{a}_2$ direction. The eigenvalues and eigenfunctions of Eq. (5.1) under the condition (6.2) are denoted by $\bar{\lambda}_n(\bar{\mathbf{k}}; \tau_2, \tau_3)$ and $\bar{\phi}_n(\bar{\mathbf{k}}, \mathbf{x}; \tau_2, \tau_3)$, where $n = 0, 1, 2, \ldots$.

For the eigenvalue problem defined by Eq. (5.1) and the boundary condition (6.2), we have the following theorem between $\bar{\lambda}_n(\bar{\mathbf{k}}; \tau_2, \tau_3)$ and the eigenvalue $\hat{\Lambda}_n(\hat{\mathbf{k}}; \tau_3)$ in (5.30) of $\hat{\psi}_n(\hat{\mathbf{k}}, \mathbf{x}; \tau_3)$.

Theorem 6.1

$$\bar{\lambda}_n(\bar{\mathbf{k}}; \tau_2, \tau_3) \geq \hat{\Lambda}_n(\hat{\mathbf{k}}; \tau_3), \quad \textit{for } (\bar{\mathbf{k}} - \hat{\mathbf{k}}) \cdot \mathbf{a}_1 = 0. \tag{6.3}$$

Note that in (6.2) and (6.3), $\hat{\mathbf{k}}$ is a wave vector in the film plane, and $\bar{\mathbf{k}}$ is a wave vector in the wire direction. In (6.3), $\hat{\mathbf{k}}$ and $\bar{\mathbf{k}}$ have the same component in the wire ($\mathbf{a}_1$) direction. In this book, we say that a two-dimensional Bloch wavevector $\hat{\mathbf{k}}$ is related to the one-dimensional Bloch wavevector $\bar{\mathbf{k}}$ in the $\mathbf{a}_1$ direction if $(\hat{\mathbf{k}} - \bar{\mathbf{k}}) \cdot \mathbf{a}_1 = 0$, or briefly, $\hat{\mathbf{k}}$ is related to $\bar{\mathbf{k}}$.

[1] For the further quantum confinement of these two-dimensional Bloch waves, it is assumed that in Sects. 6.2 and 6.3, the existence of boundary τ_2 does not change the one-dimensional translational symmetry in the $\mathbf{a}_1$ direction.

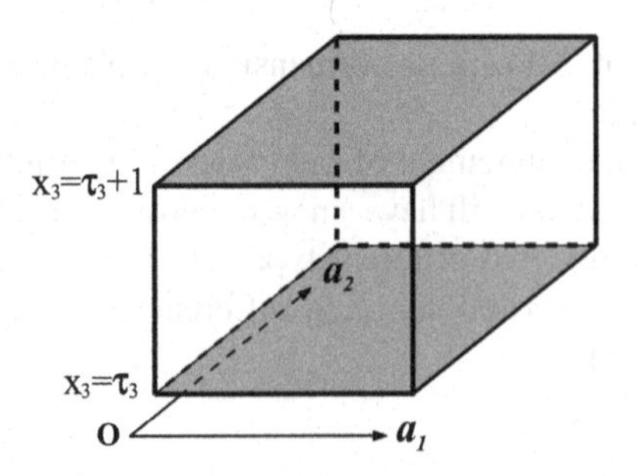

Fig. 6.1 The rectangular parallelepiped B for the eigenvalue problem of (5.1) under the boundary condition (6.2). The two shadowed faces of ∂B_3 are the two faces of which each $\hat{\psi}_n(\hat{\mathbf{k}}, \mathbf{x}; \tau_3)$ is zero (thus the specific direction of $\mathbf{a}_3$ no longer matters). The two *thick-lined* faces of ∂B_2 are the two faces on which each function $\bar{\phi}(\bar{\mathbf{k}}, \mathbf{x}; \tau_2, \tau_3)$ is further required to be zero

Since the two-dimensional Bloch wave $\hat{\psi}_n(\hat{\mathbf{k}}, \mathbf{x}; \tau_3)$ satisfies

$$\begin{cases} \hat{\psi}(\hat{\mathbf{k}}, \mathbf{x}+\mathbf{a}_i; \tau_3) = \mathrm{e}^{\mathrm{i}k_i}\hat{\psi}(\hat{\mathbf{k}}, \mathbf{x}; \tau_3), & -\pi < k_i \le \pi, \quad i = 1, 2 \\ \hat{\psi}(\hat{\mathbf{k}}, \mathbf{x}; \tau_3) = 0, & \mathbf{x} \in \partial B_3, \end{cases} \tag{6.4}$$

Theorem 6.1 can be proved similar to Theorem 5.1. The major difference is in the Dirichlet integral

$$\begin{aligned} J(f, g) &= \int_B \{\nabla f(\mathbf{x}) \cdot \nabla g^*(\mathbf{x}) + v(\mathbf{x}) f(\mathbf{x}) g^*(\mathbf{x})\}\, \mathrm{d}\mathbf{x} \\ &= \int_B f(\mathbf{x})\{-\nabla^2 g^*(\mathbf{x}) + v(\mathbf{x}) g^*(\mathbf{x})\}\, \mathrm{d}\mathbf{x} + \int_{\partial B} f(\mathbf{x}) \frac{\partial g^*(\mathbf{x})}{\partial n}\, \mathrm{d}S. \end{aligned} \tag{6.5}$$

If both $f(\mathbf{x})$ and $g(\mathbf{x})$ satisfy the conditions (6.4), the integral over ∂B in (6.5) is zero because that the integrals over two opposite faces of ∂B_1 and ∂B_2 cancel out and $f(\mathbf{x}) = 0$ when $\mathbf{x} \in \partial B_3$. If $f(\mathbf{x}) = \bar{\phi}(\bar{\mathbf{k}}, \mathbf{x}; \tau_2, \tau_3)$ and $g(\mathbf{x}) = \hat{\psi}(\hat{\mathbf{k}}, \mathbf{x}; \tau_3)$, the integral over ∂B in (6.5) is also zero because the integrals over two opposite faces of ∂B_1 cancel out since $(\bar{\mathbf{k}} - \hat{\mathbf{k}}) \cdot \mathbf{a}_1 = 0$ and the integral over each face of ∂B_2 and ∂B_3 is zero since $f(\mathbf{x}) = 0$ when $\mathbf{x} \in \partial B_2$ or $\mathbf{x} \in \partial B_3$.

Theorem 6.1 is similar to Theorem 5.1; the consequences of the quantum confinement of three-dimensional Bloch waves $\phi_n(\mathbf{k}, \mathbf{x})$ in the $\mathbf{a}_3$ direction due to Theorem 5.1 can be similarly used to the quantum confinement of two-dimensional Bloch waves $\hat{\psi}_n(\hat{\mathbf{k}}, \mathbf{x}; \tau_3)$ in the $\mathbf{a}_2$ direction.

For each bulk energy band n and each $\bar{\mathbf{k}}$, there is one $\bar{\phi}_n(\bar{\mathbf{k}}, \mathbf{x}; \tau_2, \tau_3)$.

Because $v(\mathbf{x} + \mathbf{a}_2) = v(\mathbf{x})$, the function $\bar{\phi}_n(\bar{\mathbf{k}}, \mathbf{x}; \tau_2, \tau_3)$ has the form

$$\bar{\phi}_n(\bar{\mathbf{k}}, \mathbf{x} + \mathbf{a}_2; \tau_2, \tau_3) = \mathrm{e}^{\mathrm{i}k_2}\bar{\phi}_n(\bar{\mathbf{k}}, \mathbf{x}; \tau_2, \tau_3), \tag{6.6}$$

and, here, k_2 can be real or complex. If k_2 in (6.6) is real, then $\bar{\phi}_n(\bar{\mathbf{k}}, \mathbf{x}; \tau_2, \tau_3)$ is a $\hat{\psi}_{n'}(\hat{\mathbf{k}}, \mathbf{x}; \tau_3)$. By Theorem 6.1, a $\hat{\psi}_{n'}(\hat{\mathbf{k}}, \mathbf{x}; \tau_3)$ cannot be a $\bar{\phi}_n(\bar{\mathbf{k}}, \mathbf{x}; \tau_2, \tau_3)$ except in some special cases where $\hat{\psi}_{n'}(\hat{\mathbf{k}}, \mathbf{x}; \tau_3)$ has a nodal plane perpendicularly intersecting the $\mathbf{a}_2$ axis at $x_2 = \tau_2$. Therefore, k_2 in (6.6) can be real only in such particular cases; in most cases, k_2 in (6.6) is complex with a nonzero imaginary part.

Depending on τ_2 (and also τ_3), n, and $\bar{\mathbf{k}}$, the imaginary part of k_2 in (6.6) can be either positive or negative, corresponding to whether $\bar{\phi}_n(\bar{\mathbf{k}}, \mathbf{x}; \tau_2, \tau_3)$ decays in either the positive or the negative direction of $\mathbf{a}_2$. Such states $\bar{\phi}_n(\bar{\mathbf{k}}, \mathbf{x}; \tau_2, \tau_3)$ with a nonzero imaginary part of k_2 in (6.6) cannot exist in a film with two-dimensional translational invariance because they are divergent in either the negative or the positive direction of $\mathbf{a}_2$. However, they can play a significant role in a quantum wire with a finite size in the $\mathbf{a}_2$ direction.

The further quantum confinement of the two-dimensional Bloch waves $\hat{\psi}_n(\hat{\mathbf{k}}, \mathbf{x}; \tau_3)$ in the $\mathbf{a}_2$ direction will produce two different types of electronic states in the quantum wire.

One type of nontrivial solutions of (6.1) from the quantum confinement of $\hat{\psi}_n(\hat{\mathbf{k}}, \mathbf{x}; \tau_3)$ can be obtained from (6.6) by assigning

$$\bar{\psi}_n(\bar{\mathbf{k}}, \mathbf{x}; \tau_2, \tau_3) = \begin{cases} c_{N_2,N_3}\bar{\phi}_n(\bar{\mathbf{k}}, \mathbf{x}; \tau_2, \tau_3), & \mathbf{x} \in \text{the wire}, \\ 0, & \mathbf{x} \notin \text{the wire}, \end{cases} \tag{6.7}$$

where c_{N_2,N_3} is a normalization constant. The corresponding eigenvalue

$$\bar{\Lambda}_n(\bar{\mathbf{k}}; \tau_2, \tau_3) = \bar{\lambda}_n(\bar{\mathbf{k}}; \tau_2, \tau_3) \tag{6.8}$$

is dependent on τ_2 and τ_3.

Therefore, for each bulk energy band n and each wave vector $\bar{\mathbf{k}}$, there is one electronic state $\bar{\psi}_n(\bar{\mathbf{k}}, \mathbf{x}; \tau_2, \tau_3)$. It is $\bar{\phi}_n(\bar{\mathbf{k}}, \mathbf{x}; \tau_2, \tau_3)$ inside the wire but zero otherwise, whose energy $\bar{\Lambda}_n(\bar{\mathbf{k}}; \tau_2, \tau_3)$ depends on τ_2 and τ_3. This is an edge-like state because $\bar{\phi}_n(\bar{\mathbf{k}}, \mathbf{x}; \tau_2, \tau_3)$ decays in either the positive or the negative direction of $\mathbf{a}_2$ and $\mathbf{a}_3$ in most cases.

Now, we try to find other solutions of (6.1) from the quantum confinement of $\hat{\psi}_n(\hat{\mathbf{k}}, \mathbf{x}; \tau_3)$. We can expect that there are stationary Bloch states in the $\mathbf{a}_2$ direction, formed due to the multiple reflections of $\hat{\psi}_n(\hat{\mathbf{k}}, \mathbf{x}; \tau_3)$ between two confinement boundary surfaces that perpendicularly intersect the $\mathbf{a}_2$ axis at τ_2 and $(\tau_2 + N_2)$.[2]

Since it is assumed that the existence of the boundary τ_3 does not change the two-dimensional space group symmetry of the system, in many quantum films discussed in Chap. 5

$$\hat{\Lambda}_n(k_1\bar{\mathbf{b}}_1 + k_2\hat{\mathbf{b}}_2; \tau_3) = \hat{\Lambda}_n(k_1\bar{\mathbf{b}}_1 - k_2\hat{\mathbf{b}}_2; \tau_3) \tag{6.9}$$

[2]The author does not have an enough understanding on solutions of periodic partial differential equations to assert that for what τ_2 the stationary Bloch state solutions in the $\mathbf{a}_2$ direction of (6.1) exist. However, he expects that the existence of such type of states is reasonable for a physical wire with a physical boundary τ_2.

in (5.30) is true; thus, in general,

$$f_{n,k_1,k_2}(\mathbf{x};\tau_3) = c_+\hat{\psi}_n(k_1\bar{\mathbf{b}}_1 + k_2\hat{\mathbf{b}}_2, \mathbf{x};\tau_3) \\ + c_-\hat{\psi}_n(k_1\bar{\mathbf{b}}_1 - k_2\hat{\mathbf{b}}_2, \mathbf{x};\tau_3), \quad 0 < k_2 < \pi,$$

where $c_\pm$ are not zero, is a nontrivial solution of Eq. (5.12) due to (6.9). It is easy to see that $f_{n,k_1,k_2}(\mathbf{x};\tau_3)$ is a one-dimensional Bloch wave with a wave vector $\bar{\mathbf{k}} = k_1\bar{\mathbf{b}}_1$ in the wire direction:

$$f_{n,k_1,k_2}(\mathbf{x}+\mathbf{a}_1;\tau_3) = \mathrm{e}^{ik_1} f_{n,k_1,k_2}(\mathbf{x};\tau_3), \quad -\pi < k_1 \leq \pi,$$

due to (6.4). To be a solution of Eq. (6.1), the function $f_{n,k_1,k_2}(\mathbf{x};\tau_3)$ is required to be zero at the front face and the rear face of the wire. By writing the front face equation of the wire as $x_2 = x_{2,f}(x_1, x_3)$ and the rear face equation of the wire as $x_2 = x_{2,r}(x_1, x_3)$, we should have

$$\begin{aligned} c_+\hat{\psi}_n[k_1\bar{\mathbf{b}}_1 + &k_2\hat{\mathbf{b}}_2, \mathbf{x} \in x_{2,f}(x_1,x_3);\tau_3] \\ &+ c_-\hat{\psi}_n[k_1\bar{\mathbf{b}}_1 - k_2\hat{\mathbf{b}}_2, \mathbf{x} \in x_{2,f}(x_1,x_3);\tau_3] = 0, \\ c_+\hat{\psi}_n[k_1\hat{\mathbf{b}}_1 + &k_2\hat{\mathbf{b}}_2, \mathbf{x} \in x_{2,r}(x_1,x_3);\tau_3] \\ &+ c_-\hat{\psi}_n[k_1\bar{\mathbf{b}}_1 - k_2\hat{\mathbf{b}}_2, \mathbf{x} \in x_{2,r}(x_1,x_3);\tau_3] = 0. \end{aligned} \tag{6.10}$$

Since $x_{2,r}(x_1, x_3) = x_{2,f}(x_1, x_3) + N_2$, we have

$$\hat{\psi}_n[k_1\bar{\mathbf{b}}_1 + k_2\hat{\mathbf{b}}_2, \mathbf{x} \in x_{2,r}(x_1,x_3);\tau_3] \\ = \mathrm{e}^{ik_2N_2}\hat{\psi}_n[k_1\bar{\mathbf{b}}_1 + k_2\hat{\mathbf{b}}_2, \mathbf{x} \in x_{2,f}(x_1,x_3);\tau_3],$$

and

$$\hat{\psi}_n[k_1\bar{\mathbf{b}}_1 - k_2\hat{\mathbf{b}}_2, \mathbf{x} \in x_{2,r}(x_1,x_3);\tau_3] \\ = \mathrm{e}^{-ik_2N_2}\hat{\psi}_n[k_1\bar{\mathbf{b}}_1 - k_2\hat{\mathbf{b}}_2, \mathbf{x} \in x_{2,f}(x_1,x_3);\tau_3],$$

due to (6.4). Therefore, for $c_\pm$ in (6.10) are not both zero, $\mathrm{e}^{ik_2N_2} - \mathrm{e}^{-ik_2N_2} = 0$ has to be true for each such stationary Bloch state, for such a specific τ_2.

The stationary Bloch state solutions of Eq. (6.1) obtained from the further quantum confinement of $\hat{\psi}_n(\hat{\mathbf{k}}, \mathbf{x};\tau_3)$ should have the form

$$\bar{\psi}_{n,j_2}(\bar{\mathbf{k}}, \mathbf{x};\tau_2,\tau_3) = \begin{cases} f_{n,k_1,\kappa_2}(\mathbf{x};\tau_2,\tau_3), & \mathbf{x} \in \text{the wire}, \\ 0, & \mathbf{x} \notin \text{the wire}, \end{cases} \tag{6.11}$$

where $\bar{\mathbf{k}} = k_1\bar{\mathbf{b}}_1$ and

$$f_{n,k_1,k_2}(\mathbf{x};\tau_2,\tau_3) = c_{n,k_1,k_2;\tau_2}\hat{\psi}_n(k_1\bar{\mathbf{b}}_1 + k_2\hat{\mathbf{b}}_2, \mathbf{x};\tau_3) \\ + c_{n,k_1,-k_2;\tau_2}\hat{\psi}_n(k_1\bar{\mathbf{b}}_1 - k_2\hat{\mathbf{b}}_2, \mathbf{x};\tau_3),$$

$c_{n,k_1,\pm k_2;\tau_2}$ are dependent on τ_2, and

$$\kappa_2 = j_2\,\pi/N_2, \quad j_2 = 1, 2, \ldots, N_2 - 1; \tag{6.12}$$

here, j_2 is the stationary state index in the $\mathbf{a}_2$ direction. The energy $\bar{\Lambda}$ of each such stationary Bloch state is given by

$$\bar{\Lambda}_{n,j_2}(\bar{\mathbf{k}}; \tau_3) = \hat{\Lambda}_n(\bar{\mathbf{k}} + \kappa_2\hat{\mathbf{b}}_2; \tau_3). \tag{6.13}$$

Each eigenvalue $\bar{\Lambda}_{n,j_2}(\bar{\mathbf{k}}; \tau_3)$ depends on N_2 and τ_3. These states are surface-like states in the quantum wire since $\hat{\psi}_n(\hat{\mathbf{k}}, \mathbf{x}; \tau_3)$ are surface-like states in the quantum film.

Similar to (5.33), due to (6.3), (6.8), and (6.13), in general, the energy of an edge-like state is above the energy of a relevant surface-like state in an ideal quantum wire:

$$\bar{\Lambda}_n(\bar{\mathbf{k}}; \tau_2, \tau_3) > \bar{\Lambda}_{n,j_2}(\bar{\mathbf{k}}; \tau_3). \tag{6.14}$$

6.3 Further Quantum Confinement of $\hat{\psi}_{n,j_3}(\hat{\mathbf{k}}, \mathbf{x}; \tau_3)$

The further confinement of two-dimensional Bloch waves $\hat{\psi}_{n,j_3}(\hat{\mathbf{k}}, \mathbf{x}; \tau_3)$ in the $\mathbf{a}_2$ direction can be similarly discussed. $\hat{\psi}_{n,j_3}(\hat{\mathbf{k}}, \mathbf{x}; \tau_3)$ with different stationary state index j_3 will be confined in the $\mathbf{a}_2$ direction independently.

Suppose B' is a rectangular parallelepiped with surfaces oriented in the $\mathbf{a}_1$, $\mathbf{a}_2$, or $\mathbf{b}_3$ direction[3] as shown in Fig. 6.2. It has a bottom face and a top face at $x_3 = \tau_3$ and $x_3 = \tau_3 + N_3$, a front face and a rear face perpendicularly intersecting the $\mathbf{a}_2$ axis at $x_2 = \tau_2$ and $x_2 = \tau_2 + 1$, and a left face and a right face perpendicularly intersecting the $\mathbf{a}_1$ axis and separated by a_1. The function set $\bar{\phi}_{j_3}(\bar{\mathbf{k}}, \mathbf{x}; \tau_2, \tau_3)$ is defined by the condition that each function behaves as a Bloch stationary state with the stationary index j_3 in the $\mathbf{b}_3$ direction as $\hat{\psi}_{j_3}(\hat{\mathbf{k}}, \mathbf{x}; \tau_3)$,[4] and

$$\begin{cases} \bar{\phi}_{j_3}(\bar{\mathbf{k}}, \mathbf{x} + \mathbf{a}_1; \tau_2, \tau_3) = e^{ik_1}\bar{\phi}_{j_3}(\bar{\mathbf{k}}, \mathbf{x}; \tau_2, \tau_3), & -\pi < k_1 \leq \pi, \\ \bar{\phi}_{j_3}(\bar{\mathbf{k}}, \mathbf{x}; \tau_2, \tau_3) = 0, & \mathbf{x} \in \partial B_2' \bigcup \partial B_3'. \end{cases} \tag{6.15}$$

The eigenvalues and eigenfunctions of Eq. (5.1) under this condition are denoted by $\bar{\lambda}_{n,j_3}(\bar{\mathbf{k}}; \tau_2)$ and $\bar{\phi}_{n,j_3}(\bar{\mathbf{k}}, \mathbf{x}; \tau_2, \tau_3)$, where $n = 0, 1, 2\ldots$.

For the eigenvalue problem defined by Eq. (5.1) and this condition, similar to Theorem 6.1, we expect to have the following theorem between $\bar{\lambda}_{n,j_3}(\bar{\mathbf{k}}; \tau_2)$ and the eigenvalue $\hat{\Lambda}_{n,j_3}(\hat{\mathbf{k}})$ in (5.28) of $\hat{\psi}_{n,j_3}(\hat{\mathbf{k}}, \mathbf{x}; \tau_3)$.

---

[3] That is, in the $\hat{\mathbf{b}}_1$, $\hat{\mathbf{b}}_2$, or $\mathbf{b}_3$ direction.

[4] $\hat{\psi}_{j_3}(\hat{\mathbf{k}}, \mathbf{x}; \tau_3)$ generally can be an (any) linear combination of $\hat{\psi}_{n,j_3}(\hat{\mathbf{k}}, \mathbf{x}; \tau_3)$ of different n.

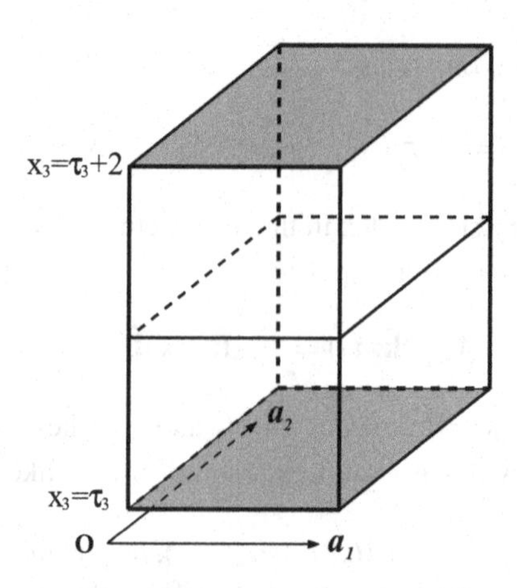

Fig. 6.2 The rectangular parallelepiped B' for the quantum confinement of $\hat{\psi}_{n,j_3}(\hat{\mathbf{k}}, \mathbf{x}; \tau_3)$. The two shadowed faces of $\partial B_3'$ determined by $x_3 = \tau_3$ and $x_3 = (\tau_3 + N_3)$ (in the figure is shown the case $N_3 = 2$) are the two faces on which each $\hat{\psi}_{n,j_3}(\hat{\mathbf{k}}, \mathbf{x}; \tau_3)$ is zero (thus, the specific direction of $\mathbf{a}_3$ no longer matters). The two *thick-lined* faces of $\partial B_2'$ perpendicularly intersecting the $\mathbf{a}_2$ axis at $\tau_2\mathbf{a}_2$ and $(\tau_2 + 1)\mathbf{a}_2$ are the two faces on which each function $\bar{\phi}_{j_3}(\bar{\mathbf{k}}, \mathbf{x}; \tau_2, \tau_3)$ is further required to be zero

Theorem 6.2

$$\bar{\lambda}_{n,j_3}(\bar{\mathbf{k}}; \tau_2) \geq \hat{\Lambda}_{n,j_3}(\hat{\mathbf{k}}) \quad for\ (\bar{\mathbf{k}} - \hat{\mathbf{k}}) \cdot \mathbf{a}_1 = 0. \tag{6.16}$$

As in (6.3), in (6.16) $\hat{\mathbf{k}}$ is a wave vector in the film plane, and $\bar{\mathbf{k}}$ is a wave vector in the wire direction. In (6.16), $\hat{\mathbf{k}}$ is related to $\bar{\mathbf{k}}$. $\hat{\psi}_{j_3}(\hat{\mathbf{k}}, \mathbf{x}; \tau_3)$ with a different stationary state index j_3 are orthogonal to each other, each of them will be confined in the $\mathbf{a}_2$ direction independently, Theorem 6.2 can be proved similar to Theorem 6.1.

Theorem 6.2 is similar to Theorem 6.1. The consequences of the quantum confinement of two-dimensional Bloch waves $\hat{\psi}_n(\hat{\mathbf{k}}, \mathbf{x}; \tau_3)$ in the $\mathbf{a}_2$ direction due to Theorem 6.1 can be similarly applied to the further quantum confinement of two-dimensional Bloch waves $\hat{\psi}_{n,j_3}(\hat{\mathbf{k}}, \mathbf{x}; \tau_3)$ in the $\mathbf{a}_2$ direction.

Because $v(\mathbf{x} + \mathbf{a}_2) = v(\mathbf{x})$, the function $\bar{\phi}_{n,j_3}(\bar{\mathbf{k}}, \mathbf{x}; \tau_2, \tau_3)$ has the form

$$\bar{\phi}_{n,j_3}(\bar{\mathbf{k}}, \mathbf{x} + \mathbf{a}_2; \tau_2, \tau_3) = \mathrm{e}^{ik_2}\bar{\phi}_{n,j_3}(\bar{\mathbf{k}}, \mathbf{x}; \tau_2, \tau_3), \tag{6.17}$$

and k_2 is either complex with a nonzero imaginary part or a real number.

If k_2 is real in (6.17), $\bar{\phi}_{n,j_3}(\bar{\mathbf{k}}, \mathbf{x}; \tau_2, \tau_3)$ is a $\hat{\psi}_{n',j_3}(\hat{\mathbf{k}}, \mathbf{x}; \tau_3)$. By Theorem 6.2, a $\hat{\psi}_{n',j_3}(\hat{\mathbf{k}}, \mathbf{x}; \tau_3)$ can be a $\bar{\phi}_{n,j_3}(\bar{\mathbf{k}}, \mathbf{x}; \tau_2, \tau_3)$ only in special cases where $\hat{\psi}_{n',j_3}(\hat{\mathbf{k}}, \mathbf{x}; \tau_3)$ has a nodal plane perpendicularly intersecting the $\mathbf{a}_2$ axis at $x_2 = \tau_2$. Therefore, k_2 in (6.17) can be real only in such special cases; in most cases, k_2 in (6.17) is complex with a nonzero imaginary part.

The imaginary part of k_2 in (6.17) can be either positive or negative; this corresponds to that $\bar{\phi}_{n,j_3}(\bar{\mathbf{k}}, \mathbf{x}; \tau_2, \tau_3)$ decays in either the positive or the negative direction of $\mathbf{a}_2$. Such states $\bar{\phi}_{n,j_3}(\bar{\mathbf{k}}, \mathbf{x}; \tau_2, \tau_3)$ with a nonzero imaginary part of k_2 cannot exist in a film with two-dimensional translational invariance since they are divergent in either the negative or the positive direction of $\mathbf{a}_2$. However, they can play a significant role in a quantum wire with a finite size in the $\mathbf{a}_2$ direction.

The quantum confinement of two-dimensional Bloch waves $\hat{\psi}_{n,j_3}(\hat{\mathbf{k}}, \mathbf{x}; \tau_3)$ in the $\mathbf{a}_2$ direction will produce two different types of solutions of (6.1) in the quantum wire.

One type of nontrivial solutions can be obtained from (6.17) by assigning

$$\bar{\psi}_{n,j_3}(\bar{\mathbf{k}}, \mathbf{x}; \tau_2, \tau_3) = \begin{cases} c_{N_2,N_3}\bar{\phi}_{n,j_3}(\bar{\mathbf{k}}, \mathbf{x}; \tau_2, \tau_3), & \mathbf{x} \in \text{the wire}, \\ 0, & \mathbf{x} \notin \text{the wire}, \end{cases} \tag{6.18}$$

where c_{N_2,N_3} is a normalization constant. The corresponding eigenvalue

$$\bar{\Lambda}_{n,j_3}(\bar{\mathbf{k}}; \tau_2) = \bar{\lambda}_{n,j_3}(\bar{\mathbf{k}}; \tau_2) \tag{6.19}$$

is dependent on τ_2 and N_3. A consequence of Theorem 6.2 is that for each bulk energy band n, each j_3, and each wave vector $\bar{\mathbf{k}}$, there is one such solution (6.18) of Eq. (6.1). This state is a surface-like state since $\bar{\phi}_{n,j_3}(\bar{\mathbf{k}}, \mathbf{x}; \tau_2, \tau_3)$ decays in either the positive or negative direction of $\mathbf{a}_2$ in most cases.

Now, we try to find other solutions of Eq. (6.1) from the further quantum confinement of $\hat{\psi}_{n,j_3}(\hat{\mathbf{k}}, \mathbf{x}; \tau_3)$. We can expect that there are stationary Bloch states in the $\mathbf{a}_2$ direction, formed due to the multiple reflections of $\hat{\psi}_{n,j_3}(\hat{\mathbf{k}}, \mathbf{x}; \tau_3)$ between two boundary surfaces of the wire perpendicular to the $\mathbf{a}_2$ axis.

For many quantum films discussed in Chap. 5, we have

$$\hat{\Lambda}_{n,j_3}(k_1\bar{\mathbf{b}}_1 + k_2\hat{\mathbf{b}}_2) = \hat{\Lambda}_{n,j_3}(k_1\bar{\mathbf{b}}_1 - k_2\hat{\mathbf{b}}_2) \tag{6.20}$$

in (5.28); thus, in general,

$$\begin{aligned} f_{n,k_1,k_2,j_3}(\mathbf{x}; \tau_3) &= c_+\hat{\psi}_{n,j_3}(k_1\bar{\mathbf{b}}_1 + k_2\hat{\mathbf{b}}_2, \mathbf{x}; \tau_3) \\ &\quad + c_-\hat{\psi}_{n,j_3}(k_1\bar{\mathbf{b}}_1 - k_2\hat{\mathbf{b}}_2, \mathbf{x}; \tau_3), \quad 0 < k_2 < \pi, \end{aligned}$$

where the $c_\pm$ are not zero, is a nontrivial solution of Eq. (5.12) due to (6.20). It is easy to see that $f_{n,k_1,k_2,j_3}(\mathbf{x}; \tau_3)$ is a one-dimensional Bloch wave with a wave vector $\bar{\mathbf{k}} = k_1\bar{\mathbf{b}}_1$ in the wire direction:

$$f_{n,k_1,k_2,j_3}(\mathbf{x} + \mathbf{a}_1; \tau_3) = \mathrm{e}^{\mathrm{i}k_1} f_{n,k_1,k_2,j_3}(\mathbf{x}; \tau_3), \quad -\pi < k_1 \leq \pi,$$

due to (5.27). To be a solution of Eq. (6.1), the function $f_{n,k_1,k_2,j_3}(\mathbf{x}; \tau_3)$ is required to be zero on the front face and the rear face of the wire. By writing the front face equation of the wire as $x_2 = x_{2,f}(x_1, x_3)$ and the rear face equation of the wire as $x_2 = x_{2,r}(x_1, x_3)$, we should have

$$c_+\hat{\psi}_{n,j_3}[k_1\bar{\mathbf{b}}_1 + k_2\hat{\mathbf{b}}_2, \mathbf{x} \in x_{2,f}(x_1, x_3); \tau_3] \\ + c_-\hat{\psi}_{n,j_3}[k_1\bar{\mathbf{b}}_1 - k_2\hat{\mathbf{b}}_2, \mathbf{x} \in x_{2,f}(x_1, x_3); \tau_3] = 0,$$
$$c_+\hat{\psi}_{n,j_3}[k_1\bar{\mathbf{b}}_1 + k_2\hat{\mathbf{b}}_2, \mathbf{x} \in x_{2,r}(x_1, x_3); \tau_3] \\ + c_-\hat{\psi}_{n,j_3}[k_1\bar{\mathbf{b}}_1 - k_2\hat{\mathbf{b}}_2, \mathbf{x} \in x_{2,r}(x_1, x_3); \tau_3] = 0.$$

Since $x_{2,r}(x_1, x_3) = x_{2,f}(x_1, x_3) + N_2$, we have

$$\hat{\psi}_{n,j_3}[k_1\bar{\mathbf{b}}_1 + k_2\hat{\mathbf{b}}_2, \mathbf{x} \in x_{2,r}(x_1, x_3); \tau_3] \\ = e^{ik_2N_2}\hat{\psi}_{n,j_3}[k_1\bar{\mathbf{b}}_1 + k_2\hat{\mathbf{b}}_2, \mathbf{x} \in x_{2,f}(x_1, x_3); \tau_3],$$

and

$$\hat{\psi}_{n,j_3}[k_1\bar{\mathbf{b}}_1 - k_2\hat{\mathbf{b}}_2, \mathbf{x} \in x_{2,r}(x_1, x_3); \tau_3] \\ = e^{-ik_2N_2}\hat{\psi}_{n,j_3}[k_1\bar{\mathbf{b}}_1 - k_2\hat{\mathbf{b}}_2, \mathbf{x} \in x_{2,f}(x_1, x_3); \tau_3],$$

due to (5.27). Therefore, for $c_\pm$ not both zero, $e^{ik_2N_2} - e^{-ik_2N_2} = 0$ has to be true for each such stationary Bloch state, for the quantum wire.

Therefore, the stationary Bloch state solutions of Eq. (6.1) from the quantum confinement of $\hat{\psi}_{n,j_3}(\hat{\mathbf{k}}, \mathbf{x}; \tau_3)$ should have the form

$$\bar{\psi}_{n,j_2,j_3}(\bar{\mathbf{k}}, \mathbf{x}; \tau_2, \tau_3) = \begin{cases} f_{n,k_1,\kappa_2,j_3}(\mathbf{x}; \tau_2, \tau_3), & \mathbf{x} \in \text{the wire}, \\ 0, & \mathbf{x} \notin \text{the wire}, \end{cases} \tag{6.21}$$

where

$$f_{n,k_1,k_2,j_3}(\mathbf{x}; \tau_2, \tau_3) = c_{n,k_1,k_2,j_3;\tau_2}\hat{\psi}_{n,j_3}(k_1\bar{\mathbf{b}}_1 + k_2\hat{\mathbf{b}}_2, \mathbf{x}; \tau_3) \\ + c_{n,k_1,-k_2,j_3;\tau_2}\hat{\psi}_{n,j_3}(k_1\bar{\mathbf{b}}_1 - k_2\hat{\mathbf{b}}_2, \mathbf{x}; \tau_3),$$

$c_{n,k_1,\pm k_2,j_3;\tau_2}$ are dependent on τ_2, and

$$\kappa_2 = j_2\pi/N_2, \quad j_2 = 1, 2, \ldots, N_2 - 1;$$

here, j_2 is the stationary state index in the $\mathbf{a}_2$ direction. Each stationary Bloch state $\bar{\psi}_{n,j_2,j_3}(\bar{\mathbf{k}}, \mathbf{x}; \tau_2, \tau_3)$ satisfying (6.1) has the energy

$$\bar{\Lambda}_{n,j_2,j_3}(\bar{\mathbf{k}}) = \hat{\Lambda}_{n,j_3}(\bar{\mathbf{k}} + \kappa_2\hat{\mathbf{b}}_2). \tag{6.22}$$

Each energy $\bar{\Lambda}_{n,j_2,j_3}(\bar{\mathbf{k}})$ for this case is dependent on N_2 and N_3, the wire size. There are $(N_2 - 1)(N_3 - 1)$ such stationary Bloch states for each n and $\bar{\mathbf{k}}$ in the quantum wire. Their energies map the $\hat{\Lambda}_{n,j_3}(\hat{\mathbf{k}})$ exactly and thus also map the corresponding bulk energy band $\varepsilon_n(\mathbf{k})$ exactly by (5.28): $\bar{\Lambda}_{n,j_2,j_3}(\bar{\mathbf{k}}) = \hat{\Lambda}_{n,j_3}(\bar{\mathbf{k}} + \kappa_2\hat{\mathbf{b}}_2) = \varepsilon_n(\bar{\mathbf{k}} +$

$\kappa_2\hat{\mathbf{b}}_2 + \kappa_3\mathbf{b}_3)$. Therefore, $\bar{\psi}_{n,j_2,j_3}(\bar{\mathbf{k}}, \mathbf{x}; \tau_2, \tau_3)$ can be considered as bulk-like states in the quantum wire.

From the quantum confinement of $\hat{\psi}_{n,j_3}(\hat{\mathbf{k}}, \mathbf{x}; \tau_3)$ in an ideal quantum wire, between a surface-like state and relevant bulk-like states obtained, in general

$$\bar{\Lambda}_{n,j_3}(\bar{\mathbf{k}}; \tau_2) > \bar{\Lambda}_{n,j_2,j_3}(\bar{\mathbf{k}}) \tag{6.23}$$

is true, due to (6.16), (6.19), and (6.22); similar to (6.14).

We have seen that for the further quantum confinement of two-dimensional Bloch waves $\hat{\psi}_n(\hat{\mathbf{k}}, \mathbf{x}; \tau_3)$ or $\hat{\psi}_{n,j_3}(\hat{\mathbf{k}}, \mathbf{x}; \tau_3)$, each one will produce two different types of one-dimensional Bloch waves in the ideal quantum wire. For the ideal rectangular quantum wire obtained from a quantum film of N_3 layers in the $\mathbf{a}_3$ direction and with the bottom face defined by $\tau_3\mathbf{a}_3$ being further confined by two boundary faces in the $\mathbf{a}_2$ direction defined by τ_2 and $N_2\mathbf{a}_2$ apart from each other, there are four sets of electronic states in the quantum wire:

The energy $\bar{\Lambda}_{n,j_2,j_3}(\bar{\mathbf{k}})$ (6.22) of each electronic state $\bar{\psi}_{n,j_2,j_3}(\bar{\mathbf{k}}, \mathbf{x}; \tau_2, \tau_3)$ in (6.21) depends on N_2 and N_3. The energies of these states map the energy band of the bulk $\varepsilon_n(\mathbf{k})$ exactly. These states are bulk-like states, and there are $(N_2 - 1)(N_3 - 1)$ such states in the quantum wire for each bulk energy band n and each $\bar{\mathbf{k}}$.

The energy $\bar{\Lambda}_{n,j_2}(\bar{\mathbf{k}}; \tau_3)$ (6.13) of each electronic state $\bar{\psi}_{n,j_2}(\bar{\mathbf{k}}, \mathbf{x}; \tau_2, \tau_3)$ in (6.11) depends on N_2 and τ_3. The energies of these states map the surface-like energy subband $\hat{\Lambda}_n(\hat{\mathbf{k}}; \tau_3)$ in the film exactly. These states are surface-like states. There are $(N_2 - 1)$ such states in the quantum wire for each bulk energy band n and each $\bar{\mathbf{k}}$.

The energy $\bar{\Lambda}_{n,j_3}(\bar{\mathbf{k}}; \tau_2)$ (6.19) of each state $\bar{\psi}_{n,j_3}(\bar{\mathbf{k}}, \mathbf{x}; \tau_2, \tau_3)$ in (6.18) depends on N_3 and τ_2. These are also surface-like states. There are $(N_3 - 1)$ such states in the quantum wire for each bulk energy band n and each $\bar{\mathbf{k}}$.

The energy $\bar{\Lambda}_n(\bar{\mathbf{k}}; \tau_2, \tau_3)$ (6.8) of each electronic state $\bar{\psi}_n(\bar{\mathbf{k}}, \mathbf{x}; \tau_2, \tau_3)$ in (6.7) depends on τ_2 and τ_3. These are edge-like states. Although a rectangular quantum wire always has *four* edges, there is only one such edge-like state for each bulk energy band n and each $\bar{\mathbf{k}}$.

Therefore, the effect of the further quantum confinement in one more direction is *to have one and only one* boundary-dependent sub-subband for each subband of the electronic states in the film obtained in Chap. 5; the energies of other states can be obtained either from the $\hat{\Lambda}_{n,j_3}(\hat{\mathbf{k}})$ (which originally is determined by the energy band structure $\varepsilon_n(\mathbf{k})$ of the bulk crystal by (5.28)) by (6.22) or from the surface-like band structure, such as $\hat{\Lambda}_n(\hat{\mathbf{k}}; \tau_3)$ by (6.13). In general, the boundary-dependent state always has a higher energy than the relevant size-dependent states.

The results in Sects. 6.2 and 6.3 were obtained by a specific quantum confinement order. To achieve a more comprehensive understanding of the electronic states in an ideal quantum wire, we need to consider two different confinement orders.

6.4 Quantum Wires of Crystals with a sc, tetr, or ortho Bravais Lattice

We expect that the simplest cases where the theory in this chapter is applicable are the rectangular quantum wires of crystals with a sc, tetr, or ortho Bravais lattice in which (5.13), (6.9), and (6.20) are true. In these crystals, the three primitive lattice vectors $\mathbf{a}_1$, $\mathbf{a}_2$, and $\mathbf{a}_3$ are perpendicular to each other and equivalent; consequently, the three primitive lattice vectors in the k space, $\mathbf{b}_1$, $\mathbf{b}_2$, and $\mathbf{b}_3$ are also perpendicular to each other and equivalent. Such a quantum wire in the direction of $\mathbf{a}_1$ can be considered as a film with the film plane defined by $\mathbf{a}_1$ and $\mathbf{a}_2$ being further confined in the $\mathbf{a}_2$ direction as we have done so far. Equivalently, it can also be considered as a film with the film plane defined by $\mathbf{a}_1$ and $\mathbf{a}_3$ being further confined in the $\mathbf{a}_3$ direction. If we consider the electronic states in the quantum wire in the latter way, we will obtain that

$$\bar{\psi}_{n,j_3}(\bar{\mathbf{k}}, \mathbf{x}; \tau_2, \tau_3) = \begin{cases} f_{n,k_1,\kappa_3}(\mathbf{x}; \tau_2, \tau_3), & \mathbf{x} \in \text{the wire}, \\ 0, & \mathbf{x} \notin \text{the wire}, \end{cases} \tag{6.24}$$

instead of (6.18), where $\bar{\mathbf{k}} = k_1\bar{\mathbf{b}}_1$ and

$$\begin{aligned} f_{n,k_1,k_3}(\mathbf{x}; \tau_2, \tau_3) &= c_{n,k_1,k_3;\tau_3}\hat{\psi}_n(k_1\bar{\mathbf{b}}_1 + k_3\mathbf{b}_3, \mathbf{x}; \tau_2) \\ &+ c_{n,k_1,-k_3;\tau_3}\hat{\psi}_n(k_1\bar{\mathbf{b}}_1 - k_3\mathbf{b}_3, \mathbf{x}; \tau_2), \end{aligned}$$

$c_{n,k_1,\pm k_3;\tau_3}$ are dependent on τ_3, and κ_3 and j_3 are given by (5.26). Equation (6.24) gives a more specific relationship between the surface-like states $\bar{\psi}_{n,j_3}(\bar{\mathbf{k}}, \mathbf{x}; \tau_2, \tau_3)$ in the quantum wire and the surface-like states $\hat{\psi}_n(\hat{\mathbf{k}}, \mathbf{x}; \tau_2)$ in the quantum film with a film plane defined by $\mathbf{a}_1$ and $\mathbf{a}_3$. Correspondingly,

$$\bar{\Lambda}_{n,j_3}(\bar{\mathbf{k}}; \tau_2) = \hat{\Lambda}_n(\bar{\mathbf{k}} + \kappa_3\mathbf{b}_3; \tau_2) \tag{6.25}$$

can be obtained instead of (6.19). Equation (6.25) gives a more specific relationship between the surface-like subbands $\bar{\Lambda}_{n,j_3}(\bar{\mathbf{k}}; \tau_2)$ in the quantum wire and the surface-like subband $\hat{\Lambda}_n(\hat{\mathbf{k}}; \tau_2)$ in the quantum film with a film plane defined by $\mathbf{a}_1$ and $\mathbf{a}_3$.

Similar to (6.14) and (6.23), we can obtain that

$$\bar{\Lambda}_n(\bar{\mathbf{k}}; \tau_2, \tau_3) > \bar{\Lambda}_{n,j_3}(\bar{\mathbf{k}}; \tau_2), \tag{6.26}$$

and

$$\bar{\Lambda}_{n,j_2}(\bar{\mathbf{k}}; \tau_3) > \bar{\Lambda}_{n,j_2,j_3}(\bar{\mathbf{k}}). \tag{6.27}$$

Therefore, for an ideal rectangular quantum wire of a crystal with a sc, tetr, or ortho Bravais lattice, for each bulk energy band n, there are the following:

$(N_2 - 1)(N_3 - 1)$ bulk-like subbands with energies

$$\bar{\Lambda}_{n,j_2,j_3}(\bar{\mathbf{k}}) = \varepsilon_n \left(\bar{\mathbf{k}} + \frac{j_2\pi}{N_2}\mathbf{b}_2 + \frac{j_3\pi}{N_3}\mathbf{b}_3 \right) \tag{6.28}$$

from (6.22) and (5.28);
$(N_3 - 1)$ surface-like subbands with energies

$$\bar{\Lambda}_{n,j_3}(\bar{\mathbf{k}}; \tau_2) = \hat{\Lambda}_n \left(\bar{\mathbf{k}} + \frac{j_3\pi}{N_3}\mathbf{b}_3; \tau_2 \right) \tag{6.29}$$

from (6.25) and (5.26);
$(N_2 - 1)$ surface-like subbands with energies

$$\bar{\Lambda}_{n,j_2}(\bar{\mathbf{k}}; \tau_3) = \hat{\Lambda}_n \left(\bar{\mathbf{k}} + \frac{j_2\pi}{N_2}\mathbf{b}_2; \tau_3 \right) \tag{6.30}$$

from (6.13) and (6.12);
one edge-like subband with energy

$$\bar{\Lambda}_n(\bar{\mathbf{k}}; \tau_2, \tau_3) = \bar{\lambda}_n(\bar{\mathbf{k}}; \tau_2, \tau_3) \tag{6.31}$$

from (6.8). Here, τ_2 and N_2 define the boundary face positions and the size of the quantum wire in the $\mathbf{a}_2$ direction, τ_3 and N_3 define the boundary face positions and the size in the $\mathbf{a}_3$ direction; $j_2 = 1, 2, \ldots, N_2 - 1$ and $j_3 = 1, 2, \ldots, N_3 - 1$; $\bar{\mathbf{k}}$ is a wave vector in the wire direction and $\hat{\Lambda}_n(\hat{\mathbf{k}}; \tau_3)$ is the surface-like band structure of a quantum film with the film plane oriented in the $\mathbf{a}_3$ direction with a wave vector $\hat{\mathbf{k}}$ in the film plane; $\hat{\Lambda}_n(\hat{\mathbf{k}}; \tau_2)$ is the surface-like band structure of a quantum film with the film plane oriented in the $\mathbf{a}_2$ direction with a wave vector $\hat{\mathbf{k}}$ in the film plane.

Between the energies of electronic states with the same energy band index n and with the same wave vector $\bar{\mathbf{k}}$ in the quantum wire, the following general relations exist:

$$\begin{aligned}
\bar{\Lambda}_n(\bar{\mathbf{k}}; \tau_2, \tau_3) &> \bar{\Lambda}_{n,j_2}(\bar{\mathbf{k}}; \tau_3), \\
\bar{\Lambda}_n(\bar{\mathbf{k}}; \tau_2, \tau_3) &> \bar{\Lambda}_{n,j_3}(\bar{\mathbf{k}}; \tau_2), \\
\bar{\Lambda}_{n,j_3}(\bar{\mathbf{k}}; \tau_2) &> \bar{\Lambda}_{n,j_2,j_3}(\bar{\mathbf{k}}), \\
\bar{\Lambda}_{n,j_2}(\bar{\mathbf{k}}; \tau_3) &> \bar{\Lambda}_{n,j_2,j_3}(\bar{\mathbf{k}}),
\end{aligned}$$

from (6.14), (6.26), (6.23), and (6.27), respectively.

However, probably the practically more interesting cases are quantum wires of crystals with an fcc or bcc Bravais lattice in which (5.20), (6.9), and (6.20) are true. For these crystals, the choice of primitive lattice vectors for films depends on the film direction, as we have seen in Sect. 5.6. In the following, we try to obtain predictions on the electronic states in several such quantum wires, based on the results obtained in Sects. 6.1–6.3.

6.5 fcc Quantum Wires with (110) and (001) Surfaces

We consider an fcc $[1\bar{1}0]$ quantum wire with (110) and (001) surfaces and having a rectangular cross section $N_{110}a/\sqrt{2} \times N_{001}a$, where a is the lattice constant, N_{110} and N_{001} are two positive integers. The electronic states in such a quantum wire can be considered as the electronic states in a (001) fcc quantum film of thickness $N_{001}a$ discussed in Sect. 5.6.1 further confined by two (110) boundary surfaces. They can also equivalently be considered as the electronic states in an (110) fcc quantum film of thickness $N_{110}a/\sqrt{2}$, discussed in Sect. 5.6.2, further confined by two (001) boundary surfaces.

6.5.1 fcc Quantum Wires Obtained from (001) Films Further Confined by Two (110) Surfaces

For an fcc quantum wire obtained from a (001) film further confined by two (110) surfaces, we begin with an fcc (001) film of thickness $N_{001}a$ and choose the primitive lattice vectors as in (5.34): $\mathbf{a}_1 = a/2(1, -1, 0)$ and $\mathbf{a}_2 = a/2(1, 1, 0)$, $\mathbf{a}_3 = a/2(1, 0, 1)$ and thus $\mathbf{b}_1 = 1/a(1, -1, -1)$, $\mathbf{b}_2 = 1/a(1, 1, -1)$, and $\mathbf{b}_3 = 1/a(0, 0, 2)$. Correspondingly, $\hat{\mathbf{b}}_1 = 1/a(1, -1, 0)$ and $\hat{\mathbf{b}}_2 = 1/a(1, 1, 0)$.

Now, we have a (001) film with $N_3 = 2N_{001}$. From the results obtained in Sect. 5.6.1, in such a film for each bulk energy band, there are $2N_{001} - 1$ bulk-like subbands with energies

$$\hat{\Lambda}_{n,j_3}(k_1\hat{\mathbf{b}}_1 + k_2\hat{\mathbf{b}}_2) = \varepsilon_n\left[k_1\hat{\mathbf{b}}_1 + k_2\hat{\mathbf{b}}_2 + \frac{j_3\pi}{N_{001}a}(0, 0, 1)\right]$$

by (5.35), where

$$j_3 = 1, 2, \ldots, 2N_{001} - 1,$$

and one surface-like subband whose energy

$$\hat{\Lambda}_n(k_1\hat{\mathbf{b}}_1 + k_2\hat{\mathbf{b}}_2;\ \tau_{001}) = \hat{\lambda}_n(k_1\hat{\mathbf{b}}_1 + k_2\hat{\mathbf{b}}_2;\ \tau_{001})$$

by (5.30) since now $\tau_3 = \tau_{001}$. $\hat{\mathbf{k}} = k_1\hat{\mathbf{b}}_1 + k_2\hat{\mathbf{b}}_2$ is a wave vector in the (001) plane.

Then we consider the (001) fcc quantum film further confined by two (110) boundary surfaces which are $N_{110}a/\sqrt{2}$ apart. The energies $\hat{\Lambda}_{n,j_3}(\hat{\mathbf{k}})$ of bulk-like states $\hat{\psi}_{n,j_3}(\hat{\mathbf{k}}, \mathbf{x}; \tau_3)$ in the (001) quantum film satisfy (6.20): $\hat{\Lambda}_{n,j_3}(k_1\bar{\mathbf{b}}_1 + k_2\hat{\mathbf{b}}_2) = \hat{\Lambda}_{n,j_3}(k_1\bar{\mathbf{b}}_1 - k_2\hat{\mathbf{b}}_2)$. We also expect that the energies $\hat{\Lambda}_n(\hat{\mathbf{k}}; \tau_{001})$ of surface-like states $\hat{\psi}_n(\hat{\mathbf{k}}, \mathbf{x}; \tau_{001})$ in the fcc (001) film satisfy (6.9): $\hat{\Lambda}_n(k_1\bar{\mathbf{b}}_1 + k_2\hat{\mathbf{b}}_2; \tau_{001}) = \hat{\Lambda}_n(k_1\bar{\mathbf{b}}_1 - k_2\hat{\mathbf{b}}_2; \tau_{001})$ (see Sect. 5.6.1). Therefore, the results obtained in Sects. 6.2 and 6.3 can be applied. We now have $N_2 = N_{110}$ and $\tau_2 = \tau_{110}$; thus, for each bulk energy band, there are four different sets of one-dimensional Bloch waves in the quantum wire.

From (6.22), there are $(N_{110} - 1)(2N_{001} - 1)$ bulk-like subbands for each bulk energy band n in the quantum wire; each subband has the energy

$$\bar{\Lambda}_{n,j_{110},j_3}(\bar{\mathbf{k}}) = \varepsilon_n \left[\bar{\mathbf{k}} + \frac{j_{110}\pi}{N_{110}a}(1, 1, 0) + \frac{j_3\pi}{N_{001}a}(0, 0, 1) \right], \tag{6.32}$$

where $\bar{\mathbf{k}}$ is a wave vector in the wire direction $\mathbf{a}_1$,

$$j_3 = 1, 2, \ldots, 2N_{001} - 1,$$

and

$$j_{110} = 1, 2, \ldots, N_{110} - 1. \tag{6.33}$$

By defining

$$j_{001} = \begin{cases} j_3, & j_3 < N_{001}, \\ 2N_{001} - j_3, & j_3 > N_{001}, \end{cases} \tag{6.34}$$

where

$$j_{001} = 1, 2, \ldots, N_{001} - 1; \tag{6.35}$$

those $(N_{110} - 1)(2N_{001} - 1)$ bulk-like subbands in (6.32) in the quantum wire can be separated into three subsets according to (6.34) and (6.35). They are as follows: $(N_{110} - 1)(N_{001} - 1)$ bulk-like subbands in the quantum wire with energies

$$\bar{\Lambda}^{bk,a}_{n,j_{110},j_{001}}(\bar{\mathbf{k}}) = \varepsilon_n \left[\bar{\mathbf{k}} + \frac{j_{110}\pi}{N_{110}a}(1, 1, 0) + \frac{j_{001}\pi}{N_{001}a}(0, 0, 1) \right], \tag{6.32a}$$

$(N_{110} - 1)$ bulk-like subbands in the quantum wire with energies

$$\bar{\Lambda}^{bk,b}_{n,j_{110}}(\bar{\mathbf{k}}) = \varepsilon_n \left[\bar{\mathbf{k}} + \frac{j_{110}\pi}{N_{110}a}(1, 1, 0) + \frac{\pi}{a}(0, 0, 1) \right], \tag{6.32b}$$

$(N_{110} - 1)(N_{001} - 1)$ bulk-like subbands in the quantum wire with energies[5]

---

[5]Since for cubic semiconductors and fcc metals, in general, $\varepsilon_n(k_x, k_y, k_z) = \varepsilon_n(k_x, k_y, -k_z)$ and $1/a(0, 0, 2)$ is a reciprocal lattice vector for crystals with an fcc Bravais lattice.

$$\begin{aligned}\bar{\Lambda}^{bk,c}_{n,j_{110},j_{001}}(\bar{\mathbf{k}}) &= \varepsilon_n\left[\bar{\mathbf{k}} + \frac{j_{110}\pi}{N_{110}a}(1,1,0) - \frac{j_{001}\pi}{N_{001}a}(0,0,1) + \frac{2\pi}{a}(0,0,1)\right] \\ &= \varepsilon_n\left[\bar{\mathbf{k}} + \frac{j_{110}\pi}{N_{110}a}(1,1,0) + \frac{j_{001}\pi}{N_{001}a}(0,0,1) + \frac{2\pi}{a}(0,0,1)\right].\end{aligned} \tag{6.32c}$$

From (6.19), for each bulk energy band n, there are $2N_{001}-1$ surface-like subbands in the quantum wire due to the existence of (110) boundary surfaces with the energies

$$\bar{\Lambda}_{n,j_3}(\bar{\mathbf{k}}; \tau_{110}) = \bar{\lambda}_{n,j_3}(\bar{\mathbf{k}}; \tau_{110}). \tag{6.36}$$

By (6.34), these $2N_{001} - 1$ surface-like bands in (6.36) can be separated into three subsets. They are as follows:
$N_{001} - 1$ surface-like subbands in the quantum wire with energies

$$\bar{\Lambda}^{sf,1}_{n,j_{001}}(\bar{\mathbf{k}}; \tau_{110}) = \bar{\lambda}_{n,j_{001}}(\bar{\mathbf{k}}; \tau_{110}), \tag{6.36a}$$

one surface-like subband in the quantum wire with energy

$$\bar{\Lambda}^{sf,2}_{n,N_{001}}(\bar{\mathbf{k}}; \tau_{110}) = \bar{\lambda}_{n,N_{001}}(\bar{\mathbf{k}}; \tau_{110}), \tag{6.36b}$$

$N_{001} - 1$ surface-like subbands in the quantum wire with energies

$$\bar{\Lambda}^{sf,3}_{n,j_{001}}(\bar{\mathbf{k}}; \tau_{110}) = \bar{\lambda}_{n,2N_{001}-j_{001}}(\bar{\mathbf{k}}; \tau_{110}). \tag{6.36c}$$

From (6.13), due to the existence of (001) boundary surfaces for each bulk energy band n, there are $N_{110} - 1$ surface-like subbands in the quantum wire with energies

$$\bar{\Lambda}^{sf,a}_{n,j_{110}}(\bar{\mathbf{k}}; \tau_{001}) = \hat{\Lambda}_n\left[\bar{\mathbf{k}} + \frac{j_{110}\pi}{N_{110}a}(1,1,0); \tau_{001}\right]. \tag{6.37}$$

From (6.8), for each bulk energy band n, there is one edge-like subband in the quantum wire with energy

$$\bar{\Lambda}^{eg}_n(\bar{\mathbf{k}}; \tau_{110}, \tau_{001}) = \bar{\lambda}_n(\bar{\mathbf{k}}; \tau_{110}, \tau_{001}). \tag{6.38}$$

6.5.2 *fcc Quantum Wires Obtained from (110) Films Further Confined by Two (001) Surfaces*

For an fcc quantum wire obtained from a (110) film further confined by two (001) surfaces, we begin with a (110) film and the primitive lattice vectors can be chosen as in (5.36): $\mathbf{a}_1 = a/2(1, -1, 0)$, $\mathbf{a}_2 = a(0, 0, -1)$, and $\mathbf{a}_3 = a/2(0, 1, 1)$ and, thus,

$\mathbf{b}_1 = 1/a(2, 0, 0)$, $\mathbf{b}_2 = 1/a(1, 1, -1)$, and $\mathbf{b}_3 = 1/a(2, 2, 0)$. Correspondingly, $\hat{\mathbf{b}}_1 = 1/a(1, -1, 0)$ and $\hat{\mathbf{b}}_2 = 1/a(0, 0, -1)$.

For an fcc quantum wire with a rectangular cross section $N_{110}a/\sqrt{2} \times N_{001}a$, we now have $N_3 = 2N_{110}$, $\tau_3 = \tau_{110}$, and $N_2 = N_{001}$, $\tau_2 = \tau_{001}$. The energies $\hat{\Lambda}_{n,j_3}(\hat{\mathbf{k}})$ of bulk-like states $\hat{\psi}_{n,j_3}(\hat{\mathbf{k}}, \mathbf{x}; \tau_{110})$ in the (110) quantum film satisfy (6.20): $\hat{\Lambda}_{n,j_3}(k_1\bar{\mathbf{b}}_1 + k_2\hat{\mathbf{b}}_2) = \hat{\Lambda}_{n,j_3}(k_1\bar{\mathbf{b}}_1 - k_2\hat{\mathbf{b}}_2)$. We also expect that the energies $\hat{\Lambda}_n(\hat{\mathbf{k}}; \tau_{110})$ of surface-like states $\hat{\psi}_n(\hat{\mathbf{k}}, \mathbf{x}; \tau_{110})$ in the fcc (110) film satisfy (6.9): $\hat{\Lambda}_n(k_1\bar{\mathbf{b}}_1 + k_2\hat{\mathbf{b}}_2; \tau_{110}) = \hat{\Lambda}_n(k_1\bar{\mathbf{b}}_1 - k_2\hat{\mathbf{b}}_2; \tau_{110})$ (see Sect. 5.6.2). Therefore, the results obtained in Sects. 6.2 and 6.3 can be applied. Similar to the results obtained in Sect. 6.5.1, for each bulk energy band n we have four different sets of one-dimensional Bloch waves in the quantum wire.

From (6.22), for each bulk energy band n, there are $(N_{001}-1)(2N_{110}-1)$ bulk-like subbands in the quantum wire; each subband has the energy

$$\bar{\Lambda}_{n,j_{001},j_3}(\bar{\mathbf{k}}) = \varepsilon_n\left[\bar{\mathbf{k}} + \frac{j_{001}\pi}{N_{001}a}(0, 0, 1) + \frac{j_3\pi}{N_{110}a}(1, 1, 0)\right], \tag{6.39}$$

where $\bar{\mathbf{k}}$ is a wave vector in the wire direction $\mathbf{a}_1$,

$$j_{001} = 1, 2, \ldots, N_{001} - 1, \tag{6.40}$$

and

$$j_3 = 1, 2, \ldots, 2N_{110} - 1.$$

By defining

$$j_{110} = \begin{cases} j_3, & j_3 < N_{110}, \\ 2N_{110} - j_3, & j_3 > N_{110}, \end{cases} \tag{6.41}$$

where

$$j_{110} = 1, 2, \ldots, N_{110} - 1; \tag{6.42}$$

those $(N_{001} - 1)(2N_{110} - 1)$ bulk-like subbands in (6.39) in the quantum wire can be separated into three subsets according to (6.41) and (6.42). They are as follows: $(N_{001} - 1)(N_{110} - 1)$ bulk-like subbands in the quantum wire with energies

$$\bar{\Lambda}^{bk,a}_{n,j_{001},j_{110}}(\bar{\mathbf{k}}) = \varepsilon_n\left[\bar{\mathbf{k}} + \frac{j_{001}\pi}{N_{001}a}(0, 0, 1) + \frac{j_{110}\pi}{N_{110}a}(1, 1, 0)\right], \tag{6.39a}$$

$(N_{001} - 1)$ bulk-like subbands in the quantum wire with energies

$$\bar{\Lambda}^{bk,b}_{n,j_{001}}(\bar{\mathbf{k}}) = \varepsilon_n\left[\bar{\mathbf{k}} + \frac{j_{001}\pi}{N_{001}a}(0, 0, 1) + \frac{\pi}{a}(1, 1, 0)\right], \tag{6.39b}$$

$(N_{001} - 1)(N_{110} - 1)$ bulk-like subbands in the quantum wire with energies[6]

$$\bar{\Lambda}^{bk,c}_{n,j_{001},j_{110}}(\bar{\mathbf{k}}) = \varepsilon_n \left[\bar{\mathbf{k}} + \frac{j_{001}\pi}{N_{001}a}(0,0,1) - \frac{j_{110}\pi}{N_{110}a}(1,1,0) + \frac{2\pi}{a}(1,1,0)\right]$$
$$= \varepsilon_n \left[\bar{\mathbf{k}} + \frac{j_{001}\pi}{N_{001}a}(0,0,1) + \frac{j_{110}\pi}{N_{110}a}(1,1,0) + \frac{2\pi}{a}(1,1,0)\right]. \quad (6.39c)$$

From (6.19), for each bulk energy band n, there are $2N_{110}-1$ surface-like subbands in the quantum wire due to the existence of two boundary surfaces in the (001) plane with energies

$$\bar{\Lambda}_{n,j_3}(\bar{\mathbf{k}}; \tau_{001}) = \bar{\lambda}_{n,j_3}(\bar{\mathbf{k}}; \tau_{001}). \quad (6.43)$$

By (6.41), these $2N_{110} - 1$ subbands in (6.43) can be separated into three subsets. They are as follows:

$N_{110} - 1$ surface-like subbands in the quantum wire with energies

$$\bar{\Lambda}^{sf,1}_{n,j_{110}}(\bar{\mathbf{k}}; \tau_{001}) = \bar{\lambda}_{n,j_{110}}(\bar{\mathbf{k}}; \tau_{001}). \quad (6.43a)$$

one surface-like subband in the quantum wire with energy

$$\bar{\Lambda}^{sf,2}_{n}(\bar{\mathbf{k}}; \tau_{001}) = \bar{\lambda}_{n,N_{110}}(\bar{\mathbf{k}}; \tau_{001}), \quad (6.43b)$$

$N_{110} - 1$ surface-like subbands in the quantum wire with energies

$$\bar{\Lambda}^{sf,3}_{n,j_{110}}(\bar{\mathbf{k}}; \tau_{001}) = \bar{\lambda}_{n,2N_{110}-j_{110}}(\bar{\mathbf{k}}; \tau_{001}). \quad (6.43c)$$

From (6.13), due to the existence of (110) boundary surfaces, for each bulk energy band n, there are $N_{001} - 1$ surface-like subbands in the quantum wire with energies

$$\bar{\Lambda}^{sf,a}_{n,j_{001}}(\bar{\mathbf{k}}; \tau_{110}) = \hat{\Lambda}_n \left[\bar{\mathbf{k}} + \frac{j_{001}\pi}{N_{001}a}(0,0,1); \tau_{110}\right]. \quad (6.44)$$

From (6.8), for each bulk energy band n, there is one edge-like subband in the quantum wire with energy

$$\bar{\Lambda}^{eg}_{n}(\bar{\mathbf{k}}; \tau_{001}, \tau_{110}) = \bar{\lambda}_n(\bar{\mathbf{k}}; \tau_{001}, \tau_{110}). \quad (6.45)$$

[6]Since for cubic semiconductors and fcc metals, in general, $\varepsilon_n(k_x, k_y, k_z) = \varepsilon_n(k_y, k_x, k_z)$ and $1/a(2,2,0)$ is a reciprocal lattice vector for crystals with an fcc Bravais lattice.

6.5.3 Results Obtained by Combining Sects. 6.5.1 and 6.5.2

For an fcc quantum wire with (001) surfaces and (110) surfaces and with a rectangular cross section $N_{110}a/\sqrt{2} \times N_{001}a$, the electronic states are one-dimensional Bloch waves with a wave vector $\bar{\mathbf{k}}$ in the $[1\bar{1}0]$ direction. We can consider it either from the confinement order in Sect. 6.5.1 or from that in Sect. 6.5.2. However, in either such confinement order, the whole symmetry of the system has not been considered, since neither such way of choosing the primary lattice vectors contains the full symmetry of the system. Nevertheless, by combining the results obtained in the two different confinement orders, a more complete and comprehensive understanding of the electronic states in the quantum wire can be achieved.

We can easily see that some expressions for the energies of the electronic states in the wire in Sects. 6.5.1 or 6.5.2 are the same, such as (6.32a) and (6.39a). Some might look different, but they are actually the same, such as (6.32c) and (6.39c).[7]

Some of them seem different. In Sect. 6.5.1, $2N_{001}$ subbands whose energies depend on τ_{110} ($2N_{001} - 1$ subbands in (6.36) and one subband in (6.38)) and N_{110} subbands whose energies depend on τ_{001} ($N_{110} - 1$ subbands in (6.37) and one subband in (6.38)) were obtained. In comparison, in Sect. 6.5.2, N_{001} subbands whose energies depend on τ_{110} ($N_{001} - 1$ subbands in (6.44) and one subband in (6.45)) and $2N_{110}$ subbands whose energies depend on τ_{001} ($2N_{110} - 1$ subbands in (6.43) and one subband in (6.45)) were obtained.

From the discussions in Sects. 5.6.1 and 5.6.2, we see that these differences come from the fact that in Sect. 6.5.1, the symmetry of the system in the (110) direction was not fully used; there is a band-folding at $\frac{\pi}{a}(1, 1, 0)$. In Sect. 6.5.2, the symmetry of the system in the (001) direction was not fully used; there is a band-folding at $\frac{\pi}{a}(0, 0, -1)$.

By considering these points, we can predict that the electronic states in such an ideal quantum wire should be as follows:

For each bulk energy band n, there are $2(N_{001} - 1)(N_{110} - 1) + (N_{001} - 1) + (N_{110} - 1) + 1$ bulk-like subbands. They include the following:
$(N_{001} - 1)(N_{110} - 1)$ subbands with energies

$$\bar{\Lambda}^{bk,a}_{n,j_{001},j_{110}}(\bar{\mathbf{k}}) = \varepsilon_n \left[\bar{\mathbf{k}} + \frac{j_{001}\pi}{N_{001}a}(0, 0, 1) + \frac{j_{110}\pi}{N_{110}a}(1, 1, 0) \right] \tag{6.46}$$

from either (6.32a) or (6.39a), $(N_{001} - 1)(N_{110} - 1)$ subbands with energies

$$\bar{\Lambda}^{bk,c}_{n,j_{001},j_{110}}(\bar{\mathbf{k}}) = \varepsilon_n \left[\bar{\mathbf{k}} + \frac{j_{001}\pi}{N_{001}a}(0, 0, 1) + \frac{j_{110}\pi}{N_{110}a}(1, 1, 0) + \frac{2\pi}{a}(1, 1, 0) \right] \tag{6.47}$$

from either (6.32c) or (6.39c), $(N_{001} - 1)$ subbands with energies

[7]Since $1/a(1, 1, -1)$ is a reciprocal lattice vector for crystals with an fcc Bravais lattice.

$$\bar{\Lambda}_{n,j_{001}}^{bk,b_1}(\bar{\mathbf{k}}) = \varepsilon_n\left[\bar{\mathbf{k}} + \frac{j_{001}\pi}{N_{001}a}(0,0,1) + \frac{\pi}{a}(1,1,0)\right] \tag{6.48}$$

from (6.39b), $(N_{110}-1)$ subbands with energies

$$\bar{\Lambda}_{n,j_{110}}^{bk,b_2}(\bar{\mathbf{k}}) = \varepsilon_n\left[\bar{\mathbf{k}} + \frac{j_{110}\pi}{N_{110}a}(1,1,0) + \frac{\pi}{a}(0,0,1)\right] \tag{6.49}$$

from (6.39b). Here the range of j_{001} is given by (6.35) or (6.40), j_{110} is given by (6.33) or (6.42).

In addition to those bulk-like subbands, for each bulk energy band n, there is one bulk-like subband in the wire with energy given by

$$\bar{\Lambda}_{n}^{bk,d}(\bar{\mathbf{k}}) = \varepsilon_n\left[\bar{\mathbf{k}} + \frac{\pi}{a}(0,0,1) + \frac{\pi}{a}(1,1,0)\right]. \tag{6.50}$$

This state is obtained from (6.36b) and ((6.43b): By (6.36b), each state in this subband is a stationary Bloch state with a $\kappa_{001} = \pi/2$ in the [001] direction; thus, its energy does not depend on τ_{001}. By (6.43b), each state in this subband is a stationary Bloch state with a $\kappa_{110} = \pi/2$ in the [110] direction; thus, its energy does not depend on τ_{110}.

For each bulk energy band n, there are $(N_{001}-1)+(N_{110}-1)$ surface-like subbands. They are $(N_{001}-1)$ subbands with energies

$$\bar{\Lambda}_{n,j_{001}}^{sf,a_1}(\bar{\mathbf{k}};\tau_{110}) = \hat{\Lambda}_n\left[\bar{\mathbf{k}} + \frac{j_{001}\pi}{N_{001}a}(0,0,1);\tau_{110}\right] \tag{6.51}$$

from (6.44) and $(N_{110}-1)$ subbands with energies

$$\bar{\Lambda}_{n,j_{110}}^{sf,a_2}(\bar{\mathbf{k}};\tau_{001}) = \hat{\Lambda}_n\left[\bar{\mathbf{k}} + \frac{j_{110}\pi}{N_{110}a}(1,1,0);\tau_{001}\right] \tag{6.52}$$

from (6.37).

For each bulk energy band n, there is one edge band in the wire with energy given by (6.38) (i.e., (6.45)).

Therefore, among $2N_{001}-1$ subbands $\bar{\Lambda}_{n,j_3}(\bar{\mathbf{k}};\tau_{110})$ in (6.36) in Sect. 6.5.1, there are in fact $N_{001}-1$ bulk-like subbands $\bar{\Lambda}_{n,j_{001}}^{bk,b_1}(\bar{\mathbf{k}})$ in (6.48), one bulk-like subband $\bar{\Lambda}_{n}^{bk,d}(\bar{\mathbf{k}})$ in (6.50), and $N_{001}-1$ surface-like subbands $\bar{\Lambda}_{n,j_{001}}^{sf,a_1}(\bar{\mathbf{k}};\tau_{110})$ in (6.51). Thus, there are totally N_{001} subbands in the quantum wire whose energies are dependent on τ_{110}: $N_{001}-1$ surface-like subbands $\bar{\Lambda}_{n,j_{001}}^{sf,a_1}(\bar{\mathbf{k}};\tau_{110})$ in (6.51) plus one edge-like subband $\bar{\Lambda}_{n}^{eg}(\bar{\mathbf{k}};\tau_{110},\tau_{001})$ in (6.38). We should also have

$$\bar{\Lambda}_{n,j_{001}}^{sf,a_1}(\bar{\mathbf{k}};\tau_{110}) > \bar{\Lambda}_{n,j_{001}}^{bk,b_1}(\bar{\mathbf{k}}), \tag{6.53}$$

and

$$\bar{\Lambda}_n^{eg}(\bar{\mathbf{k}}; \tau_{110}, \tau_{001}) > \bar{\Lambda}_n^{bk,d}(\bar{\mathbf{k}}), \tag{6.54}$$

since in our discussions on the two different ways of choosing the primitive lattice vectors in Sect. 5.6.2, the real surface-like subband has a higher energy by (5.33).

Similarly, among $2N_{110} - 1$ subbands $\bar{\Lambda}_{n,j_3}(\bar{\mathbf{k}}; \tau_{001})$ in (6.43) in Sect. 6.5.2, actually, there are $N_{110} - 1$ bulk-like subbands $\bar{\Lambda}_{n,j_{110}}^{bk,b_2}(\bar{\mathbf{k}})$ in (6.49), one bulk-like subband $\bar{\Lambda}_n^{bk,d}(\bar{\mathbf{k}})$ in (6.50), and $N_{110} - 1$ surface-like subbands $\bar{\Lambda}_{n,j_{110}}^{sf,a_2}(\bar{\mathbf{k}}; \tau_{001})$ in (6.52). There are totally N_{110} subbands in the quantum wire whose energies are dependent on τ_{001}: $N_{110} - 1$ surface-like subbands $\bar{\Lambda}_{n,j_{110}}^{sf,a_2}(\bar{\mathbf{k}}; \tau_{001})$ in (6.52) plus one edge-like subband $\bar{\Lambda}_n^{eg}(\bar{\mathbf{k}}; \tau_{110}, \tau_{001})$ in (6.45). We should also have

$$\bar{\Lambda}_{n,j_{110}}^{sf,a_2}(\bar{\mathbf{k}}; \tau_{001}) > \bar{\Lambda}_{n,j_{110}}^{bk,b_2}(\bar{\mathbf{k}}), \tag{6.55}$$

and $\bar{\Lambda}_n^{eg}(\bar{\mathbf{k}}; \tau_{110}, \tau_{001}) > \bar{\Lambda}_n^{bk,d}(\bar{\mathbf{k}})$ as in (6.54).

Since in a cubic semiconductor, one of the triply-degenerated VBM states may have a (001) nodal surface or an (110) nodal surface; there can be one state in a Si (001) film or (110) film or a GaAs (110) film with the energy of VBM. This energy does not depend on the film thickness, as observed in [1–3]. However, no one VBM state in Si or GaAs can have both a (001) nodal surface and an (110) nodal surface simultaneously. Therefore, there is not an electronic state in an ideal rectangular quantum wire of a cubic semiconductor discussed here whose energy is the energy of the VBM and does not change as the wire size changes. Consequently, it can be predicted that there are, at least, three edge-like states in such a quantum wire with (110) and (001) surfaces, whose energies are above the VBM and do not depend on the wire size and shape.

6.6 fcc Quantum Wires with (110) and (1$\bar{1}$0) Surfaces

The electronic states in an ideal rectangular quantum wire of an fcc crystal with (110) and (1$\bar{1}$0) surfaces are one-dimensional Bloch waves with a wave vector $\bar{\mathbf{k}}$ in the [001] direction. Such a quantum wire has a cross section $N_{110}a/\sqrt{2} \times N_{1\bar{1}0}a/\sqrt{2}$, where a is the lattice constant, N_{110} and $N_{1\bar{1}0}$ are positive integers. By using an approach similar to that used in Sect. 6.5, the properties of electronic states in such a quantum wire can be predicted.

For each bulk energy band n, there are $2(N_{1\bar{1}0} - 1)(N_{110} - 1) + (N_{1\bar{1}0} - 1) + (N_{110} - 1) + 1$ bulk-like subbands in the quantum wire. They include $(N_{1\bar{1}0} - 1)(N_{110} - 1)$ subbands with energies

$$\bar{\Lambda}_{n,j_{1\bar{1}0},j_{110}}^{bk,a}(\bar{\mathbf{k}}) = \varepsilon_n\left[\bar{\mathbf{k}} + \frac{j_{1\bar{1}0}\pi}{N_{1\bar{1}0}a}(1,-1,0) + \frac{j_{110}\pi}{N_{110}a}(1,1,0)\right], \tag{6.56}$$

$(N_{1\bar{1}0} - 1)(N_{110} - 1)$ subbands with energies

$$\bar{\Lambda}^{bk,c}_{n,j_{1\bar{1}0},j_{110}}(\bar{\mathbf{k}}) = \varepsilon_n \left[\bar{\mathbf{k}} + \frac{j_{1\bar{1}0}\pi}{N_{1\bar{1}0}a}(1,-1,0) + \frac{j_{110}\pi}{N_{110}a}(1,1,0) + \frac{2\pi}{a}(1,1,0) \right], \tag{6.57}$$

$(N_{1\bar{1}0} - 1)$ subbands with energies

$$\bar{\Lambda}^{bk,b_1}_{n,j_{1\bar{1}0}}(\bar{\mathbf{k}}) = \varepsilon_n \left[\bar{\mathbf{k}} + \frac{j_{1\bar{1}0}\pi}{N_{1\bar{1}0}a}(1,-1,0) + \frac{\pi}{a}(1,1,0) \right], \tag{6.58}$$

and $(N_{110} - 1)$ subbands with energies

$$\bar{\Lambda}^{bk,b_2}_{n,j_{110}}(\bar{\mathbf{k}}) = \varepsilon_n \left[\bar{\mathbf{k}} + \frac{j_{110}\pi}{N_{110}a}(1,1,0) + \frac{\pi}{a}(1,-1,0) \right]. \tag{6.59}$$

Here, $j_{1\bar{1}0} = 1, 2, \ldots, N_{1\bar{1}0} - 1$ and $j_{110} = 1, 2, \ldots, N_{110} - 1$.

For each bulk energy band n, there is one bulk-like subband in the wire with energy

$$\bar{\Lambda}^{bk,d}_{n}(\bar{\mathbf{k}}) = \varepsilon_n \left[\bar{\mathbf{k}} + \frac{\pi}{a}(1,1,0) + \frac{\pi}{a}(1,-1,0) \right]. \tag{6.60}$$

For each bulk energy band n, there are $(N_{1\bar{1}0} - 1) + (N_{110} - 1)$ surface-like subbands in the quantum wire. They are $(N_{1\bar{1}0} - 1)$ subbands with energies

$$\bar{\Lambda}^{sf,a_1}_{n,j_{1\bar{1}0}}(\bar{\mathbf{k}}; \tau_{110}) = \hat{\Lambda}_n \left[\bar{\mathbf{k}} + \frac{j_{1\bar{1}0}\pi}{N_{1\bar{1}0}a}(1,-1,0); \tau_{110} \right], \tag{6.61}$$

and $(N_{110} - 1)$ subbands with energies

$$\bar{\Lambda}^{sf,a_2}_{n,j_{110}}(\bar{\mathbf{k}}; \tau_{1\bar{1}0}) = \hat{\Lambda}_n \left[\bar{\mathbf{k}} + \frac{j_{110}\pi}{N_{110}a}(1,1,0); \tau_{1\bar{1}0} \right]. \tag{6.62}$$

Here, τ_{110} and $\tau_{1\bar{1}0}$ define the boundary faces of the quantum wire in the [110] and $[1\bar{1}0]$ directions.

For each bulk energy band n, there is one edge-like subband in the quantum wire with energy

$$\bar{\Lambda}^{eg}_{n}(\bar{\mathbf{k}}; \tau_{110}, \tau_{1\bar{1}0}) = \bar{\lambda}_n(\bar{\mathbf{k}}; \tau_{110}, \tau_{1\bar{1}0}). \tag{6.63}$$

In a cubic semiconductor, no one VBM state can have both an (110) nodal plane and an $(1\bar{1}0)$ nodal plane simultaneously. There is not an electronic state in an ideal quantum wire of a cubic semiconductor discussed here whose energy is the energy of the VBM and does not change as the wire size or shape changes, as observed in the numerical calculations of Franceschetti and Zunger [3] on GaAs free-standing quantum wires shown in Fig. 5.4b. Consequently, it is also predicted that there are at least three edge states in such a rectangular quantum wire whose energies are above the VBM and do not depend on the wire size and shape.

6.7 bcc Quantum Wires with (001) and (010) Surfaces

For a bcc quantum wire with (010) and (001) surfaces and having a rectangular cross section $N_{010}a \times N_{001}a$, where N_{010} and N_{001} are positive integers, the electronic states are one-dimensional Bloch waves with a wave vector $\bar{\mathbf{k}}$ in the [100] direction. They can be similarly obtained as in Sect. 6.5.

For each bulk energy band n, there are $2(N_{010} - 1)(N_{001} - 1) + (N_{010} - 1) + (N_{001} - 1) + 1$ bulk-like subbands in the quantum wire. They include $(N_{010} - 1)(N_{001} - 1)$ subbands with energies

$$\bar{\Lambda}^{bk,a}_{n,j_{010},j_{001}}(\bar{\mathbf{k}}) = \varepsilon_n \left[\bar{\mathbf{k}} + \frac{j_{010}\pi}{N_{010}a}(0,1,0) + \frac{j_{001}\pi}{N_{001}a}(0,0,1) \right], \tag{6.64}$$

$(N_{001} - 1)(N_{010} - 1)$ subbands with energies

$$\bar{\Lambda}^{bk,c}_{n,j_{010},j_{001}}(\bar{\mathbf{k}}) = \varepsilon_n \left[\bar{\mathbf{k}} + \frac{j_{010}\pi}{N_{010}a}(0,1,0) + \frac{j_{001}\pi}{N_{001}a}(0,0,1) + \frac{2\pi}{a}(0,1,0) \right], \tag{6.65}$$

$(N_{001} - 1)$ subbands with energies

$$\bar{\Lambda}^{bk,b_1}_{n,j_{001}}(\bar{\mathbf{k}}) = \varepsilon_n \left[\bar{\mathbf{k}} + \frac{j_{001}\pi}{N_{001}a}(0,0,1) + \frac{\pi}{a}(0,1,0) \right], \tag{6.66}$$

and $(N_{010} - 1)$ subbands with energies

$$\bar{\Lambda}^{bk,b_2}_{n,j_{010}}(\bar{\mathbf{k}}) = \varepsilon_n \left[\bar{\mathbf{k}} + \frac{j_{010}\pi}{N_{010}a}(0,1,0) + \frac{\pi}{a}(0,0,1) \right]. \tag{6.67}$$

Here $j_{001} = 1, 2, \ldots, N_{001} - 1$ and $j_{010} = 1, 2, \ldots, N_{010} - 1$.

For each bulk energy band n, there is one bulk-like subband with energy

$$\bar{\Lambda}^{bk,d}_{n}(\bar{\mathbf{k}}) = \varepsilon_n \left[\bar{\mathbf{k}} + \frac{\pi}{a}(0,1,0) + \frac{\pi}{a}(0,0,1) \right]. \tag{6.68}$$

For each bulk energy band n, there are $(N_{001} - 1) + (N_{010} - 1)$ surface-like subbands in the quantum wire. They are $(N_{001} - 1)$ subbands with energies

$$\bar{\Lambda}^{sf,a_1}_{n,j_{001}}(\bar{\mathbf{k}}; \tau_{010}) = \hat{\Lambda}_n \left[\bar{\mathbf{k}} + \frac{j_{001}\pi}{N_{001}a}(0,0,1); \tau_{010} \right], \tag{6.69}$$

and $(N_{010} - 1)$ subbands with energies

$$\bar{\Lambda}^{sf,a_2}_{n,j_{010}}(\bar{\mathbf{k}}; \tau_{001}) = \hat{\Lambda}_n \left[\bar{\mathbf{k}} + \frac{j_{010}\pi}{N_{010}a}(0, 1, 0); \tau_{001} \right]. \tag{6.70}$$

Here, τ_{010} and τ_{001} define the boundary faces of the quantum wire in the [010] and [001] directions.

For each bulk energy band n, there is one edge-like subband in the quantum wire with energy

$$\bar{\Lambda}^{eg}_n(\bar{\mathbf{k}}; \tau_{001}, \tau_{010}) = \bar{\lambda}_n(\bar{\mathbf{k}}; \tau_{001}, \tau_{010}). \tag{6.71}$$

6.8 Summary and Discussions

By considering the further quantum confinement of two-dimensional Bloch waves $\hat{\psi}_n(\hat{\mathbf{k}}, \mathbf{x}; \tau_3)$ and $\hat{\psi}_{n,j_3}(\hat{\mathbf{k}}, \mathbf{x}; \tau_3)$ in one more direction and considering two different confinement orders, the properties of electronic states in ideal quantum wires discussed here can generally and analytically be understood. There are three different types of electronic states in an ideal quantum wire: bulk-like states, surface-like states, and edge-like states.

Similar to a surface-like subband, the physics origin of an edge-like subband is also related to a bulk energy band. An edge-like electronic state in a quantum wire is better understood as an electronic state whose properties and energy are determined by the edge position, rather than that the state is localized near a specific edge. Only when a Bloch function has two nodal surfaces which are the surfaces of the quantum wire, might the edge state be a Bloch state. It seems that such cases rarely happen in most quantum wires of cubic semiconductors of general interest.

Due to the different crystal structure of an fcc crystal or a bcc crystal, the numbers of each type of electronic states in a quantum wire discussed in Sects. 6.5–6.7 are somewhat different from what in a quantum wire of a crystal with a sc, tetr, or ortho Bravais lattice discussed in Sect. 6.4.

Nevertheless, since the results in Sects. 6.5–6.7 were also obtained from understandings of the further quantum confinement effects of $\hat{\psi}_n(\hat{\mathbf{k}}, \mathbf{x}; \tau_3)$ and $\hat{\psi}_{n,j_3}(\hat{\mathbf{k}}, \mathbf{x}; \tau_3)$ in Sects. 6.1–6.3, there are similar relationships between the three different types of electronic states. For example, for an ideal fcc quantum wire with (110) and (001) surfaces, we should have

$$\bar{\Lambda}^{eg}_n(\bar{\mathbf{k}}; \tau_{001}, \tau_{110}) > \bar{\Lambda}^{sf,a_1}_{n,j_{001}}(\bar{\mathbf{k}}; \tau_{110}) \tag{6.72}$$

and

$$\bar{\Lambda}^{eg}_n(\bar{\mathbf{k}}; \tau_{001}, \tau_{110}) > \bar{\Lambda}^{sf,a_2}_{n,j_{110}}(\bar{\mathbf{k}}; \tau_{001}) \tag{6.73}$$

between the energy of an edge-like state in (6.38) or (6.45) and the energies of relevant surface-like states in (6.51) and (6.52). These two equations are obtained from the relationship (6.14) between (6.38) and (6.37) (i.e., (6.52)), or from the relationship (6.14) between (6.45) and (6.44) (i.e., (6.51)). We also have

$$\bar{\Lambda}_n^{eg}(\bar{\mathbf{k}}; \tau_{001}, \tau_{110}) > \bar{\Lambda}_n^{bk,d}(\bar{\mathbf{k}}) \tag{6.74}$$

between the energy of an edge-like state in (6.38) or (6.45) and the energy of a relevant bulk-like state in (6.50). This equation comes from the relationship (6.54) between (6.38) and (6.50).

We have

$$\bar{\Lambda}_{n,j_{001}}^{sf,a_1}(\bar{\mathbf{k}}; \tau_{110}) > \bar{\Lambda}_{n,j_{001}}^{bk,b_1}(\bar{\mathbf{k}}), \tag{6.75}$$

$$\bar{\Lambda}_{n,j_{001}}^{sf,a_1}(\bar{\mathbf{k}}; \tau_{110}) > \bar{\Lambda}_{n,j_{001},j_{110}}^{bk,a}(\bar{\mathbf{k}}), \tag{6.76}$$

and

$$\bar{\Lambda}_{n,j_{001}}^{sf,a_1}(\bar{\mathbf{k}}; \tau_{110}) > \bar{\Lambda}_{n,j_{001},j_{110}}^{bk,c}(\bar{\mathbf{k}}) \tag{6.77}$$

between the energy of a surface-like state in (6.51) and the energies of relevant bulk-like states in (6.48), (6.46), and (6.47). These three equations come from the relationship (6.23) between (6.36) and (6.32) and/or from (6.53). Similarly, we have

$$\bar{\Lambda}_{n,j_{110}}^{sf,a_2}(\bar{\mathbf{k}}; \tau_{001}) > \bar{\Lambda}_{n,j_{110}}^{bk,b_2}(\bar{\mathbf{k}}), \tag{6.78}$$

$$\bar{\Lambda}_{n,j_{110}}^{sf,a_2}(\bar{\mathbf{k}}; \tau_{001}) > \bar{\Lambda}_{n,j_{001},j_{110}}^{bk,a}(\bar{\mathbf{k}}), \tag{6.79}$$

and

$$\bar{\Lambda}_{n,j_{110}}^{sf,a_2}(\bar{\mathbf{k}}; \tau_{001}) > \bar{\Lambda}_{n,j_{001},j_{110}}^{bk,c}(\bar{\mathbf{k}}) \tag{6.80}$$

between the energy of a surface-like state in (6.52) and the energies of relevant bulk-like states in (6.49), (6.46), and (6.47). These three equations come from the relationship (6.23) between (6.43) and (6.39) and/or from (6.55).

Corresponding relationships between the electronic states in an ideal rectangular fcc quantum wire with (110) and $(1\bar{1}0)$ surfaces or in an ideal rectangular bcc quantum wire with (001) and (010) surfaces can be similarly obtained.

Therefore, in an ideal rectangular quantum wire discussed in Sects. 6.5–6.7, by (6.72)–(6.80) or similar equations, we can understand that for the electronic states with the same energy band index n and the same wave vector $\bar{\mathbf{k}}$ in the quantum wire, the following general relations exist:

The energy of the edge-like state

> The energy of every surface-like state

> The energy of every relevant bulk-like state.

In an everywhere neutral semiconductor quantum wire, the edge-like subbands should be as equally occupied as the bulk-like subbands. The edge-like subbands are usually even higher in energy than the relevant surface-like subbands. Thus, in a quantum wire of a cubic semiconductor, the edge-like subbands originating from the valence bands might partly be even higher in energy than some bulk-like subbands originating from a conduction band. If such a case happens, the equality of the Fermi energy in the quantum wire must force some electrons to flow from those edge-like subbands originating from the valence bands into the bulk-like subbands originating from that conduction band. Such a redistribution of electrons will make a quantum wire of a semiconductor crystal have an electrical conductivity of the metal.

Based on similar reasonings, the edges of an alkali metal quantum wire could be even more positively charged than the surfaces.

There are also edge-like subbands originating from bulk conduction bands in a semiconductor quantum wire. These edge-like subbands will be even higher in energy than the surface-like subbands originating from the bulk conduction bands and, thus, will usually not be occupied. It seems unlikely that these edge-like subbands will have a significant effect on the properties of a semiconductor quantum wire.

Although there is only one edge-like subband for each bulk energy band in a quantum wire, it does not mean that all electronic states in the edge-like subband are localized on the same edge of the wire. A clearer understanding of the electronic states in an edge-like subband in a quantum wire requires a clearer understanding of solutions of the partial differential equation (5.1), in particular, the non-Bloch state solutions in different energy ranges.

The electronic states in an ideal quantum wire discussed here are solutions of the *partial* differential equation (5.1) under the boundary condition

$$\bar{\psi}(\bar{\mathbf{k}}, \mathbf{x}) = 0, \qquad \mathbf{x} \notin \text{ the wire.}$$

The one-dimensional Bloch waves obtained are fundamentally different from the cases treated in Chap. 4, where all electronic states are solutions of the *ordinary* differential equation (4.1). Therefore, for the further quantum confinement of these one-dimensional Bloch waves in an ideal finite crystal or quantum dot, we should use the approach used in Chap. 5 and this chapter, rather than the results obtained in Chap. 4.

References

1. S.B. Zhang, A. Zunger, Appl. Phys. Lett. **63**, 1399 (1993)
2. S.B. Zhang, C.-Y. Yeh, A. Zunger, Phys. Rev. B **48**, 11204 (1993)
3. A. Franceschetti, A. Zunger, Appl. Phys. Lett. **68**, 3455 (1996)

Chapter 7
Electronic States in Ideal Finite Crystals or Quantum Dots

In this chapter, we are interested in electronic states in a finite crystal or quantum dot of rectangular cuboid shape obtained by a further truncation of a quantum wire. They can be seen as the one-dimensional Bloch waves in an ideal rectangular quantum wire discussed in Chap. 6 further confined by two boundary surfaces perpendicularly intersecting the $\mathbf{a}_1$ axis. By applying an approach similar to that used in last two chapters, we can understand that the further quantum confinement of each set of one-dimensional Bloch waves in the quantum wire will produce two different types of electronic states in the ideal finite crystal or quantum dot.

A finite crystal or quantum dot of rectangular cuboid shape always has six boundary faces: two faces in the $(h_1k_1l_1)$ plane, two faces in the $(h_2k_2l_2)$ plane, and two faces in the $(h_3k_3l_3)$ plane. The electronic states in such a finite crystal or quantum dot can be considered as the electronic states in a quantum film with two faces in the $(h_3k_3l_3)$ plane further confined by two faces in the $(h_2k_2l_2)$ plane and, finally, further confined by two faces in the $(h_1k_1l_1)$ plane. They can also be considered as three-dimensional Bloch waves $\phi_n(\mathbf{k}, \mathbf{x})$ to be confined in the three directions in other confinement orders. There are altogether six different orders. The results obtained in these six different confinement orders are equally valid and can be complementary to each other. By combining the results obtained from the six different confinement orders, we can achieve a more comprehensive understanding of the electronic states in the finite crystal or quantum dot.

The simplest cases where the results obtained in this chapter are applicable are the electronic states in a finite crystal or quantum dot of rectangular cuboid shape with a sc, tetr, or ortho Bravais lattice in which (5.13), (6.9), and (6.20) are true. In those crystals, the three primitive lattice vectors $\mathbf{a}_1$, $\mathbf{a}_2$, and $\mathbf{a}_3$ are perpendicular to each other and are essentially equivalent. Properties of the electronic states in such an ideal finite crystal or quantum dot can generally and analytically be predicted.

Many cubic semiconductors and metals have an fcc or bcc Bravais lattice, the electronic states in a finite crystal or quantum dot with an fcc or bcc Bravais lattice often are practically more interesting. Based on the general theory developed in this chapter, we can also give predictions of the electronic states in some ideal finite

S.Y. Ren, *Electronic States in Crystals of Finite Size*, Springer Tracts in Modern Physics 270, DOI 10.1007/978-981-10-4718-3_7

crystals or quantum dots of rectangular cuboid shape with an fcc or a bcc Bravais lattice.

In Sects. 7.1–7.5, we consider the cases where the each set of one-dimensional waves in an ideal quantum wire obtained in Sects. 6.1–6.3 is further confined in one more direction. That is, we try to understand the consequences of three-dimensional Bloch waves being confined in three directions in a specific confinement order. In Sects. 7.6–7.8, we obtain predictions on the electronic states in several finite crystals or quantum dots from the understanding obtained in Sects. 7.1–7.5 and by considering different quantum confinement orders. Section 7.9 is a summary and discussions.

7.1 Basic Considerations

We are interested in electronic states in a finite crystal of rectangular cuboid shape. The crystal has a bottom face and a top face at $x_3 = \tau_3$, and $x_3 = \tau_3 + N_3$, a front face and a rear face perpendicularly intersecting the $\mathbf{a}_2$ axis at $x_2 = \tau_2$ and $x_2 = \tau_2 + N_2$, and a left face and a right face perpendicularly intersecting the $\mathbf{a}_1$ axis at $x_1 = \tau_1$ and $x_1 = \tau_1 + N_1$. Here, τ_1, τ_2, and τ_3 define the boundary face locations and N_1, N_2, and N_3 are three positive integers indicating the size and shape of the crystal. We look for the eigenvalues Λ and eigenfunctions $\psi(\mathbf{x})$ of the following two equations:

$$\begin{cases} -\nabla^2\psi(\mathbf{x}) + [v(\mathbf{x}) - \Lambda]\psi(\mathbf{x}) = 0, & \mathbf{x} \in \text{ the crystal}, \\ \psi(\mathbf{x}) = 0, & \mathbf{x} \notin \text{ the crystal}, \end{cases} \tag{7.1}$$

where $v(\mathbf{x})$ is a periodic potential:

$$v(\mathbf{x} + \mathbf{a}_1) = v(\mathbf{x} + \mathbf{a}_2) = v(\mathbf{x} + \mathbf{a}_3) = v(\mathbf{x}).$$

For the further quantum confinement of one-dimensional Bloch waves in the quantum wire, $\bar{\psi}_n(\bar{\mathbf{k}}, \mathbf{x}; \tau_2, \tau_3)$, $\bar{\psi}_{n,j_3}(\bar{\mathbf{k}}, \mathbf{x}; \tau_2, \tau_3)$, $\bar{\psi}_{n,j_2}(\bar{\mathbf{k}}, \mathbf{x}; \tau_2, \tau_3)$, and $\bar{\psi}_{n,j_2,j_3}(\bar{\mathbf{k}}, \mathbf{x}; \tau_2, \tau_3)$, each set has a new corresponding eigenvalue problem and a new theorem. The theorem is similar to Theorem 6.1 or 6.2 for the electronic states in a quantum wire. The confinement of each set will produce two types of confined electronic states in the finite crystal or quantum dot.

7.2 Further Quantum Confinement of $\bar{\psi}_n(\bar{\mathbf{k}}, \mathbf{x}; \tau_2, \tau_3)$

For the quantum confinement of the edge-like states $\bar{\psi}_n(\bar{\mathbf{k}}, \mathbf{x}; \tau_2, \tau_3)$, we consider a rectangular parallelepiped C as shown in Fig. 7.1 with a bottom face $x_3 = \tau_3$ and a top face $x_3 = \tau_3 + 1$, a front face and a rear face perpendicularly intersecting the $\mathbf{a}_2$ axis at $x_2 = \tau_2$ and $x_2 = \tau_2 + 1$, a left face and a right face perpendicularly

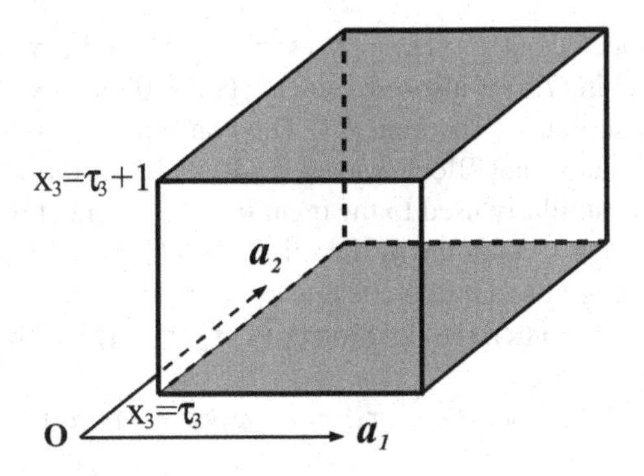

Fig. 7.1 The rectangular parallelepiped C for the eigenvalue problem of (5.1) under the boundary condition (7.2). The two shadowed faces of ∂C_3 and the two *thick-lined* faces of ∂C_2 are the four faces on which each $\bar{\psi}_n(\bar{\mathbf{k}}, \mathbf{x}; \tau_2, \tau_3)$ is zero. The *left* face and the *right* face are the two faces on which each function $\phi(\mathbf{x}; \tau_1, \tau_2, \tau_3)$ is further required to be zero

intersecting the $\mathbf{a}_1$ axis at $x_1 = \tau_1$ and $x_1 = \tau_1 + 1$. The function set $\phi(\mathbf{x}; \tau_1, \tau_2, \tau_3)$ is defined by (5.1) and the boundary condition

$$\phi(\mathbf{x}; \tau_1, \tau_2, \tau_3) = 0, \qquad \mathbf{x} \in \partial C. \tag{7.2}$$

Here, ∂C is the boundary of C. The eigenvalues and eigenfunctions of (5.1) under the condition (7.2) are denoted by $\lambda_n(\tau_1, \tau_2, \tau_3)$ and $\phi_n(\mathbf{x}; \tau_1, \tau_2, \tau_3)$, respectively.

For the eigenvalue problem defined by (5.1) and (7.2), we have the following theorem between $\lambda_n(\tau_1, \tau_2, \tau_3)$ and the eigenvalue $\bar{\Lambda}_n(\bar{\mathbf{k}}; \tau_2, \tau_3)$ of $\bar{\psi}_n(\bar{\mathbf{k}}, \mathbf{x}; \tau_2, \tau_3)$ in (6.8).

Theorem 7.1

$$\lambda_n(\tau_1, \tau_2, \tau_3) \geq \bar{\Lambda}_n(\bar{\mathbf{k}}; \tau_2, \tau_3). \tag{7.3}$$

Since each $\bar{\psi}_n(\bar{\mathbf{k}}, \mathbf{x}; \tau_2, \tau_3)$ satisfies

$$\begin{cases} \bar{\psi}(\bar{\mathbf{k}}, \mathbf{x} + \mathbf{a}_1; \tau_2, \tau_3) = \mathrm{e}^{\mathrm{i}k_1} \bar{\psi}(\bar{\mathbf{k}}, \mathbf{x}; \tau_2, \tau_3), & -\pi < k_1 \leq \pi, \\ \bar{\psi}(\bar{\mathbf{k}}, \mathbf{x}; \tau_2, \tau_3) = 0, & \mathbf{x} \in \partial C_2 \text{ or } \mathbf{x} \in \partial C_3, \end{cases} \tag{7.4}$$

Theorem 7.1 can be proven similar to Theorem 5.1. The major difference is in the Dirichlet integral

$$\begin{aligned} J(f, g) &= \int_C \{\nabla f(\mathbf{x}) \cdot \nabla g^*(\mathbf{x}) + v(\mathbf{x}) f(\mathbf{x}) g^*(\mathbf{x})\} \, \mathrm{d}\mathbf{x} \\ &= \int_C f(\mathbf{x}) \{-\nabla^2 g^*(\mathbf{x}) + v(\mathbf{x}) g^*(\mathbf{x})\} \, \mathrm{d}\mathbf{x} + \int_{\partial C} f(\mathbf{x}) \frac{\partial g^*(\mathbf{x})}{\partial n} \, \mathrm{d}S, \end{aligned} \tag{7.5}$$

if both $f(\mathbf{x})$ and $g(\mathbf{x})$ satisfy the conditions (7.4), the integral over ∂C in (7.5) is zero since $f(\mathbf{x}) = 0$ when $\mathbf{x} \in \partial C_2$ or $\mathbf{x} \in \partial C_3$ and the integrals over the two opposite faces of ∂C_1 are canceled out. If $f(\mathbf{x}) = \phi_n(\mathbf{x}; \tau_1, \tau_2, \tau_3)$ and $g(\mathbf{x}) = \bar{\psi}(\bar{\mathbf{k}}, \mathbf{x}; \tau_2, \tau_3)$, the integral over ∂C in (7.5) is also zero since $f(\mathbf{x}) = 0$ when $\mathbf{x} \in \partial C$.

Theorem 7.1 is similar to Theorem 6.1. The consequences of the quantum confinement of two-dimensional Bloch waves $\hat{\psi}_n(\hat{\mathbf{k}}, \mathbf{x}; \tau_3)$ in the $\mathbf{a}_2$ direction due to Theorem 6.1 can be similarly used to the quantum confinement of one-dimensional Bloch waves $\bar{\psi}_n(\bar{\mathbf{k}}, \mathbf{x}; \tau_2, \tau_3)$ in the $\mathbf{a}_1$ direction.

For each bulk energy band n, there is one $\phi_n(\mathbf{x}; \tau_1, \tau_2, \tau_3)$.

Because $v(\mathbf{x} + \mathbf{a}_1) = v(\mathbf{x})$, the function $\phi_n(\mathbf{x}; \tau_1, \tau_2, \tau_3)$ has the form

$$\phi_n(\mathbf{x} + \mathbf{a}_1; \tau_1, \tau_2, \tau_3) = \mathrm{e}^{\mathrm{i}k_1}\phi_n(\mathbf{x}; \tau_1, \tau_2, \tau_3), \tag{7.6}$$

and, here, k_1 can be complex with a nonzero imaginary part or a real number.

If k_1 is real in (7.6), then $\phi_n(\mathbf{x}; \tau_1, \tau_2, \tau_3)$ is a $\bar{\psi}_{n'}(\bar{\mathbf{k}}, \mathbf{x}; \tau_2, \tau_3)$. By Theorem 7.1, a $\bar{\psi}_{n'}(\bar{\mathbf{k}}, \mathbf{x}; \tau_2, \tau_3)$ cannot be a $\phi_n(\mathbf{x}; \tau_1, \tau_2, \tau_3)$ except in some special cases where the $\bar{\psi}_{n'}(\bar{\mathbf{k}}, \mathbf{x}; \tau_2, \tau_3)$ has a nodal plane perpendicularly intersecting the $\mathbf{a}_1$ axis at $x_1 = \tau_1$. Thus, k_1 in (7.6) can be real only in such special cases; in most cases, k_1 in (7.6) is complex with a nonzero imaginary part.

The imaginary part of k_1 in (7.6) can be either positive or negative, corresponding to that $\phi_n(\mathbf{x}; \tau_1, \tau_2, \tau_3)$ decays in the either positive or negative direction of $\mathbf{a}_1$. Such states $\phi_n(\mathbf{x}; \tau_1, \tau_2, \tau_3)$ with a nonzero imaginary part of k_1 in (7.6) cannot exist in a quantum wire with translational invariance in the $\mathbf{a}_1$ direction since they are divergent in either the negative or the positive direction of $\mathbf{a}_1$. However, they can play a significant role in the electronic states in a finite crystal or quantum dot without translational invariance.

The further quantum confinement of one-dimensional Bloch waves $\bar{\psi}_n(\bar{\mathbf{k}}, \mathbf{x}; \tau_2, \tau_3)$ in the $\mathbf{a}_1$ direction will produce two different types of electronic states in the finite crystal or quantum dot.

One type of nontrivial solutions of (7.1) can be obtained from (7.6) by assigning

$$\psi_n(\mathbf{x}; \tau_1, \tau_2, \tau_3) = \begin{cases} c_{N_1,N_2,N_3}\phi_n(\mathbf{x}; \tau_1, \tau_2, \tau_3), & \mathbf{x} \in \text{the crystal}, \\ 0, & \mathbf{x} \notin \text{the crystal}, \end{cases} \tag{7.7}$$

where c_{N_1,N_2,N_3} is a normalization constant. The corresponding eigenvalue

$$\Lambda_n(\tau_1, \tau_2, \tau_3) = \lambda_n(\tau_1, \tau_2, \tau_3) \tag{7.8}$$

is dependent on τ_1, τ_2, and τ_3. A consequence of Theorem 7.1 is that for each energy band index n, there is only one such solution (7.7) of (7.1). This state is a vertex-like state in the finite crystal or the quantum dot, since, in most cases $\phi_n(\mathbf{x}; \tau_1, \tau_2, \tau_3)$ decays in either the positive or the negative direction of $\mathbf{a}_1$, $\mathbf{a}_2$, and $\mathbf{a}_3$.

Now, we try to find other solutions of (7.1) from the further quantum confinement of $\bar{\psi}_n(\bar{\mathbf{k}}, \mathbf{x}; \tau_2, \tau_3)$. We can expect that there are stationary Bloch states in the $\mathbf{a}_1$

direction, formed due to multiple reflections of $\bar{\psi}_n(\bar{\mathbf{k}}, \mathbf{x}; \tau_2, \tau_3)$ between the two boundary surfaces intersecting the $\mathbf{a}_1$ axis at $\tau_1\mathbf{a}_1$ and $(\tau_1 + N_1)\mathbf{a}_1$.[1]

Since[2]

$$\bar{\Lambda}_n(\bar{\mathbf{k}}; \tau_2, \tau_3) = \bar{\Lambda}_n(-\bar{\mathbf{k}}; \tau_2, \tau_3),$$

in general

$$f_{n,k_1}(\mathbf{x}; \tau_2, \tau_3) = c_+\bar{\psi}_n(k_1\bar{\mathbf{b}}_1, \mathbf{x}; \tau_2, \tau_3) + c_-\bar{\psi}_n(-k_1\bar{\mathbf{b}}_1, \mathbf{x}; \tau_2, \tau_3), \quad 0 < k_1 < \pi,$$

where $c_\pm$ are not zero, is a nontrivial solution of Eq. (6.1).

To be a solution of (7.1), the function $f_{n,k_1}(\mathbf{x}; \tau_2, \tau_3)$ is further required to be zero at the left and the right faces of the finite crystal or quantum dot. By writing the left face equation of the finite crystal as $x_1 = x_{1,l}(x_2, x_3)$ and the right face equation of the finite crystal as $x_1 = x_{1,r}(x_2, x_3)$, we have

$$\begin{aligned} c_+\bar{\psi}_n[k_1\bar{\mathbf{b}}_1, \mathbf{x} \in \; & x_{1,l}(x_2, x_3); \tau_2, \tau_3] \\ & + c_-\bar{\psi}_n[-k_1\bar{\mathbf{b}}_1, \mathbf{x} \in x_{1,l}(x_2, x_3); \tau_2, \tau_3] = 0, \\ c_+\bar{\psi}_n[k_1\bar{\mathbf{b}}_1, \mathbf{x} \in \; & x_{1,r}(x_2, x_3); \tau_2, \tau_3] \\ & + c_-\bar{\psi}_n[-k_1\bar{\mathbf{b}}_1, \mathbf{x} \in x_{1,r}(x_2, x_3); \tau_2, \tau_3] = 0. \end{aligned} \tag{7.9}$$

Since $x_{1,r}(x_2, x_3) = x_{1,l}(x_2, x_3) + N_1$, we have

$$\bar{\psi}_n[k_1\bar{\mathbf{b}}_1, \mathbf{x} \in x_{1,r}(x_2, x_3); \tau_2, \tau_3] = \mathrm{e}^{\mathrm{i}k_1N_1}\bar{\psi}_n[k_1\bar{\mathbf{b}}_1, \mathbf{x} \in x_{1,l}(x_2, x_3); \tau_2, \tau_3]$$

and

$$\hat{\psi}_n[-k_1\bar{\mathbf{b}}_1, \mathbf{x} \in x_{1,r}(x_2, x_3); \tau_2, \tau_3] = \mathrm{e}^{-\mathrm{i}k_1N_1}\hat{\psi}_n[-k_1\bar{\mathbf{b}}_1, \mathbf{x} \in x_{1,l}(x_2, x_3); \tau_2, \tau_3]$$

due to (7.4). Therefore, if $c_\pm$ in (7.9) are not both zero, $\mathrm{e}^{\mathrm{i}k_1N_1} - \mathrm{e}^{-\mathrm{i}k_1N_1} = 0$ has to be true for each such stationary Bloch state, for this specific τ_1.

Stationary Bloch state solutions of (7.1) from further confinement of $\bar{\psi}_n(\bar{\mathbf{k}}, \mathbf{x}; \tau_2, \tau_3)$ should have the form

$$\psi_{n,j_1}(\mathbf{x}; \tau_1, \tau_2, \tau_3) = \begin{cases} f_{n,\kappa_1}(\mathbf{x}; \tau_1, \tau_2, \tau_3), & \mathbf{x} \in \text{the crystal}, \\ 0, & \mathbf{x} \notin \text{the crystal}, \end{cases} \tag{7.10}$$

[1]The author does not have enough understanding on solutions of periodic partial differential equations to assert that for what τ_1 the stationary Bloch state solutions in the $\mathbf{a}_1$ direction of (7.1) exist. However, he expects that the existence of such type of states is reasonable for a physical dot or finite crystal with a physical boundary τ_1.

[2]As solutions of (6.1), $\bar{\psi}(\bar{\mathbf{k}}, \mathbf{x})$ and $\bar{\psi}^*(\bar{\mathbf{k}}, \mathbf{x})$ have the same energy. $\bar{\psi}_n^*(\bar{\mathbf{k}}, \mathbf{x}; \tau_2, \tau_3) = \bar{\psi}_n(-\bar{\mathbf{k}}, \mathbf{x}; \tau_2, \tau_3)$ leads to that $\bar{\Lambda}_n(-\bar{\mathbf{k}}; \tau_2, \tau_3) = \bar{\Lambda}_n(\bar{\mathbf{k}}; \tau_2, \tau_3)$.

where

$$f_{n,k_1}(\mathbf{x}; \tau_1, \tau_2, \tau_3) = c_{n,k_1;\tau_1} \bar{\psi}_n(k_1 \bar{\mathbf{b}}_1, \mathbf{x}; \tau_2, \tau_3) + c_{n,-k_1;\tau_1} \bar{\psi}_n(-k_1 \bar{\mathbf{b}}_1, \mathbf{x}; \tau_2, \tau_3);$$

here, $c_{n,\pm k_1;\tau_1}$ are dependent on τ_1 and

$$\kappa_1 = j_1 \, \pi / N_1, \quad j_1 = 1, 2, \ldots, N_1 - 1, \tag{7.11}$$

where j_1 is the stationary Bloch state index in the $\mathbf{a}_1$ direction. These solutions $\psi_{n,j_1}(\mathbf{x}; \tau_1, \tau_2, \tau_3)$ satisfying (7.1) have energies Λ given by

$$\Lambda_{n,j_1}(\tau_2, \tau_3) = \bar{\Lambda}_n(\kappa_1 \bar{\mathbf{b}}_1; \tau_2, \tau_3). \tag{7.12}$$

Each energy given in (7.12) is dependent on N_1 and τ_2, τ_3. For each bulk energy band n, there are $N_1 - 1$ such states. These states are edge-like states in the finite crystal or quantum dot because $\bar{\psi}_n(\bar{\mathbf{k}}, \mathbf{x}; \tau_2, \tau_3)$ are edge-like states in the quantum wire.

Due to (7.3), (7.8), and (7.12), for the further quantum confinement of one-dimensional Bloch wave $\bar{\psi}_n(\bar{\mathbf{k}}, \mathbf{x}; \tau_2, \tau_3)$, in general, the energy of the vertex-like state is always above the energy of every relevant edge-like state:

$$\Lambda_n(\tau_1, \tau_2, \tau_3) > \Lambda_{n,j_1}(\tau_2, \tau_3). \tag{7.13}$$

7.3 Further Quantum Confinement of $\bar{\psi}_{n,j_3}(\bar{\mathbf{k}}, \mathbf{x}; \tau_2, \tau_3)$

For the quantum confinement of surface-like states $\bar{\psi}_{n,j_3}(\bar{\mathbf{k}}, \mathbf{x}; \tau_2, \tau_3)$, we consider a rectangular parallelepiped C' as shown in Fig. 7.2 with a bottom face and a top face at $x_3 = \tau_3$ and $x_3 = \tau_3 + N_3$, a front face and a rear face perpendicularly intersecting the $\mathbf{a}_2$ axis at $x_2 = \tau_2$ and $x_2 = \tau_2 + 1$, a left face and right face perpendicularly intersecting the $\mathbf{a}_1$ axis at $x_1 = \tau_1$ and $x_1 = \tau_1 + 1$.

We define a function set $\phi_{j_3}(\mathbf{x}; \tau_1, \tau_2, \tau_3)$ by the condition that each function is zero at the boundary $\partial C'$ of C' and behaves like a Bloch stationary state $\bar{\psi}_{j_3}(\bar{\mathbf{k}}, \mathbf{x}; \tau_2, \tau_3)$[3] with a stationary state index j_3 in the $\mathbf{b}_3$ direction. The eigenvalues and eigenfunctions of (5.1) with this condition are denoted by $\lambda_{n,j_3}(\tau_1, \tau_2)$ and $\phi_{n,j_3}(\mathbf{x}; \tau_1, \tau_2, \tau_3)$, where $n = 0, 1, 2, \ldots$. For each eigenvalue problem defined by (5.1) and this condition, we expect to have the following theorem between $\lambda_{n,j_3}(\tau_1, \tau_2)$ and the eigenvalue $\bar{\Lambda}_{n,j_3}(\bar{\mathbf{k}}; \tau_2)$ (6.19) of $\bar{\psi}_{n,j_3}(\bar{\mathbf{k}}, \mathbf{x}; \tau_2, \tau_3)$ in (6.18).

[3] $\bar{\psi}_{j_3}(\bar{\mathbf{k}}, \mathbf{x}; \tau_2, \tau_3)$ generally can be an (any) linear combination of $\bar{\psi}_{n,j_3}(\bar{\mathbf{k}}, \mathbf{x}; \tau_2, \tau_3)$ with different n.

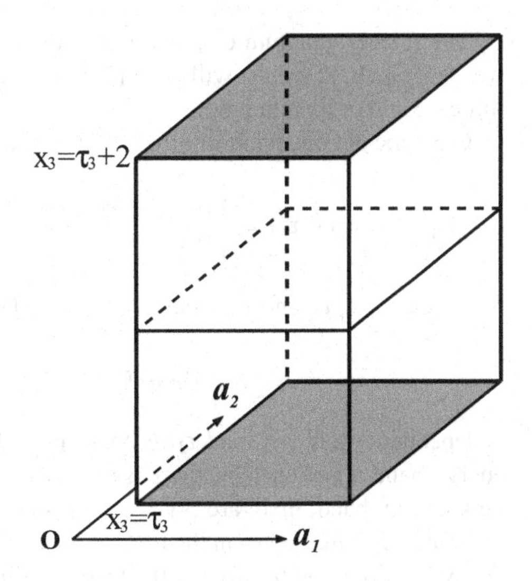

Fig. 7.2 The rectangular parallelepiped C' for the further quantum confinement of $\bar{\psi}_{n,j_3}(\bar{\mathbf{k}}, \mathbf{x}; \tau_2, \tau_3)$. The two shadowed faces of $\partial C'_3$ determined by $x_3 = \tau_3$ and $x_3 = \tau_3 + N_3$ (in the figure is shown the case $N_3 = 2$) and the two *thick-lined* faces of $\partial C'_2$ are the four faces on which each function $\bar{\psi}_{n,j_3}(\bar{\mathbf{k}}, \mathbf{x}; \tau_2, \tau_3)$ is zero. The *left face* and the *right face* are the two faces on which each function $\phi_{j_3}(\mathbf{x}; \tau_1, \tau_2, \tau_3)$ is further required to be zero

Theorem 7.2

$$\lambda_{n,j_3}(\tau_1, \tau_2) \geq \bar{\Lambda}_{n,j_3}(\bar{\mathbf{k}}; \tau_2). \tag{7.14}$$

$\bar{\psi}_{j_3}(\bar{\mathbf{k}}, \mathbf{x}; \tau_2, \tau_3)$ with a different j_3 are orthogonal to each other, each of them will be confined in the $\mathbf{a}_1$ direction independently. Theorem 7.2 can be proven similar to Theorem 7.1.

Theorem 7.2 is similar to Theorem 6.2. The consequences of the quantum confinement of two-dimensional Bloch waves $\hat{\psi}_{n,j_3}(\hat{\mathbf{k}}, \mathbf{x}; \tau_3)$ in the $\mathbf{a}_2$ direction due to Theorem 6.2 can be similarly used to the quantum confinement of one-dimensional Bloch waves $\bar{\psi}_{n,j_3}(\bar{\mathbf{k}}, \mathbf{x}; \tau_2, \tau_3)$ in the $\mathbf{a}_1$ direction.

Because $v(\mathbf{x} + \mathbf{a}_1) = v(\mathbf{x})$, the function $\phi_{n,j_3}(\mathbf{x}; \tau_1, \tau_2, \tau_3)$ has the form

$$\phi_{n,j_3}(\mathbf{x} + \mathbf{a}_1; \tau_1, \tau_2, \tau_3) = \mathrm{e}^{\mathrm{i}k_1} \phi_{n,j_3}(\mathbf{x}; \tau_1, \tau_2, \tau_3). \tag{7.15}$$

k_1 in (7.15) can be complex with a nonzero imaginary part or a real number. If k_1 is real in (7.15), then $\phi_{n,j_3}(\mathbf{x}; \tau_1, \tau_2, \tau_3)$ is a $\bar{\psi}_{n',j_3}(\bar{\mathbf{k}}, \mathbf{x}; \tau_2, \tau_3)$. By Theorem 7.2, a $\bar{\psi}_{n',j_3}(\bar{\mathbf{k}}, \mathbf{x}; \tau_2, \tau_3)$ cannot be a $\phi_{n,j_3}(\mathbf{x}; \tau_1, \tau_2, \tau_3)$ except in some special cases where the $\bar{\psi}_{n',j_3}(\bar{\mathbf{k}}, \mathbf{x}; \tau_2, \tau_3)$ has a nodal plane perpendicularly intersecting the $\mathbf{a}_1$ axis at $x_1 = \tau_1$. Therefore, k_1 in (7.15) can be real only in such special cases; in most cases, k_1 in (7.15) is complex with a nonzero imaginary part.

The imaginary part of k_1 in (7.15) can be either positive or negative, corresponding to that $\phi_{n,j_3}(\mathbf{x}; \tau_1, \tau_2, \tau_3)$ decays in either the positive or the negative direction of $\mathbf{a}_1$. Such states $\phi_{n,j_3}(\mathbf{x}; \tau_1, \tau_2, \tau_3)$ with a nonzero imaginary part of k_1 in (7.15) cannot exist in a quantum wire with translational invariance in the $\mathbf{a}_1$ direction since they are divergent in either the negative or the positive direction of $\mathbf{a}_1$. However, they can play a significant role in the electronic states in a finite crystal or quantum dot without translational invariance.

The further quantum confinement in the $\mathbf{a}_1$ direction of one-dimensional Bloch waves $\bar{\psi}_{n,j_3}(\bar{\mathbf{k}}, \mathbf{x}; \tau_2, \tau_3)$ will produce two different types of electronic states in the finite crystal or quantum dot.

One type of nontrivial solutions of (7.1) can be obtained from (7.15) by assigning

$$\psi_{n,j_3}(\mathbf{x}; \tau_1, \tau_2, \tau_3) = \begin{cases} c_{N_1,N_2,N_3}\phi_{n,j_3}(\mathbf{x}; \tau_1, \tau_2, \tau_3), & \mathbf{x} \in \text{the crystal}, \\ 0, & \mathbf{x} \notin \text{the crystal}, \end{cases} \tag{7.16}$$

where c_{N_1,N_2,N_3} is a normalization constant. The corresponding eigenvalue

$$\Lambda_{n,j_3}(\tau_1, \tau_2) = \lambda_{n,j_3}(\tau_1, \tau_2) \tag{7.17}$$

is dependent on N_3, τ_1, and τ_2. A consequence of Theorem 7.2 is that for each bulk energy band n and each j_3, there is only one such solution (7.16) of (7.1). For each bulk energy band, there are $N_3 - 1$ such states in the finite crystal or quantum dot. They are edge-like states in the finite crystal or quantum dot since $\psi_{n,j_3}(\mathbf{x}; \tau_1, \tau_2, \tau_3)$ decays in either the positive or the negative direction of $\mathbf{a}_1$ and $\mathbf{a}_2$ in most cases.

Now, we try to find other solutions of (7.1) from the further quantum confinement of $\bar{\psi}_{n,j_3}(\bar{\mathbf{k}}, \mathbf{x}; \tau_2, \tau_3)$. We can expect that there are stationary Bloch states in the $\mathbf{a}_1$ direction, formed due to the multiple reflections of the one-dimensional Bloch waves $\bar{\psi}_{n,j_3}(\bar{\mathbf{k}}, \mathbf{x}; \tau_2, \tau_3)$ between two boundary surfaces intersecting the $\mathbf{a}_1$ axis at $\tau_1\mathbf{a}_1$ and $(\tau_1 + N_1)\mathbf{a}_1$.

Since the energies (6.19) of $\bar{\psi}_{n,j_3}(\bar{\mathbf{k}}, \mathbf{x}; \tau_2, \tau_3)$ satisfy[4]

$$\bar{\Lambda}_{n,j_3}(\bar{\mathbf{k}}; \tau_2) = \bar{\Lambda}_{n,j_3}(-\bar{\mathbf{k}}; \tau_2), \tag{7.18}$$

in general

$$\begin{aligned} f_{n,k_1,j_3}(\mathbf{x}; \tau_2, \tau_3) = {} & c_+\bar{\psi}_{n,j_3}(k_1\bar{\mathbf{b}}_1, \mathbf{x}; \tau_2, \tau_3) \\ & + c_-\bar{\psi}_{n,j_3}(-k_1\bar{\mathbf{b}}_1, \mathbf{x}; \tau_2, \tau_3), \quad 0 < k_1 < \pi, \end{aligned}$$

where $c_\pm$ are not zero, is a nontrivial solution of Eq. (6.1) because of (7.18). Very similarly to what we did in Sect. 7.2, we can expect that the stationary Bloch state solutions of (7.1) from the further confinement of $\bar{\psi}_{n,j_3}(\bar{\mathbf{k}}, \mathbf{x}; \tau_2, \tau_3)$ should have the form

$$\psi_{n,j_1,j_3}(\mathbf{x}; \tau_1, \tau_2, \tau_3) = \begin{cases} f_{n,\kappa_1,j_3}(\mathbf{x}; \tau_1, \tau_2, \tau_3), & \mathbf{x} \in \text{the crystal}, \\ 0, & \mathbf{x} \notin \text{the crystal}, \end{cases} \tag{7.19}$$

[4] $\bar{\psi}^*_{n,j_3}(\bar{\mathbf{k}}, \mathbf{x}; \tau_2, \tau_3) = \bar{\psi}_{n,j_3}(-\bar{\mathbf{k}}, \mathbf{x}; \tau_2, \tau_3)$ leads to (7.18).

where

$$f_{n,k_1,j_3}(\mathbf{x}; \tau_1, \tau_2, \tau_3) = c_{n,k_1,j_3;\tau_1} \bar{\psi}_{n,j_3}(k_1\bar{\mathbf{b}}_1, \mathbf{x}; \tau_2, \tau_3) \\ + c_{n,-k_1,j_3;\tau_1} \bar{\psi}_{n,j_3}(-k_1\bar{\mathbf{b}}_1, \mathbf{x}; \tau_2, \tau_3);$$

$c_{n,\pm k_1,j_3;\tau_1}$ are dependent on $\tau_1, \kappa_1 = j_1\pi/N_1$, and $j_1 = 1, 2, \ldots, N_1 - 1$ as in (7.11). Stationary Bloch state solutions $\psi_{n,j_1,j_3}(\mathbf{x}; \tau_1, \tau_2, \tau_3)$ satisfying (7.1) have energies Λ given by

$$\Lambda_{n,j_1,j_3}(\tau_2) = \bar{\Lambda}_{n,j_3}(\kappa_1\bar{\mathbf{b}}_1; \tau_2). \tag{7.20}$$

Each energy in (7.20) for this case is dependent on N_1, N_3, and τ_2. Those are surface-like states in the finite crystal or quantum dot because $\bar{\psi}_{n,j_3}(\bar{\mathbf{k}}, \mathbf{x}; \tau_2, \tau_3)$ are surface-like states in the quantum wire. For each bulk energy band n, there are $(N_1-1)(N_3-1)$ such states in the finite crystal or quantum dot.

Similar to (7.13), because of (7.14), (7.17), and (7.20), for the further quantum confinement of $\bar{\psi}_{n,j_3}(\bar{\mathbf{k}}, \mathbf{x}; \tau_2, \tau_3)$, in general, the energy of an edge-like state is above the energy of a relevant surface-like state:

$$\Lambda_{n,j_3}(\tau_1, \tau_2) > \Lambda_{n,j_1,j_3}(\tau_2). \tag{7.21}$$

7.4 Further Quantum Confinement of $\bar{\psi}_{n,j_2}(\bar{\mathbf{k}}, \mathbf{x}; \tau_2, \tau_3)$

The further quantum confinement of the surface-like states $\bar{\psi}_{n,j_2}(\bar{\mathbf{k}}, \mathbf{x}; \tau_2, \tau_3)$ in the quantum wire $\mathbf{a}_1$ direction can be similarly discussed. For each energy band n, we will have $N_2 - 1$ edge-like states and $(N_1 - 1)(N_2 - 1)$ surface-like states in the finite crystal or quantum dot.

For the quantum confinement of $\bar{\psi}_{n,j_2}(\bar{\mathbf{k}}, \mathbf{x}; \tau_2, \tau_3)$, we consider a rectangular parallelepiped C'' as shown in Fig. 7.3 with a bottom face $x_3 = \tau_3$, a top face $x_3 = \tau_3 + 1$, a front face and a rear face perpendicularly intersecting the $\mathbf{a}_2$ axis at $x_2 = \tau_2$ and $x_2 = \tau_2 + N_2$, a left face and a right face perpendicularly intersecting the $\mathbf{a}_1$ axis at $x_1 = \tau_1$ and $x_1 = \tau_1 + 1$. We define a function set $\phi_{j_2}(\mathbf{x}; \tau_1, \tau_2, \tau_3)$ by the condition that each function is zero at the boundary $\partial C''$ of C'' and behaves as a Bloch stationary state $\bar{\psi}_{j_2}(\bar{\mathbf{k}}, \mathbf{x}; \tau_2, \tau_3)$[5] with a stationary state index j_2 in the $\hat{\mathbf{b}}_2$ direction. The eigenvalues and eigenfunctions of (5.1) with this condition are denoted by $\lambda_{n,j_2}(\tau_1, \tau_3)$ and $\phi_{n,j_2}(\mathbf{x}; \tau_1, \tau_2, \tau_3)$, where $n = 0, 1, 2, \ldots$. For each eigenvalue problem defined by (5.1) and this condition, we expect to have the following theorem between $\lambda_{n,j_2}(\tau_1, \tau_3)$ and the eigenvalue $\bar{\Lambda}_{n,j_2}(\bar{\mathbf{k}}; \tau_3)$ (6.13) of $\bar{\psi}_{n,j_2}(\bar{\mathbf{k}}, \mathbf{x}; \tau_2, \tau_3)$ in (6.11).

[5] $\bar{\psi}_{j_2}(\bar{\mathbf{k}}, \mathbf{x}; \tau_2, \tau_3)$ generally can be an (any) linear combination of $\bar{\psi}_{n,j_2}(\bar{\mathbf{k}}, \mathbf{x}; \tau_2, \tau_3)$ of different n.

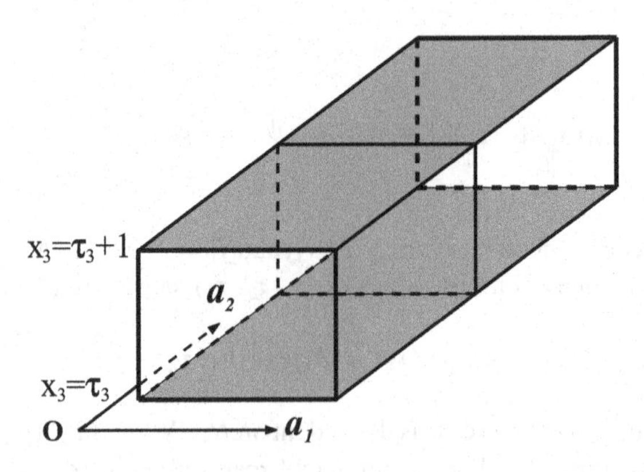

Fig. 7.3 The rectangular parallelepiped C'' for the quantum confinement of $\bar{\psi}_{n,j_2}(\bar{\mathbf{k}}, \mathbf{x}; \tau_2, \tau_3)$. The two shadowed faces of $\partial C''_3$ determined by $x_3 = \tau_3$ and $x_3 = \tau_3 + 1$ and the two *thick-lined* faces of $\partial C''_2$ determined by $\tau_2\mathbf{a}_2$ and $(\tau_2 + N_2)\mathbf{a}_2$ (in the figure is shown the case $N_2 = 2$) are the four faces on which each function $\bar{\psi}_{n,j_2}(\bar{\mathbf{k}}, \mathbf{x}; \tau_2, \tau_3)$ is zero. The *left face* and the *right face* of $\partial C''_1$ are the two faces on which each function $\phi_{j_2}(\mathbf{x}; \tau_1, \tau_2, \tau_3)$ is further required to be zero

Theorem 7.3

$$\lambda_{n,j_2}(\tau_1, \tau_3) \geq \bar{\Lambda}_{n,j_2}(\bar{\mathbf{k}}; \tau_3). \tag{7.22}$$

$\bar{\psi}_{j_2}(\bar{\mathbf{k}}, \mathbf{x}; \tau_2, \tau_3)$ with a different j_2 are orthogonal to each other, each of them will be confined in the $\mathbf{a}_1$ direction independently. Theorem 7.3 can be proven similar to Theorem 7.1.

Theorem 7.3 is similar to Theorem 7.2. The consequences of the quantum confinement of one-dimensional Bloch waves $\bar{\psi}_{n,j_3}(\bar{\mathbf{k}}, \mathbf{x}; \tau_2, \tau_3)$ in the $\mathbf{a}_1$ direction due to Theorem 7.2 can be similarly applied to the quantum confinement of one-dimensional Bloch waves $\bar{\psi}_{n,j_2}(\bar{\mathbf{k}}, \mathbf{x}; \tau_2, \tau_3)$ in the $\mathbf{a}_1$ direction.

Because $v(\mathbf{x} + \mathbf{a}_1) = v(\mathbf{x})$, the function $\phi_{n,j_2}(\mathbf{x}; \tau_1, \tau_2, \tau_3)$ has the form

$$\phi_{n,j_2}(\mathbf{x} + \mathbf{a}_1; \tau_1, \tau_2, \tau_3) = \mathrm{e}^{ik_1}\phi_{n,j_2}(\mathbf{x}; \tau_1, \tau_2, \tau_3). \tag{7.23}$$

k_1 in (7.23) can be complex with a nonzero imaginary part or a real number. If k_1 is real in (7.23), then $\phi_{n,j_2}(\mathbf{x}; \tau_1, \tau_2, \tau_3)$ is a $\bar{\psi}_{n',j_2}(\bar{\mathbf{k}}, \mathbf{x}; \tau_2, \tau_3)$. By Theorem 7.3, only in particular cases where a $\bar{\psi}_{n',j_2}(\bar{\mathbf{k}}, \mathbf{x}; \tau_2, \tau_3)$ has a nodal plane perpendicularly intersecting the $\mathbf{a}_1$ axis at $x_1 = \tau_1$, the $\bar{\psi}_{n',j_2}(\bar{\mathbf{k}}, \mathbf{x}; \tau_2, \tau_3)$ can be a $\phi_{n,j_2}(\mathbf{x}; \tau_1, \tau_2, \tau_3)$. Therefore, k_1 in (7.23) can be real only in such particular cases; in most cases, k_1 in (7.23) is complex with a nonzero imaginary part.

The imaginary part of k_1 in (7.23) can be either positive or negative; this corresponds to that $\phi_{n,j_2}(\mathbf{x}; \tau_1, \tau_2, \tau_3)$ decays in either the positive or the negative direction of $\mathbf{a}_1$. Such states $\phi_{n,j_2}(\mathbf{x}; \tau_1, \tau_2, \tau_3)$ with a nonzero imaginary part of k_1 in (7.23) cannot exist in a quantum wire with translational invariance in the $\mathbf{a}_1$ direction, because they are divergent in either the negative or the positive direction of $\mathbf{a}_1$. However, these states can play a significant role in the electronic states in a finite crystal or quantum dot without translational invariance.

The further quantum confinement in the $\mathbf{a}_1$ direction of one-dimensional Bloch waves $\bar{\psi}_{n,j_2}(\bar{\mathbf{k}}, \mathbf{x}; \tau_2, \tau_3)$ will produce two different types of electronic states in the finite crystal or quantum dot.

One type of nontrivial solutions of (7.1) can be obtained from (7.23) by assigning

$$\psi_{n,j_2}(\mathbf{x}; \tau_1, \tau_2, \tau_3) = \begin{cases} c_{N_1,N_2,N_3}\phi_{n,j_2}(\mathbf{x}; \tau_1, \tau_2, \tau_3), & \mathbf{x} \in \text{the crystal}, \\ 0, & \mathbf{x} \notin \text{the crystal}, \end{cases} \tag{7.24}$$

where c_{N_1,N_2,N_3} is a normalization constant. The corresponding eigenvalue is

$$\Lambda_{n,j_2}(\tau_1, \tau_3) = \lambda_{n,j_2}(\tau_1, \tau_3). \tag{7.25}$$

For each band n and each j_2, there is one electronic state $\psi_{n,j_2}(\mathbf{x}; \tau_1, \tau_2, \tau_3)$ which is $\phi_{n,j_2}(\mathbf{x}; \tau_1, \tau_2, \tau_3)$ inside the finite crystal or dot but zero otherwise. Its energy $\Lambda_{n,j_2}(\tau_1, \tau_3)$ depends on τ_1, τ_3, and N_2. For each bulk energy band n, there are $N_2 - 1$ such states in the finite crystal or quantum dot. They are edge-like states in the finite crystal or quantum dot since $\phi_{n,j_2}(\mathbf{x}; \tau_1, \tau_2, \tau_3)$ decays in either the positive or the negative direction of $\mathbf{a}_1$ and $\mathbf{a}_3$ in most cases.

Now, we try to find other solutions of (7.1) from the quantum confinement of $\bar{\psi}_{n,j_2}(\bar{\mathbf{k}}, \mathbf{x}; \tau_2, \tau_3)$. We can expect that there are stationary Bloch states in the $\mathbf{a}_1$ direction, formed due to the multiple reflections of $\bar{\psi}_{n,j_2}(\bar{\mathbf{k}}, \mathbf{x}; \tau_2, \tau_3)$ between two boundary surfaces perpendicularly intersecting the $\mathbf{a}_1$ axis at $x_1 = \tau_1$ and $x_1 = \tau_1 + N_1$.

Since for the energies of the electronic states $\bar{\psi}_{n,j_2}(\bar{\mathbf{k}}, \mathbf{x}; \tau_2, \tau_3)$ in the quantum wire we have[6]

$$\bar{\Lambda}_{n,j_2}(\bar{\mathbf{k}}; \tau_3) = \bar{\Lambda}_{n,j_2}(-\bar{\mathbf{k}}; \tau_3), \tag{7.26}$$

in general

$$\begin{aligned} f_{n,k_1,j_2}(\mathbf{x}; \tau_2, \tau_3) &= c_+\bar{\psi}_{n,j_2}(k_1\bar{\mathbf{b}}_1, \mathbf{x}; \tau_2, \tau_3) \\ &+ c_-\bar{\psi}_{n,j_2}(-k_1\bar{\mathbf{b}}_1, \mathbf{x}; \tau_2, \tau_3), \qquad 0 < k_1 < \pi, \end{aligned}$$

where $c_\pm$ are not zero, is a nontrivial solution of Eq. (6.1) due to (7.26). Similar to the cases in Sects. 7.2 and 7.3, we can expect that the stationary Bloch state solutions of (7.1) originating from the quantum confinement of $\bar{\psi}_{n,j_2}(\bar{\mathbf{k}}, \mathbf{x}; \tau_2, \tau_3)$ should have the form

$$\psi_{n,j_1,j_2}(\mathbf{x}; \tau_1, \tau_2, \tau_3) = \begin{cases} f_{n,\kappa_1,j_2}(\mathbf{x}; \tau_1, \tau_2, \tau_3), & \mathbf{x} \in \text{the crystal}, \\ 0, & \mathbf{x} \notin \text{the crystal}, \end{cases} \tag{7.27}$$

---

[6] $\bar{\psi}^*_{n,j_2}(\bar{\mathbf{k}}, \mathbf{x}; \tau_2, \tau_3) = \bar{\psi}_{n,j_2}(-\bar{\mathbf{k}}, \mathbf{x}; \tau_2, \tau_3)$ leads to (7.26).

where

$$f_{n,k_1,j_2}(\mathbf{x};\tau_1,\tau_2,\tau_3) = c_{n,k_1,j_2;\tau_1}\bar{\psi}_{n,j_2}(k_1\bar{\mathbf{b}}_1,\mathbf{x};\tau_2,\tau_3) + c_{n,-k_1,j_2;\tau_1}\bar{\psi}_{n,j_2}(-k_1\bar{\mathbf{b}}_1,\mathbf{x};\tau_2,\tau_3),$$

$c_{n,\pm k_1,j_2;\tau_1}$ are dependent on τ_1, $\kappa_1 = j_1\pi/N_1$, and $j_1 = 1, 2, \ldots, N_1 - 1$ as in (7.11). Stationary Bloch state solutions $\psi_{n,j_1,j_2}(\mathbf{x};\tau_1,\tau_2,\tau_3)$ satisfying (7.1) have energies Λ given by

$$\Lambda_{n,j_1,j_2}(\tau_3) = \bar{\Lambda}_{n,j_2}(\kappa_1\bar{\mathbf{b}}_1;\tau_3). \tag{7.28}$$

Each energy $\Lambda_{n,j_1,j_2}(\tau_3)$ for this case is dependent on N_1, N_2, and τ_3. These are surface-like states in the finite crystal or quantum dot since $\bar{\psi}_{n,j_2}(\bar{\mathbf{k}},\mathbf{x};\tau_2,\tau_3)$ are surface-like states in the quantum wire. For each bulk energy band n, there are $(N_1-1)(N_2-1)$ such states in the finite crystal or quantum dot.

Similar to (7.21), because of (7.22), (7.25), and (7.28), for the further quantum confinement of $\bar{\psi}_{n,j_2}(\bar{\mathbf{k}},\mathbf{x};\tau_2,\tau_3)$ in general the energy of an edge-like state is above the energy of a relevant surface-like state:

$$\Lambda_{n,j_2}(\tau_1,\tau_3) > \Lambda_{n,j_1,j_2}(\tau_3). \tag{7.29}$$

7.5 Further Quantum Confinement of $\bar{\psi}_{n,j_2,j_3}(\bar{\mathbf{k}},\mathbf{x};\tau_2,\tau_3)$

For the quantum confinement of bulk-like states $\bar{\psi}_{n,j_2,j_3}(\bar{\mathbf{k}},\mathbf{x};\tau_2,\tau_3)$, we consider a rectangular parallelepiped C''' as shown in Fig. 7.4 with a bottom face at $x_3 = \tau_3$, a top face at $x_3 = \tau_3 + N_3$, a front face and a rear face perpendicularly intersecting the $\mathbf{a}_2$ axis at $x_2 = \tau_2$ and $x_2 = \tau_2 + N_2$, a left face and a right face perpendicularly intersecting the $\mathbf{a}_1$ axis at $x_1 = \tau_1$ and at $x_1 = \tau_1 + 1$.

We define a function set $\phi_{j_2,j_3}(\mathbf{x};\tau_1,\tau_2,\tau_3)$ by the condition that each function is zero at the boundary $\partial C'''$ of C''', and behaves as a Bloch stationary state with a stationary state index j_2 in the $\hat{\mathbf{b}}_2$ direction, as a Bloch stationary state with a stationary state index j_3 in the $\mathbf{b}_3$ direction.[7] The eigenvalues and eigenfunctions of (5.1) under this condition are denoted by $\lambda_{n,j_2,j_3}(\tau_1)$ and $\phi_{n,j_2,j_3}(\mathbf{x};\tau_1,\tau_2,\tau_3)$, where $n = 0, 1, 2, \ldots$. For the eigenvalue problem defined by (5.1) and this condition, we expect to have the following theorem between $\lambda_{n,j_2,j_3}(\tau_1)$ and the eigenvalues $\bar{\Lambda}_{n,j_2,j_3}(\bar{\mathbf{k}})$ (6.22) of $\bar{\psi}_{n,j_2,j_3}(\bar{\mathbf{k}},\mathbf{x};\tau_2,\tau_3)$ in (6.21).

Theorem 7.4

$$\lambda_{n,j_2,j_3}(\tau_1) \geq \bar{\Lambda}_{n,j_2,j_3}(\bar{\mathbf{k}}). \tag{7.30}$$

[7] $\bar{\psi}_{j_2,j_3}(\bar{\mathbf{k}},\mathbf{x};\tau_2,\tau_3)$ can, in general, be an (any) linear combination of $\bar{\psi}_{n,j_2,j_3}(\bar{\mathbf{k}},\mathbf{x};\tau_2,\tau_3)$ with different n.

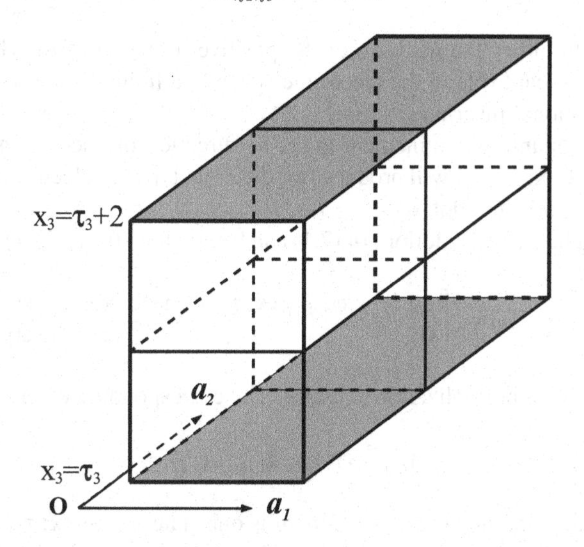

Fig. 7.4 The rectangular parallelepiped C''' for the quantum confinement of $\bar{\psi}_{n,j_2,j_3}(\bar{\mathbf{k}}, \mathbf{x}; \tau_2, \tau_3)$. The two shadowed faces of $\partial C_3'''$ determined by $x_3 = \tau_3$ and $x_3 = (\tau_3 + N_3)$ (in the figure is shown the case $N_3 = 2$) and the two *thick-lined* faces of $\partial C_2'''$ determined by $\tau_2 \mathbf{a}_2$ and $(\tau_2 + N_2)\mathbf{a}_2$ (in the figure is shown the case $N_2 = 2$) are the four faces on which each function $\bar{\psi}_{n,j_2,j_3}(\bar{\mathbf{k}}, \mathbf{x}; \tau_2, \tau_3)$ is zero. The *left face* and the *right face* of $\partial C_1'''$ are the two faces on which each function $\phi_{j_2,j_3}(\mathbf{x}; \tau_1, \tau_2, \tau_3)$ is further required to be zero

$\bar{\psi}_{j_2,j_3}(\bar{\mathbf{k}}, \mathbf{x}; \tau_2, \tau_3)$ with a different j_2 or j_3 are orthogonal to each other, each of them will be confined in the $\mathbf{a}_1$ direction independently. Theorem 7.4 can be proven similar to Theorem 7.2.

Theorem 7.4 is similar to Theorem 7.2; the consequences of the further quantum confinement of one-dimensional Bloch waves $\bar{\psi}_{n,j_3}(\bar{\mathbf{k}}, \mathbf{x}; \tau_2, \tau_3)$ in the $\mathbf{a}_1$ direction due to Theorem 7.2 can be similarly applied to the further quantum confinement of one-dimensional Bloch waves $\bar{\psi}_{n,j_2,j_3}(\bar{\mathbf{k}}, \mathbf{x}; \tau_2, \tau_3)$ in the $\mathbf{a}_1$ direction.

Because $v(\mathbf{x} + \mathbf{a}_1) = v(\mathbf{x})$, the function $\phi_{n,j_2,j_3}(\mathbf{x}; \tau_1, \tau_2, \tau_3)$ has the form

$$\phi_{n,j_2,j_3}(\mathbf{x} + \mathbf{a}_1; \tau_1, \tau_2, \tau_3) = \mathrm{e}^{\mathrm{i}k_1} \phi_{n,j_2,j_3}(\mathbf{x}; \tau_1, \tau_2, \tau_3). \tag{7.31}$$

k_1 in (7.31) can be complex with a nonzero imaginary part or a real number. If k_1 is real in (7.31), then $\phi_{n,j_2,j_3}(\mathbf{x}; \tau_1, \tau_2, \tau_3)$ is a $\bar{\psi}_{n',j_2,j_3}(\bar{\mathbf{k}}, \mathbf{x}; \tau_2, \tau_3)$. By Theorem 7.4, a $\bar{\psi}_{n',j_2,j_3}(\bar{\mathbf{k}}, \mathbf{x}; \tau_2, \tau_3)$ can not be a $\phi_{n,j_2,j_3}(\mathbf{x}; \tau_1, \tau_2, \tau_3)$ except in some special cases where $\bar{\psi}_{n',j_2,j_3}(\bar{\mathbf{k}}, \mathbf{x}; \tau_2, \tau_3)$ has a nodal plane perpendicularly intersecting the $\mathbf{a}_1$ axis at $x_1 = \tau_1$. Therefore, k_1 in (7.31) can be real only in such special cases; in most cases, it is complex with a nonzero imaginary part.

The imaginary part of k_1 in (7.31) can be either positive or negative, corresponding to that $\phi_{n,j_2,j_3}(\mathbf{x}; \tau_1, \tau_2, \tau_3)$ decays in either the positive or the negative direction of $\mathbf{a}_1$. Such states $\phi_{n,j_2,j_3}(\mathbf{x}; \tau_1, \tau_2, \tau_3)$ with a nonzero imaginary part of k_1 in (7.31) cannot exist in a quantum wire with translational invariance in the $\mathbf{a}_1$ direction because they are divergent in either the negative or the positive direction of $\mathbf{a}_1$. However, they can play a significant role in the electronic states in a finite crystal or quantum dot without translational invariance.

The further quantum confinement in the $\mathbf{a}_1$ direction of one-dimensional Bloch waves $\bar{\psi}_{n,j_2,j_3}(\bar{\mathbf{k}}, \mathbf{x}; \tau_2, \tau_3)$ will produce two different types of electronic states in the finite crystal or quantum dot.

One type of nontrivial solutions of (7.1) can be obtained from (7.31) by assigning

$$\psi_{n,j_2,j_3}(\mathbf{x}; \tau_1, \tau_2, \tau_3) = \begin{cases} c_{N_1,N_2,N_3}\phi_{n,j_2,j_3}(\mathbf{x}; \tau_1, \tau_2, \tau_3), & \mathbf{x} \in \text{the crystal}, \\ 0, & \mathbf{x} \notin \text{the crystal}, \end{cases} \tag{7.32}$$

where c_{N_1,N_2,N_3} is a normalization constant. The corresponding eigenvalue is

$$\Lambda_{n,j_2,j_3}(\tau_1) = \lambda_{n,j_2,j_3}(\tau_1). \tag{7.33}$$

For each energy band and each j_2, j_3, there is one state $\psi_{n,j_2,j_3}(\mathbf{x}; \tau_1, \tau_2, \tau_3)$ which is $\phi_{n,j_2,j_3}(\mathbf{x}; \tau_1, \tau_2, \tau_3)$ inside the finite crystal or quantum dot but zero otherwise. Its energy $\Lambda_{n,j_2,j_3}(\tau_1)$ depends on τ_1, N_2, and N_3. For each bulk energy band n, there are $(N_2-1)(N_3-1)$ such states in the finite crystal or quantum dot. They are surface-like states in the finite crystal or quantum dot since $\phi_{n,j_2,j_3}(\mathbf{x}; \tau_1, \tau_2, \tau_3)$ decays in either the positive or the negative direction of $\mathbf{a}_1$ in most cases.

Now, we try to find out other solutions of (7.1) from the quantum confinement of $\bar{\psi}_{n,j_2,j_3}(\bar{\mathbf{k}}, \mathbf{x}; \tau_2, \tau_3)$. We can expect that there are stationary Bloch states in the $\mathbf{a}_1$ direction, formed due to multiple reflections of $\bar{\psi}_{n,j_2,j_3}(\bar{\mathbf{k}}, \mathbf{x}; \tau_2, \tau_3)$ between two boundary surfaces perpendicularly intersecting the $\mathbf{a}_1$ axis at $x_1 = \tau_1$ and $x_1 = \tau_1 + N_1$.

Since we have[8]

$$\bar{\Lambda}_{n,j_2,j_3}(\bar{\mathbf{k}}) = \bar{\Lambda}_{n,j_2,j_3}(-\bar{\mathbf{k}}), \tag{7.34}$$

in general

$$\begin{aligned} f_{n,k_1,j_2,j_3}(\mathbf{x}; \tau_2, \tau_3) &= c_+\bar{\psi}_{n,j_2,j_3}(k_1\bar{\mathbf{b}}_1, \mathbf{x}; \tau_2, \tau_3) \\ &+ c_-\bar{\psi}_{n,j_2,j_3}(-k_1\bar{\mathbf{b}}_1, \mathbf{x}; \tau_2, \tau_3), \quad 0 < k_1 < \pi, \end{aligned}$$

where $c_\pm$ are not both zero, is a nontrivial solution of Eq. (6.1) due to (7.34). Similar to what was done in Sects. 7.2–7.4, we can see that the stationary Bloch state solutions of (7.1) originating from the further quantum confinement of $\bar{\psi}_{n,j_2,j_3}(\bar{\mathbf{k}}, \mathbf{x}; \tau_2, \tau_3)$ should have the form

$$\psi_{n,j_1,j_2,j_3}(\mathbf{x}; \tau_1, \tau_2, \tau_3) = \begin{cases} f_{n,\kappa_1,j_2,j_3}(\mathbf{x}; \tau_1, \tau_2, \tau_3), & \mathbf{x} \in \text{the crystal}, \\ 0, & \mathbf{x} \notin \text{the crystal}, \end{cases} \tag{7.35}$$

[8] $\bar{\psi}^*_{n,j_2,j_3}(\bar{\mathbf{k}}, \mathbf{x}; \tau_2, \tau_3) = \bar{\psi}_{n,j_2,j_3}(-\bar{\mathbf{k}}, \mathbf{x}; \tau_2, \tau_3)$ leads to (7.34).

where

$$f_{n,k_1,j_2,j_3}(\mathbf{x}; \tau_1, \tau_2, \tau_3) = c_{n,k_1,j_2,j_3;\tau_1} \bar{\psi}_{n,j_2,j_3}(k_1\bar{\mathbf{b}}_1, \mathbf{x}; \tau_2, \tau_3) + c_{n,-k_1,j_2,j_3;\tau_1} \bar{\psi}_{n,j_2,j_3}(-k_1\bar{\mathbf{b}}_1, \mathbf{x}; \tau_2, \tau_3),$$

$c_{n,\pm k_1,j_2,j_3;\tau_1}$ are dependent on τ_1, $\kappa_1 = j_1\pi/N_1$, and $j_1 = 1, 2, \ldots, N_1 - 1$, as in (7.11). These stationary Bloch state solutions $\psi_{n,j_1,j_2,j_3}(\mathbf{x}; \tau_1, \tau_2, \tau_3)$ satisfying (7.1) have energies Λ given by

$$\Lambda_{n,j_1,j_2,j_3} = \bar{\Lambda}_{n,j_2,j_3}(\kappa_1\bar{\mathbf{b}}_1). \tag{7.36}$$

Each energy Λ_{n,j_1,j_2,j_3} in (7.36) for this case is dependent on N_1, N_2, and N_3. The energies Λ_{n,j_1,j_2,j_3} map the bulk energy band $\varepsilon_n(\mathbf{k})$ exactly: From (7.36), (6.22), and (5.28), one obtains that $\Lambda_{n,j_1,j_2,j_3} = \bar{\Lambda}_{n,j_2,j_3}(\kappa_1\bar{\mathbf{b}}_1) = \hat{\Lambda}_{n,j_3}(\kappa_1\bar{\mathbf{b}}_1 + \kappa_2\hat{\mathbf{b}}_2) = \varepsilon_n(\kappa_1\bar{\mathbf{b}}_1 + \kappa_2\hat{\mathbf{b}}_2 + \kappa_3\mathbf{b}_3)$. These stationary states can be considered as bulk-like states in the finite crystal or quantum dot. For each band index n, there are $(N_1 - 1)(N_2 - 1)(N_3 - 1)$ such bulk-like states in the finite crystal or quantum dot.

Because of (7.30), (7.33), and (7.36), for the further quantum confinement of one-dimensional Bloch waves $\bar{\psi}_{n,j_2,j_3}(\bar{\mathbf{k}}, \mathbf{x}; \tau_2, \tau_3)$ in general the energy of a surface-like state is above the energy of every relevant bulk-like state:

$$\Lambda_{n,j_2,j_3}(\tau_1) > \Lambda_{n,j_1,j_2,j_3}. \tag{7.37}$$

Thus, the effects of the further quantum confinement of one-dimensional Bloch waves $\bar{\psi}_n(\bar{\mathbf{k}}, \mathbf{x}; \tau_2, \tau_3)$, $\bar{\psi}_{n,j_3}(\bar{\mathbf{k}}, \mathbf{x}; \tau_2, \tau_3)$, $\bar{\psi}_{n,j_2}(\bar{\mathbf{k}}, \mathbf{x}; \tau_2, \tau_3)$, and $\bar{\psi}_{n,j_2,j_3}(\bar{\mathbf{k}}, \mathbf{x}; \tau_2, \tau_3)$ are similar to what we have seen in Chaps. 5 and 6: Each set will produce two different types of electronic states in the ideal finite crystal or quantum dot. They can be grouped into eight sets and have different behaviors. For each bulk energy band n, there are the following:

$(N_1 - 1)(N_2 - 1)(N_3 - 1)$ bulk-like states $\psi_{n,j_1,j_2,j_3}(\mathbf{x}; \tau_1, \tau_2, \tau_3)$ in (7.35), their energies Λ_{n,j_1,j_2,j_3} in (7.36) depend on N_1, N_2, and N_3;

$(N_1 - 1)(N_2 - 1)$ surface-like states $\psi_{n,j_1,j_2}(\mathbf{x}; \tau_1, \tau_2, \tau_3)$ in (7.27), their energies $\Lambda_{n,j_1,j_2}(\tau_3)$ in (7.28) depend on N_1, N_2, and τ_3;

$(N_2 - 1)(N_3 - 1)$ surface-like states $\psi_{n,j_2,j_3}(\mathbf{x}; \tau_1, \tau_2, \tau_3)$ in (7.32), their energies $\Lambda_{n,j_2,j_3}(\tau_1)$ in (7.33) depend on N_2, N_3, and τ_1;

$(N_1 - 1)(N_3 - 1)$ surface-like states $\psi_{n,j_1,j_3}(\mathbf{x}; \tau_1, \tau_2, \tau_3)$ in (7.19), their energy $\Lambda_{n,j_1,j_3}(\tau_2)$ in (7.20) depend on N_1, N_3, and τ_2;

$(N_1 - 1)$ edge-like states $\psi_{n,j_1}(\mathbf{x}; \tau_1, \tau_2, \tau_3)$ in (7.10), their energies $\Lambda_{n,j_1}(\tau_2, \tau_3)$ in (7.12) depend on N_1, τ_2, and τ_3;

$(N_2 - 1)$ edge-like states $\psi_{n,j_2}(\mathbf{x}; \tau_1, \tau_2, \tau_3)$ in (7.24), their energies $\Lambda_{n,j_2}(\tau_1, \tau_3)$ in (7.25) depend on N_2, τ_1, and τ_3;

$(N_3 - 1)$ edge-like states $\psi_{n,j_3}(\mathbf{x}; \tau_1, \tau_2, \tau_3)$ in (7.16), their energies $\Lambda_{n,j_3}(\tau_1, \tau_2)$ in (7.17) depend on N_3, τ_1, and τ_2;

one vertex-like state $\psi_n(\mathbf{x}; \tau_1, \tau_2, \tau_3)$ in (7.7), its energy $\Lambda_n(\tau_1, \tau_2, \tau_3)$ in (7.8) depends on τ_1, τ_2, and τ_3.

We have seen again that in the cases discussed here, the effect of the further quantum confinement in one more direction is *to have one and only one* boundary-dependent state for each sub-band of the electronic states in the quantum wire investigated in Chap. 6. The other states are size-dependent states, whose energies can be obtained from either the edge-like sub-band structure $\bar{\Lambda}_n(\bar{\mathbf{k}}; \tau_2, \tau_3)$ by (7.12), the surface-like subband structure $\bar{\Lambda}_{n,j_3}(\bar{\mathbf{k}}; \tau_2)$ by (7.20), the surface-like sub-band structure $\bar{\Lambda}_{n,j_2}(\bar{\mathbf{k}}; \tau_3)$ by (7.28), or the bulk-like sub-band structure $\bar{\Lambda}_{n,j_2,j_3}(\bar{\mathbf{k}})$ by (7.36) in the quantum wire. In general, the energy of the boundary-dependent state is always above the energy of every size-dependent state if they are obtained from the quantum confinement of one subband of one-dimensional Bloch waves.

The electronic states in a finite crystal or quantum dot can be seen as the Bloch electronic states confined to a film in one specific direction first, then confined to a wire in the second direction, and, finally, confined in the third direction. Equivalently, they can also be seen as the three-dimensional Bloch waves are confined in three directions in other different confinement orders. By considering the results obtained from all six different quantum confinement orders, we could achieve a more comprehensive understanding of the electronic states in an ideal finite crystal or quantum dot and a more specific expression for each set of electronic states and energies.

7.6 Finite Crystals or Quantum Dots with a sc, tetr, or ortho Bravais Lattice

We expect that the simplest cases, where the theory in this chapter can be applied are finite crystals or quantum dots of rectangular cuboid shape with a sc, tetr, or ortho Bravais lattice in which (5.13), (6.9), and (6.20) are true. In such a crystal, the three primitive lattice vectors $\mathbf{a}_1$, $\mathbf{a}_2$, and $\mathbf{a}_3$ are perpendicular to each other and equivalent; consequently, the three primitive lattice vectors in the k space, $\mathbf{b}_1$, $\mathbf{b}_2$, and $\mathbf{b}_3$, are also perpendicular to each other and equivalent. By considering the quantum confinement in six different orders, we can obtain the electronic states in such a finite crystal or quantum dot with a size $N_1 a_1$ in the $\mathbf{a}_1$ direction, a size $N_2 a_2$ in the $\mathbf{a}_2$ direction, and a size $N_3 a_3$ in the $\mathbf{a}_3$ direction. That is, for each bulk energy band, there are $(N_1 - 1)(N_2 - 1)(N_3 - 1)$ bulk-like states, $(N_1 - 1)(N_2 - 1) + (N_2 - 1)(N_3 - 1) + (N_3 - 1)(N_1 - 1)$ surface-like states, $(N_1 - 1) + (N_2 - 1) + (N_3 - 1)$ edge-like states, and one vertex-like state. They are as follows:
$(N_{001} - 1)(N_{1\bar{1}0} - 1)(N_{110} - 1)$ bulk-like states with energies

$$\Lambda_{n,j_1,j_2,j_3} = \varepsilon_n \left[\frac{j_1 \pi}{N_1} \mathbf{b}_1 + \frac{j_2 \pi}{N_2} \mathbf{b}_2 + \frac{j_3 \pi}{N_3} \mathbf{b}_3 \right] \tag{7.38}$$

from (7.36), (6.22), and (5.28); $(N_1 - 1)(N_2 - 1)$ surface-like states with energies

$$\Lambda_{n,j_1,j_2}(\tau_3) = \hat{\Lambda}_n \left[\frac{j_1\pi}{N_1}\mathbf{b}_1 + \frac{j_2\pi}{N_2}\mathbf{b}_2; \tau_3 \right] \tag{7.39}$$

from (7.28) and (6.13); $(N_2 - 1)(N_3 - 1)$ surface-like states with energies

$$\Lambda_{n,j_2,j_3}(\tau_1) = \hat{\Lambda}_n \left[\frac{j_2\pi}{N_2}\mathbf{b}_2 + \frac{j_3\pi}{N_3}\mathbf{b}_3; \tau_1 \right] \tag{7.40}$$

from equations similar to (7.28) or (6.13) obtained by considering different confinement orders; $(N_3 - 1)(N_1 - 1)$ surface-like states with energies

$$\Lambda_{n,j_3,j_1}(\tau_2) = \hat{\Lambda}_n \left[\frac{j_3\pi}{N_3}\mathbf{b}_3 + \frac{j_1\pi}{N_1}\mathbf{b}_1; \tau_2 \right] \tag{7.41}$$

from (7.20) and (6.25); $(N_1 - 1)$ edge-like states with energies

$$\Lambda_{n,j_1}(\tau_2, \tau_3) = \bar{\Lambda}_n \left[\frac{j_1\pi}{N_1}\mathbf{b}_1; \tau_2, \tau_3 \right] \tag{7.42}$$

from (7.12); $(N_2 - 1)$ edge-like states with energies

$$\Lambda_{n,j_2}(\tau_3, \tau_1) = \bar{\Lambda}_n \left[\frac{j_2\pi}{N_2}\mathbf{b}_2; \tau_3, \tau_1 \right] \tag{7.43}$$

$(N_3 - 1)$ edge-like states with energies

$$\Lambda_{n,j_3}(\tau_1, \tau_2) = \bar{\Lambda}_n \left[\frac{j_3\pi}{N_3}\mathbf{b}_3; \tau_1, \tau_2 \right] \tag{7.44}$$

from equations similar to (7.12) obtained by considering different confinement orders; and one vertex-like state with energy

$$\Lambda_n(\tau_1, \tau_2, \tau_3) = \lambda_n(\tau_1, \tau_2, \tau_3) \tag{7.45}$$

from (7.8). Here, $j_1 = 1, 2, \ldots, N_1 - 1$, $j_2 = 1, 2, \ldots, N_2 - 1$, and $j_3 = 1, 2, \ldots, N_3 - 1$. τ_1, τ_2, and τ_3 define the boundary surface locations of the finite crystal or quantum dot in the $\mathbf{a}_1$, $\mathbf{a}_2$, and $\mathbf{a}_3$ directions. $\hat{\Lambda}_n[\hat{\mathbf{k}}; \tau_l]$ is the surface-like band structure of a quantum film with film plane oriented in the $\mathbf{a}_l$ direction with a wave vector $\hat{\mathbf{k}}$ in the film plane. $\bar{\Lambda}_n[\bar{\mathbf{k}}; \tau_l, \tau_m]$ is the edge-like band structure of a rectangular quantum wire with the wire faces oriented in the $\mathbf{a}_l$ or the $\mathbf{a}_m$ direction with a wave vector $\bar{\mathbf{k}}$ in the wire direction.

Furthermore, from (7.13), (7.21), (7.29), and (7.37), and similar equations obtained from other confinement orders, we can obtain that in general,

$$\Lambda_n(\tau_1, \tau_2, \tau_3) > \Lambda_{n,j_l}(\tau_m, \tau_n) > \Lambda_{n,j_l,j_m}(\tau_n) > \Lambda_{n,j_l,j_m,j_n}; \tag{7.46}$$

here, each one of $l, m,$ and n can be any one of $1, 2,$ and 3, but no two of them are equal ($l, m,$ and n are combinations of $1, 2,$ and 3). Relation (7.46) indicates that in such an ideal finite crystal or quantum dot, for the electronic states with the same energy band index n, the vertex-like state has the highest energy, above the energy of every edge-like state. An edge-like state has an energy above the energy of every relevant surface-like state. A surface-like state has an energy above the energy of every relevant bulk-like state.

Probably, the practically more interesting cases are finite crystals or quantum dots of crystals with an fcc or bcc Bravais lattice in which (5.20), (6.9), and (6.20) are true. The choosing of the primitive lattice vectors in those crystals depends on the film direction. Based on the results obtained in Sects. 5.1–5.5, 6.1–6.3, and 7.1–7.5, we can understand the consequences of the three-dimensional Bloch waves of such a crystal being confined to a finite crystal or quantum dot in a specific order. Similar to what was done in Chap. 6, by combining the results obtained from different confinement orders, a more comprehensive understanding of the electronic states in such a finite crystal and quantum dot can be achieved.

7.7 fcc Finite Crystals with (001), (110), and $(1\bar{1}0)$ Surfaces

For an fcc finite crystal with (001), (110), and $(1\bar{1}0)$ surfaces and having a rectangular cuboid size $N_{001}a \times N_{110}a/\sqrt{2} \times N_{1\bar{1}0}a/\sqrt{2}$, for each bulk energy band n, there are $2N_{001}N_{110}N_{1\bar{1}0}$ electronic states. By combining the results of the further quantum confinement in the $[1\bar{1}0]$ direction of one-dimensional Bloch waves in Sect. 6.5 and the results of a further quantum confinement in the [001] direction of one-dimensional Bloch waves in Sect. 6.6, and similar results of the further quantum confinement in the [110] direction of one-dimensional Bloch waves in a quantum wire with $(1\bar{1}0)$ and (001) faces, the electronic states in such a finite crystal or quantum dot can be obtained as follows.

For each bulk energy band n, there are $2(N_{001} - 1)(N_{1\bar{1}0} - 1)(N_{110} - 1) + (N_{001} - 1)(N_{1\bar{1}0} - 1) + (N_{110} - 1)(N_{001} - 1) + (N_{1\bar{1}0} - 1)(N_{110} - 1) + (N_{1\bar{1}0} - 1) + (N_{110} - 1) + (N_{001} - 1) + 1$ bulk-like states in the finite crystal or quantum dot. They are as follows:

$(N_{001} - 1)(N_{1\bar{1}0} - 1)(N_{110} - 1)$ states with energies

$$\Lambda^{bk,a}_{n,j_{001},j_{1\bar{1}0},j_{110}} = \varepsilon_n \left[\frac{j_{001}\pi}{N_{001}a}(0, 0, 1) + \frac{j_{1\bar{1}0}\pi}{N_{1\bar{1}0}a}(1, -1, 0) + \frac{j_{110}\pi}{N_{110}a}(1, 1, 0) \right]; \tag{7.47}$$

$(N_{001} - 1)(N_{1\bar{1}0} - 1)(N_{110} - 1)$ states with energies

$$\Lambda^{bk,c}_{n,j_{001},j_{1\bar{1}0},j_{110}} = \varepsilon_n \left[\frac{j_{001}\pi}{N_{001}a}(0,0,1) + \frac{j_{1\bar{1}0}\pi}{N_{1\bar{1}0}a}(1,-1,0) + \frac{j_{110}\pi}{N_{110}a}(1,1,0) + \frac{2\pi}{a}(1,1,0) \right]; \quad (7.48)$$

$(N_{001} - 1)(N_{1\bar{1}0} - 1)$ states with energies

$$\Lambda^{bk,b_1}_{n,j_{001},j_{1\bar{1}0}} = \varepsilon_n \left[\frac{j_{001}\pi}{N_{001}a}(0,0,1) + \frac{j_{1\bar{1}0}\pi}{N_{1\bar{1}0}a}(1,-1,0) + \frac{\pi}{a}(1,1,0) \right]; \quad (7.49)$$

$(N_{110} - 1)(N_{001} - 1)$ states with energies

$$\Lambda^{bk,b_2}_{n,j_{110},j_{001}} = \varepsilon_n \left[\frac{j_{110}\pi}{N_{110}a}(1,1,0) + \frac{j_{001}\pi}{N_{001}a}(0,0,1) + \frac{\pi}{a}(1,-1,0) \right]; \quad (7.50)$$

$(N_{1\bar{1}0} - 1)(N_{110} - 1)$ states with energies

$$\Lambda^{bk,b_3}_{n,j_{1\bar{1}0},j_{110}} = \varepsilon_n \left[\frac{j_{1\bar{1}0}\pi}{N_{1\bar{1}0}a}(1,-1,0) + \frac{j_{110}\pi}{N_{110}a}(1,1,0) + \frac{\pi}{a}(0,0,1) \right]; \quad (7.51)$$

$(N_{001} - 1)$ states with energies

$$\Lambda^{bk,d_1}_{n,j_{001}} = \varepsilon_n \left[\frac{j_{001}\pi}{N_{001}a}(0,0,1) + \frac{\pi}{a}(1,-1,0) + \frac{\pi}{a}(1,1,0) \right]; \quad (7.52)$$

$(N_{110} - 1)$ states with energies

$$\Lambda^{bk,d_2}_{n,j_{110}} = \varepsilon_n \left[\frac{j_{110}\pi}{N_{110}a}(1,1,0) + \frac{\pi}{a}(0,0,1) + \frac{\pi}{a}(1,-1,0) \right]; \quad (7.53)$$

$(N_{1\bar{1}0} - 1)$ states with energies

$$\Lambda^{bk,d_3}_{n,j_{1\bar{1}0}} = \varepsilon_n \left[\frac{j_{1\bar{1}0}\pi}{N_{1\bar{1}0}a}(1,-1,0) + \frac{\pi}{a}(1,1,0) + \frac{\pi}{a}(0,0,1) \right]; \quad (7.54)$$

one state with energy

$$\Lambda^{bk,e}_{n} = \varepsilon_n \left[\frac{\pi}{a}(1,-1,0) + \frac{\pi}{a}(1,1,0) + \frac{\pi}{a}(0,0,1) \right]. \quad (7.55)$$

Here, $j_{001} = 1, 2, \ldots, N_{001}-1$, $j_{1\bar{1}0} = 1, 2, \ldots, N_{1\bar{1}0}-1$, $j_{110} = 1, 2, \ldots, N_{110}-1$, and $\varepsilon(k_x, k_y, k_z)$ is the bulk energy band structure in the Cartesian system. Equation (7.47) comes from the size-dependent states of the further quantum confinement of the bulk-like sub-bands (6.46), (6.56), or a similar equation for a quantum wire in the [110] direction. Equation (7.48) comes from the size-dependent states of the further quantum confinement of the bulk-like sub-bands (6.47), (6.57), or a similar equation for a quantum wire in the [110] direction. Equations (7.49)–(7.51) come from the size-dependent states of further quantum confinement of the bulk-like sub-bands (6.48) and (6.49), and (6.58) and (6.59) or two similar equations for a quantum wire in the [110] direction. Equations (7.52)–(7.54) come from the size-dependent states of the further quantum confinement of the bulk-like sub-bands (6.50) and (6.60), and a similar equation for a quantum wire in the [110] direction. Equation (7.55) comes from the boundary-dependent state of the further quantum confinement of the bulk-like sub-bands (6.50), (6.60), and a similar equation for a quantum wire in the [110] direction. Similarly to that $\bar{\Lambda}_n^{bk,d}(\bar{\mathbf{k}})$ in (6.50), (6.60) or in a similar equation for a quantum wire in the [110] direction is a bulk-like sub-band in the quantum wire, $\Lambda_n^{bk,e}$ in (7.55) is a bulk-like state in the finite crystal or quantum dot.

The energy of each such bulk-like state can be directly obtained from the energy band structure $\varepsilon_n(\mathbf{k})$ of the corresponding bulk crystal.

For each bulk energy band n, there are $(N_{001}-1)(N_{1\bar{1}0}-1)+(N_{110}-1)(N_{001}-1)+(N_{1\bar{1}0}-1)(N_{110}-1)$ surface-like states in the crystal. They are as follows:
$(N_{001}-1)(N_{1\bar{1}0}-1)$ states with energies

$$\Lambda_{n,j_{001},j_{1\bar{1}0}}^{sf,a_1}(\tau_{110}) = \hat{\Lambda}_n\left[\frac{j_{001}\pi}{N_{001}a}(0,0,1)+\frac{j_{1\bar{1}0}\pi}{N_{1\bar{1}0}a}(1,-1,0);\tau_{110}\right]; \tag{7.56}$$

$(N_{110}-1)(N_{001}-1)$ states with energies

$$\Lambda_{n,j_{110},j_{001}}^{sf,a_2}(\tau_{1\bar{1}0}) = \hat{\Lambda}_n\left[\frac{j_{110}\pi}{N_{110}a}(1,1,0)+\frac{j_{001}\pi}{N_{001}a}(0,0,1);\tau_{1\bar{1}0}\right]; \tag{7.57}$$

$(N_{1\bar{1}0}-1)(N_{110}-1)$ states with energies

$$\Lambda_{n,j_{1\bar{1}0},j_{110}}^{sf,a_3}(\tau_{001}) = \hat{\Lambda}_n\left[\frac{j_{1\bar{1}0}\pi}{N_{1\bar{1}0}a}(1,-1,0)+\frac{j_{110}\pi}{N_{110}a}(1,1,0);\tau_{001}\right]. \tag{7.58}$$

Here, τ_{110}, $\tau_{1\bar{1}0}$, or τ_{001} define the boundary surface locations of the finite crystal or quantum dot in the [110], [1$\bar{1}$0], or [001] direction, $\hat{\Lambda}_n[\hat{\mathbf{k}};\tau_l]$ is the surface-like band structure of a quantum film with the film plane oriented in the [l] direction with a wave vector $\hat{\mathbf{k}}$ in the film plane. l can be either one of 001, 110, or 1$\bar{1}$0. Equations (7.56)–(7.58) come from the size-dependent states of the further quantum confinement of the surface-like sub-bands (6.51) and (6.52), or (6.61) and (6.62), and/or two similar equations for a quantum wire in the [110] direction.

For each bulk energy band n, there are $(N_{001}-1)+(N_{110}-1)+(N_{1\bar{1}0}-1)$ edge-like states in the crystal. They are as follows:
$(N_{001}-1)$ states with energies

$$\Lambda_{n,j_{001}}^{eg,a_1}(\tau_{1\bar{1}0},\tau_{110}) = \bar{\Lambda}_n\left[\frac{j_{001}\pi}{N_{001}a}(0,0,1);\tau_{1\bar{1}0},\tau_{110}\right]; \tag{7.59}$$

$(N_{110} - 1)$ states with energies

$$\Lambda^{eg,a_2}_{n,j_{110}}(\tau_{1\bar{1}0}, \tau_{001}) = \bar{\Lambda}_n \left[\frac{j_{110}\pi}{N_{110}a}(1, 1, 0); \tau_{1\bar{1}0}, \tau_{001} \right]; \tag{7.60}$$

$(N_{1\bar{1}0} - 1)$ states with energies

$$\Lambda^{eg,a_3}_{n,j_{1\bar{1}0}}(\tau_{001}, \tau_{110}) = \bar{\Lambda}_n \left[\frac{j_{1\bar{1}0}\pi}{N_{1\bar{1}0}a}(1, -1, 0); \tau_{001}, \tau_{110} \right]. \tag{7.61}$$

Here, $\bar{\Lambda}_n[\bar{\mathbf{k}}; \tau_l, \tau_m]$ is the edge-like band structure of a rectangular quantum wire with the wire faces oriented in the $[l]$ or the $[m]$ direction with a wave vector $\bar{\mathbf{k}}$ in the wire direction. l and m can be two of 001, 110, and 1$\bar{1}$0.

For each bulk energy band n, there is one vertex-like state in the finite crystal with energy

$$\Lambda^{vt}_n(\tau_{001}, \tau_{1\bar{1}0}, \tau_{110}) = \lambda_n(\tau_{001}, \tau_{1\bar{1}0}, \tau_{110}). \tag{7.62}$$

Equations (7.59)–(7.61) come from the size-dependent states of the further quantum confinement of the edge-like subband (6.38) or (6.45), (6.63), and a similar equation for a quantum wire in the [110] direction. Equation (7.62) comes from the boundary-dependent state of the further quantum confinement of the edge-like subband (6.38) or (6.45), or (6.63), or a similar equation for a quantum wire in the [110] direction.

No one VBM state in a cubic semiconductor can have three (001), (110), and (1$\bar{1}$0) nodal planes simultaneously. Consequently, there is no electronic state in such a quantum dot whose energy is the energy of VBM and does not change as the dot size changes, as observed in the numerical calculations of Franceschetti and Zunger [1] on free-standing GaAs quantum dots, as shown in Fig. 5.4c.

7.8 bcc Finite Crystals with (100), (010), and (001) Surfaces

For a bcc finite crystal or quantum dot with (100), (010), and (001) surfaces and having a rectangular size $N_{100}a \times N_{010}a \times N_{001}a$, for each bulk energy band n, there are $2N_{100}N_{010}N_{001}$ electronic states. They can be obtained by combining the results of a further quantum confinement in the [100] direction of one-dimensional Bloch waves in the quantum wire in Sect. 6.7 and the results of a further quantum confinement of one-dimensional Bloch waves in two quantum wires in the [010] or [001] direction.

For each bulk energy band n, there are $2(N_{100}-1)(N_{010}-1)(N_{001}-1)+(N_{001}-1)(N_{010}-1)+(N_{100}-1)(N_{001}-1)+(N_{010}-1)(N_{100}-1)+(N_{100}-1)+(N_{010}-1)+(N_{001}-1)+1$ bulk-like states in the finite crystal or quantum dot. They are as follows:

$(N_{100}-1)(N_{010}-1)(N_{001}-1)$ states with energies

$$\Lambda^{bk,a}_{n,j_{100},j_{010},j_{001}} = \varepsilon_n \left[\frac{j_{100}\pi}{N_{100}a}(1,0,0) + \frac{j_{010}\pi}{N_{010}a}(0,1,0) + \frac{j_{001}\pi}{N_{001}a}(0,0,1) \right]; \quad (7.63)$$

$(N_{100}-1)(N_{010}-1)(N_{001}-1)$ states with energies

$$\begin{aligned}\Lambda^{bk,c}_{n,j_{100},j_{010},j_{001}} =& \varepsilon_n \left[\frac{j_{100}\pi}{N_{100}a}(1,0,0) + \frac{j_{010}\pi}{N_{010}a}(0,1,0) \right. \\ &+ \left. \frac{j_{001}\pi}{N_{001}a}(0,0,1) + \frac{2\pi}{a}(1,0,0) \right]; \end{aligned} \quad (7.64)$$

$(N_{010}-1)(N_{001}-1)$ states with energies

$$\Lambda^{bk,b_1}_{n,j_{010},j_{001}} = \varepsilon_n \left[\frac{j_{010}\pi}{N_{010}a}(0,1,0) + \frac{j_{001}\pi}{N_{001}a}(0,0,1) + \frac{\pi}{a}(1,0,0) \right]; \quad (7.65)$$

$(N_{001}-1)(N_{100}-1)$ states with energies

$$\Lambda^{bk,b_2}_{n,j_{001},j_{100}} = \varepsilon_n \left[\frac{j_{001}\pi}{N_{001}a}(0,0,1) + \frac{j_{100}\pi}{N_{100}a}(1,0,0) + \frac{\pi}{a}(0,1,0) \right]; \quad (7.66)$$

$(N_{100}-1)(N_{010}-1)$ states with energies

$$\Lambda^{bk,b_3}_{n,j_{100},j_{010}} = \varepsilon_n \left[\frac{j_{100}\pi}{N_{100}a}(1,0,0) + \frac{j_{010}\pi}{N_{010}a}(0,1,0) + \frac{\pi}{a}(0,0,1) \right]; \quad (7.67)$$

$(N_{100}-1)$ states with energies

$$\Lambda^{bk,d_1}_{n,j_{100}} = \varepsilon_n \left[\frac{j_{100}\pi}{N_{100}a}(1,0,0) + \frac{\pi}{a}(0,1,0) + \frac{\pi}{a}(0,0,1) \right]; \quad (7.68)$$

$(N_{010}-1)$ states with energies

$$\Lambda^{bk,d_2}_{n,j_{010}} = \varepsilon_n \left[\frac{j_{010}\pi}{N_{010}a}(0,1,0) + \frac{\pi}{a}(0,0,1) + \frac{\pi}{a}(1,0,0) \right]; \quad (7.69)$$

$(N_{001}-1)$ states with energies

$$\Lambda^{bk,d_3}_{n,j_{001}} = \varepsilon_n \left[\frac{j_{001}\pi}{N_{001}a}(0,0,1) + \frac{\pi}{a}(1,0,0) + \frac{\pi}{a}(0,1,0) \right]; \quad (7.70)$$

one state with energy

$$\Lambda_n^{bk,e} = \varepsilon_n \left[\frac{\pi}{a}(1,0,0) + \frac{\pi}{a}(0,1,0) + \frac{\pi}{a}(0,0,1) \right]. \tag{7.71}$$

Here, $j_{100} = 1, 2, \ldots, N_{100}-1$, $j_{010} = 1, 2, \ldots, N_{010}-1$, $j_{001} = 1, 2, \ldots, N_{001}-1$, and $\varepsilon(k_x, k_y, k_z)$ is the bulk energy band structure in the Cartesian system. Equation (7.63) comes from the size-dependent states of the further quantum confinement of the bulk-like sub-bands (6.64) or two similar equations for quantum wires in the [010] or [001] directions. Equation (7.64) comes from the size-dependent states of the further quantum confinement of the bulk-like sub-bands (6.65) or two similar equations for quantum wires in the [010] or [001] directions. Equations (7.65)–(7.67) come from the size-dependent states of the further quantum confinement of the bulk-like sub-bands (6.66), (6.67), and/or four similar equations for quantum wires in the [010] or [001] directions. Equations (7.68)–(7.70) come from the size-dependent states of the further quantum confinement of the bulk-like subband (6.68) and two similar equations for quantum wires in the [010] or [001] directions. Equation (7.71) come from the boundary-dependent state of the further quantum confinement of the bulk-like sub-band (6.68) and two similar equations for quantum wires in the [010] or the [001] directions. Similar to the bulk-like state $\Lambda_n^{bk,e}$ in (7.55), $\Lambda_n^{bk,e}$ in (7.71) is also a bulk-like state. The energies of all these bulk-like states can be directly obtained from the energy band structure $\varepsilon_n(\mathbf{k})$ of the corresponding bulk crystal.

For each bulk energy band n, there are $(N_{010} - 1)(N_{001} - 1) + (N_{001} - 1)(N_{100} - 1) + (N_{100} - 1)(N_{010} - 1)$ surface-like states in the finite crystal or quantum dot. They are as follows:

$(N_{010} - 1)(N_{001} - 1)$ states with energies

$$\Lambda_{n,j_{010},j_{001}}^{sf,a_1}(\tau_{100}) = \hat{\Lambda}_n \left[\frac{j_{010}\pi}{N_{010}a}(0,1,0) + \frac{j_{001}\pi}{N_{001}a}(0,0,1); \tau_{100} \right]; \tag{7.72}$$

$(N_{001} - 1)(N_{100} - 1)$ states with energies

$$\Lambda_{n,j_{001},j_{100}}^{sf,a_2}(\tau_{010}) = \hat{\Lambda}_n \left[\frac{j_{001}\pi}{N_{001}a}(0,0,1) + \frac{j_{100}\pi}{N_{100}a}(1,0,0); \tau_{010} \right]; \tag{7.73}$$

$(N_{100} - 1)(N_{010} - 1)$ states with energies

$$\Lambda_{n,j_{100},j_{010}}^{sf,a_3}(\tau_{001}) = \hat{\Lambda}_n \left[\frac{j_{100}\pi}{N_{100}a}(1,0,0) + \frac{j_{010}\pi}{N_{010}a}(0,1,0); \tau_{001} \right]. \tag{7.74}$$

Here, τ_{100}, τ_{010}, or τ_{001} defines the boundary surface locations of the finite crystal or quantum dot in the [100], [010], or [001] direction, respectively; $\hat{\Lambda}_n[\hat{\mathbf{k}}; \tau_l]$ is the surface-like band structure of a quantum film with the film plane oriented in the $[l]$ direction with a wave vector $\hat{\mathbf{k}}$ in the film plane. l can be either one of 100, 010 or 001. Equations (7.72)–(7.74) come from the size-dependent states of the further quantum confinement of the surface-like sub-bands (6.69), (6.70), and/or four similar equations for the quantum wires in the [010] or the [001] direction.

For each bulk energy band n, there are $(N_{100} - 1) + (N_{010} - 1) + (N_{001} - 1)$ edge-like states in the finite crystal or quantum dot. They are as follows:
$(N_{100} - 1)$ edge-like states with energies

$$\Lambda^{eg,a_1}_{n,j_{100}}(\tau_{010}, \tau_{001}) = \bar{\Lambda}_n \left[\frac{j_{100}\pi}{N_{100}a}(1, 0, 0); \tau_{010}, \tau_{001} \right]; \tag{7.75}$$

$(N_{010} - 1)$ edge-like states with energies

$$\Lambda^{eg,a_2}_{n,j_{010}}(\tau_{001}, \tau_{100}) = \bar{\Lambda}_n \left[\frac{j_{010}\pi}{N_{010}a}(0, 1, 0); \tau_{001}, \tau_{100} \right]; \tag{7.76}$$

$(N_{001} - 1)$ edge-like states with energies

$$\Lambda^{eg,a_3}_{n,j_{001}}(\tau_{100}, \tau_{010}) = \bar{\Lambda}_n \left[\frac{j_{001}\pi}{N_{001}a}(0, 0, 1); \tau_{100}, \tau_{010} \right]. \tag{7.77}$$

Here, $\bar{\Lambda}_n[\bar{\mathbf{k}}; \tau_l, \tau_m]$ is the edge-like band structure of a rectangular quantum wire with the wire faces oriented in the $[l]$ or the $[m]$ direction with a wave vector $\bar{\mathbf{k}}$ in the wire direction. l and m can be two of 100, 010, and 001.

For each bulk energy band n, there is one vertex-like state in the finite crystal or quantum dot with energy

$$\Lambda^{vt}_n(\tau_{100}, \tau_{010}, \tau_{001}) = \lambda_n(\tau_{100}, \tau_{010}, \tau_{001}). \tag{7.78}$$

Equations (7.75)–(7.77) come from the size-dependent states of the further quantum confinement of the edge-like subband (6.71) and two similar equations for the quantum wires in the [001] or the [010] direction. Equation (7.78) comes from the boundary-dependent state of the further quantum confinement of the edge-like subband (6.71) or two similar equations for the quantum wires in the [010] or the [001] direction.

7.9 Summary and Discussions

We have seen that in an ideal finite crystal or quantum dot of rectangular cuboid shape discussed in Sects. 7.6–7.8, there are four different types of electronic states: bulk-like states, surface-like states, edge-like states, and vertex-like states. The crystal structure has an effect on that how the numbers of each type of electronic states in a finite crystal or quantum dot depend on the sizes in the three dimensions: The simplest cases discussed in Sect. 7.6 are somewhat different from the cases of crystals with an fcc or a bcc Bravais lattice discussed in Sects. 7.7 and 7.8.

Since the results in Sects. 7.7 and 7.8 were also essentially obtained from an understanding of the further quantum confinement of one-dimensional Bloch waves discussed in Sects. 7.2–7.5, there are similar relationships among the four different types of electronic states. For example, for an ideal fcc finite crystal or quantum dot with $(1\bar{1}0)$, (110), and (001) surfaces, we should have

$$\Lambda_n^{vt}(\tau_{1\bar{1}0}, \tau_{110}, \tau_{001}) > \Lambda_{n,j_{001}}^{eg,a_1}(\tau_{1\bar{1}0}, \tau_{110}), \tag{7.79}$$

$$\Lambda_n^{vt}(\tau_{1\bar{1}0}, \tau_{110}, \tau_{001}) > \Lambda_{n,j_{110}}^{eg,a_2}(\tau_{1\bar{1}0}, \tau_{001}), \tag{7.80}$$

and

$$\Lambda_n^{vt}(\tau_{1\bar{1}0}, \tau_{110}, \tau_{001}) > \Lambda_{n,j_{1\bar{1}0}}^{eg,a_3}(\tau_{110}, \tau_{001}) \tag{7.81}$$

between the energy of a vertex-like state in (7.62) and the energies of edge-like states in (7.59), (7.60), and (7.61), and

$$\Lambda_n^{vt}(\tau_{1\bar{1}0}, \tau_{110}, \tau_{001}) > \Lambda_n^{bk,e} \tag{7.82}$$

between the energy of a vertex-like state in (7.62) and the energy of the bulk-like state in (7.55).

We have

$$\Lambda_{n,j_{001}}^{eg,a_1}(\tau_{1\bar{1}0}, \tau_{110}) > \Lambda_{n,j_{001},j_{1\bar{1}0}}^{sf,a_1}(\tau_{110}) \tag{7.83}$$

and

$$\Lambda_{n,j_{001}}^{eg,a_1}(\tau_{1\bar{1}0}, \tau_{110}) > \Lambda_{n,j_{110},j_{001}}^{sf,a_2}(\tau_{1\bar{1}0}) \tag{7.84}$$

between the energy of an edge-like state in (7.59) and the energies of relevant surface-like states in (7.56) and (7.57), and

$$\Lambda_{n,j_{001}}^{eg,a_1}(\tau_{1\bar{1}0}, \tau_{110}) > \Lambda_{n,j_{001}}^{bk,d_1} \tag{7.85}$$

between the energy of an edge-like state in (7.59) and the energy of a relevant bulk-like state in (7.52).

We have

$$\Lambda_{n,j_{110}}^{eg,a_2}(\tau_{1\bar{1}0}, \tau_{001}) > \Lambda_{n,j_{110},j_{001}}^{sf,a_2}(\tau_{1\bar{1}0}) \tag{7.86}$$

and

$$\Lambda_{n,j_{110}}^{eg,a_2}(\tau_{1\bar{1}0}, \tau_{001}) > \Lambda_{n,j_{1\bar{1}0},j_{110}}^{sf,a_3}(\tau_{001}) \tag{7.87}$$

between the energy of an edge-like state in (7.60) and the energies of relevant surface-like states in (7.57) and (7.58), and

$$\Lambda^{eg,a_2}_{n,j_{110}}(\tau_{1\bar{1}0}, \tau_{001}) > \Lambda^{bk,d_2}_{n,j_{110}} \tag{7.88}$$

between the energy of an edge-like state in (7.60) and the energy of a relevant bulk-like state in (7.53).

We have

$$\Lambda^{eg,a_3}_{n,j_{1\bar{1}0}}(\tau_{001}, \tau_{110}) > \Lambda^{sf,a_1}_{n,j_{001},j_{1\bar{1}0}}(\tau_{110}) \tag{7.89}$$

and

$$\Lambda^{eg,a_3}_{n,j_{1\bar{1}0}}(\tau_{001}, \tau_{110}) > \Lambda^{sf,a_3}_{n,j_{1\bar{1}0},j_{110}}(\tau_{001}) \tag{7.90}$$

between the energy of an edge-like state in (7.61) and the energies of relevant surface-like states in (7.56) and (7.58), and

$$\Lambda^{eg,a_3}_{n,j_{1\bar{1}0}}(\tau_{001}, \tau_{110}) > \Lambda^{bk,d_3}_{n,j_{1\bar{1}0}} \tag{7.91}$$

between the energy of an edge-like state in (7.61) and the energy of the relevant bulk-like state in (7.54).

We have

$$\Lambda^{sf,a_1}_{n,j_{001},j_{1\bar{1}0}}(\tau_{110}) > \Lambda^{bk,a}_{n,j_{001},j_{1\bar{1}0},j_{110}} \tag{7.92}$$

and

$$\Lambda^{sf,a_1}_{n,j_{001},j_{1\bar{1}0}}(\tau_{110}) > \Lambda^{bk,c}_{n,j_{001},j_{1\bar{1}0},j_{110}} \tag{7.93}$$

between the energy of a surface-like state in (7.56) and the energies of relevant bulk-like states in (7.47) and in (7.48), and

$$\Lambda^{sf,a_1}_{n,j_{001},j_{1\bar{1}0}}(\tau_{110}) > \Lambda^{bk,b_1}_{n,j_{001},j_{1\bar{1}0}} \tag{7.94}$$

between the energy of a surface-like state in (7.56) and the energy of the relevant bulk-like state in (7.49).

We have

$$\Lambda^{sf,a_2}_{n,j_{110},j_{001}}(\tau_{1\bar{1}0}) > \Lambda^{bk,a}_{n,j_{001},j_{1\bar{1}0},j_{110}} \tag{7.95}$$

and

$$\Lambda^{sf,a_2}_{n,j_{110},j_{001}}(\tau_{1\bar{1}0}) > \Lambda^{bk,c}_{n,j_{001},j_{1\bar{1}0},j_{110}} \tag{7.96}$$

between the energy of a surface-like state in (7.57) and the energies of relevant bulk-like states in (7.47) and in (7.48), and

$$\Lambda^{sf,a_2}_{n,j_{110},j_{001}}(\tau_{1\bar{1}0}) > \Lambda^{bk,b_2}_{n,j_{110},j_{001}} \tag{7.97}$$

between the energy of a surface-like state in (7.57) and the energy of the relevant bulk-like state in (7.50).

We have

$$\Lambda^{sf,a_3}_{n,j_{1\bar{1}0},j_{110}}(\tau_{001}) > \Lambda^{bk,a}_{n,j_{001},j_{1\bar{1}0},j_{110}} \tag{7.98}$$

and

$$\Lambda^{sf,a_3}_{n,j_{1\bar{1}0},j_{110}}(\tau_{001}) > \Lambda^{bk,c}_{n,j_{001},j_{1\bar{1}0},j_{110}} \tag{7.99}$$

between the energy of a surface-like state in (7.58) and the energies of relevant bulk-like states in (7.47) and in (7.48), and

$$\Lambda^{sf,a_3}_{n,j_{1\bar{1}0},j_{110}}(\tau_{001}) > \Lambda^{bk,b_3}_{n,j_{1\bar{1}0},j_{110}} \tag{7.100}$$

between the energy of a surface-like state in (7.58) and the energy of the relevant bulk-like state in (7.51). These relationships can be obtained just as the relationships in Sect. 6.8 were obtained, using reasoning based on the relationships (7.13), (7.21), (7.29), or (7.37) obtained in Sects. 7.2–7.5.

Corresponding relationships between the four different types of electronic states in an ideal bcc finite crystal or quantum dot with three (100), (010), and (001) surfaces can be similarly obtained.

As a surface-like electronic state is better understood as an electronic state whose properties and energy are determined by the surface location and an edge-like electronic state is better understood as an electronic state whose properties and energy are determined by the edge location, a vertex-like electronic state is better understood as an electronic state whose properties and energy are determined by the vertex location, rather than a state located near a specific vertex. To better understand the properties of vertex-like states, it is necessary to have a better understanding of the properties of solutions of (5.1), in particular, the non-Bloch state solutions in different energy ranges.

Similar to surface-like states and edge-like states, the physics origin of a vertex-like state is also related to a bulk energy band. The energy of a vertex-like state is above the energies of relevant edge-like states and surface-like states; thus, the vertexes of an ideal alkali metal finite crystal or quantum dot of rectangular cuboid shape might be even more positively charged than the edges and the surfaces.

Only when a Bloch function has three different nodal planes that are the surfaces of the quantum dot might the vertex-like state be a Bloch state. It seems that such kind of cases rarely happens in most finite crystals or quantum dots of current interest.

For the electronic states in an ideal finite crystal or quantum dot of rectangular cuboid shape with the same energy band index n, summarizing (7.46), (7.79)–(7.100), and similar equations for a bcc finite crystal or quantum dot discussed in Sect. 7.8, the following general relations exist:

The energy of the vertex-like state

$>$ The energy of every edge-like state

$>$ The energy of every relevant surface-like state

$>$ The energy of every relevant bulk-like state.

Therefore, we have seen that in many simple and interesting cases, the properties of electronic states in an ideal finite crystal or quantum dot of rectangular cuboid shape such as shown in Fig. 1.1 can be understood, how the energies of those electronic states depend on the size and shape can be analytically predicted, and the energies of many electronic states can be obtained from the energy band structures of the bulk crystal. Again, the major obstacle due to the lack of translational invariance can be essentially circumvented.

It was mentioned in Chap. 1 that in a semiconductor low-dimensional system, its optical properties change dramatically as its size decreases: The measured optical band gap increases as the system size decreases; an indirect semiconductor such as Si may become luminescent, and a direct semiconductor such as GaAs may develop into an indirect one. These facts can be well understood if the optical transitions between the bulk-like stationary Bloch states contribute more significantly to the optical properties of a semiconductor low-dimensional system.[9] Furthermore, it can also be well understood that why in comparison with the experimental results, the theoretical predictions from various forms of effective mass approximation overestimate the optical band gap increase in general as the system size decreases: The relevant transition states are *stationary Bloch states* with their energies determined by the equations similar to such as Eqs. (7.47)–(7.55) rather than stationary plane wave states with their energies given by equations similar to Eq. (1.13) with the corresponding effective masses.[10]

The results obtained here provided a more concrete and comprehensive understanding of the boundary effects than the previous discussions in [2].

---

[9] In many experimental investigations, the surface-like states, edge-like states, vertex-like states, etc., of the samples are likely to be passivated.

[10] A physical picture based on the effective mass approximation such as shown in Fig. 1.2 violates a fundamental in quantum mechanics that the Hamiltonian operator is bounded from below but not above.

References

1. A. Franceschetti, A. Zunger, Appl. Phys. Lett. **68**, 3455 (1996)
2. M. Born, K. Huang, *Dynamical Theory of Crystal Lattices* (Clarendon Press, Oxford, 1954) (Appendix IV)

Part IV
Epilogue

Chapter 8
Concluding Remarks

We have presented a single-electron non-spin analytical theory on the electronic states in some simple ideal low-dimensional systems and finite crystals, based on the mathematical theory of periodic differential equations and physical intuition. By ideal, it is assumed that (i) the potential $v(\mathbf{x})$ inside the low-dimensional system or the finite crystal is the same as in a crystal with translational invariance and (ii) the electronic states are completely confined to the limited size of the low-dimensional system or the finite crystal.

8.1 Summary and Brief Discussions

The most essential results obtained in this book can be summarized as follows:

1. In *a unified theoretical frame*, it is understood that due to the existence of boundaries and the finite size in one, two, or three directions, the electronic states in some simple low-dimensional systems or finite crystals are not progressive Bloch waves as indicated by the Bloch theorem. Instead, they are either (i) stationary Bloch states in the one, two, or three directions due to the finite size or (ii) another type(s) of electronic states closely related to the very existence of the boundary. Therefore, the two fundamental difficulties mentioned in Sect. 1.3 are overcome in the unified theoretical frame for those simple ideal low-dimensional systems and finite crystals.
2. Due to the very existence of the boundary-dependent electronic states, the properties of electronic states in simple low-dimensional systems and finite crystals may be substantially different from that in crystals with translational invariance as understood in conventional solid-state physics; they may also be significantly different from what is traditionally believed in the solid-state physics community

S.Y. Ren, *Electronic States in Crystals of Finite Size*, Springer Tracts in Modern Physics 270, DOI 10.1007/978-981-10-4718-3_8

regarding the properties of electronic states of low-dimensional systems and finite crystals.

These results were obtained by trying to understand the quantum confinement effects of Bloch waves.

There are similarities and differences between the quantum confinement of Bloch waves and the well-known quantum confinement of plane waves: The most significant feature in the quantum confinement of Bloch waves is the very existence of the boundary(τ in this book)-dependent states. There are also similarities and differences between the quantum confinement effects of Bloch waves in three-dimensional space and in one-dimensional space: Only in one-dimensional space, a boundary-dependent state from the quantum confinement of Bloch waves is always either inside a band gap or at a band-edge.

It is well known that in the quantum confinement of plane waves, all permitted states are stationary waves. The differences between the quantum confinement effects of plane waves and Bloch waves come from a simple origin. In a case of plane waves, the potential is everywhere equal – the potential has a *continuous* translational invariance: there is no minimum translation unit of the potential. On the contrary, in a case of Bloch waves, the potential is not everywhere equal – the potential has a *discrete* translational invariance: there is a nonzero minimum translation unit of the potential. Consequently, the confinement effects of the plane waves and the Bloch waves will be different: The former will not depend on the boundary locations since the unconfined potential is equal everywhere; the latter will depend on the boundary locations since the unconfined potential is not equal everywhere. The very existence of boundary-dependent states in this book is a natural consequence of such a fundamental difference.

In one-dimensional cases, the existence of $N-1$ size-dependent states in a confined length $L = Na$ is a general and natural consequence of the Schrödinger differential equation, the boundary condition (4.4), and Theorem 2.8. In simplest multidimensional cases, the quantum confinement of Bloch waves can be understood by reasonings based on physical intuition if a specific condition such as (5.20), (6.9), or (6.20) is satisfied. Suppose a specific subbranch of Bloch waves (with a specific $\hat{\mathbf{k}}$ or $\bar{\mathbf{k}}$) is completely confined in a specific direction and in a specific length Na—where a is the minimum translational unit in the specific direction and N is a positive integer. $N-1$ stationary Bloch states could be formed under such a specific condition, each stationary Bloch state consists of two Bloch waves from this subbranch with wave vector components k and $-k$ in the specific direction, as a result of multiple reflections of the Bloch waves at the two boundaries. The wave vector components $\pm k$ and the energies of these $N-1$ stationary Bloch states are determined by the confinement size Na, since each stationary Bloch state can only have an integer number of half-wave lengths of the Bloch wave in the confinement length. The summation of such $N-1$ size-dependent stationary Bloch states plus one boundary-dependent state gives a total of N confined electronic states from the quantum confinement of this specific subbranch of Bloch waves as one expects.

Such an understanding on the complete quantum confinement of Bloch waves in simplest cases can be summarized in a mathematical form as

$$N = \mathbf{1} + (N - 1). \quad (8.1)$$

Equation (8.1) indicates that in a truncated periodic structure contains N periods (left), corresponding to *each permitted band* of the Bloch wave, there are two different types of states in the truncated structure. There is **1** state or subband whose energy depends on the boundary, and there are $(N - 1)$ stationary Bloch wave states or subbands whose energy depends on the structure size N. It is the boldfaced **1** state or subband in Eq. (8.1) that indicates the very point where the quantum confinement of Bloch waves is different from the results of the widely used periodic boundary condition(s) and the well-known quantum confinement of plane waves.Consequently, it is due to the existence of this boundary-dependent **1** state or subband in Eq. (8.1) that leads to the results that properties of electronic states in a simple low-dimensional system or finite crystal are substantially different from the properties of electronic states in a crystal with translational invariance as understood in the conventional solid-state physics, and also significantly different from what is traditionally believed in the solid-state physics community regarding the properties of electronic states in a low-dimensional system or finite crystal, such as ideas based on the effective mass approximation.

Equation (8.1) is for the quantum confinement of a permitted band of the Bloch waves. The equation indicates that the physical origin of a boundary-dependent electronic state or subband—the boldfaced **1** state or subband in Eq. (8.1)—is closely related to a permitted band of the Bloch wave rather than a band gap. This point is essential for the understanding of the fundamental physics of surface states and other boundary-related states. As mentioned in Chap. 5, such an understanding is not easy to achieve from either a one-dimensional analysis or a widely used semi-infinite crystal analysis.

Equation (8.1) is a summary of current understandings obtained in the book on the simplest complete quantum confinement of Bloch waves.[1] Theoretically, these results can be an interesting and substantial supplement to the well-known quantum confinement of plane waves and thus could improve our understanding of the fundamental quantum confinement effects. Practically, the results may also find valuable applications in relevant problems in modern solid-state physics and related fields such as material science. If the well-known quantum confinement of plane waves has provided many interesting and valuable insights into the physics of low-dimensional systems, we have reasons to expect that a further understanding of the quantum

[1] A periodic boundary condition gives progressive Bloch waves with N discrete wave vectors in a space size of N periods for each permitted band. It corresponds to

$$N = N.$$

A complete quantum confinement of plain waves gives only stationary waves.

confinement of Bloch waves could be a substantial step toward a more thorough and in-depth understanding of the physics of low-dimensional systems and finite crystals.

The Schrödinger differential equation in a one-dimensional crystal is an ordinary periodic differential equation. The properties of solutions of periodic ordinary differential equations have been well understood mathematically. It is from those mathematical understandings—in particular, those presented in Refs. [1–4]—as summarized in Chap. 2 that most results presented in Chaps. 3 and 4 can be rigorously proven. It is due to the alternative existence of each permitted band and each band gap in a one-dimensional crystal that Theorem 2.8 limits that a boundary-dependent state in a one-dimensional crystal must be in a band gap or at a band-edge.

The differences between the quantum confinement effects of multidimensional Bloch waves and that of one-dimensional Bloch waves naturally are closely related to the differences between the unconfined multidimensional Bloch waves and unconfined one-dimensional Bloch waves. As a comparison to one-dimensional crystals, the permitted bands in a three-dimensional crystal are overlapped. The number of band gaps in a multidimensional crystal is finite; there is not always a band gap above each permitted energy band [5–10]. Correspondingly, Theorem 5.1 limits that the energy of a boundary-dependent state in a three-dimensional crystal must be higher or equal to the upper band-edge of the relevant permitted band but does not give an upper limit. Therefore, a boundary-dependent state decaying in a specific direction may have energy in a range of permitted energy bands of the bulk. Such cases, actually, might be general in multidimensional crystals.

Nevertheless, generally speaking, the properties of solutions of the Schrödinger differential equation in a multidimensional crystal—the second-order elliptic partial differential equation with periodic coefficients—are still much less understood mathematically than in one-dimensional cases. As a result, the quantum confinement of multidimensional Bloch waves can only be understood in simplest cases. Even when a specific condition such as (5.20), (6.9), or (6.20) is satisfied, reasonings used in obtaining results on stationary Bloch states in ideal quantum films, wires, dots, and finite crystals had to be based, to a large extent, on physical concepts borrowed from the well-known stationary plane wave states rather than on rigorous mathematical inferences. A mathematically more rigorous treatment and comprehensive understanding of the problems treated in Chaps. 5–7 will probably have to wait for further progress in relevant mathematical fields.

We have seen that the analytical theory in Chap. 5 is consistent with many previously published numerical results and, therefore, it might provide a more in-depth understanding of those numerical results. The author is not aware of numerical results that can be directly compared with the general predictions in Chaps. 6 and 7, except the numerical calculations of Franceschetti and Zunger [11] on GaAs free-standing quantum wires and dots, shown in Fig. 5.4b, c, mentioned in Sects. 6.6 and 7.7. We have also seen that there are cases where there are differences between the general theory in Chap. 5 and the numerical results presented in [12, 13]. The author hopes that the analytical theory presented in this book will stimulate further numerical calculations to verify the general analytical predictions obtained here. Either the general analytical predictions will be confirmed or be negated somewhere, it could improve our current understandings of this interesting and fundamental problem on the electronic states in low-dimensional systems and finite crystals and the quantum confinement of Bloch waves, including to clarify those previously mentioned differences. In case the general predictions presented are incorrect in places, the defects in reasonings in this book could be relatively easily and straightforwardly traced, and hopefully, a corrected theory can then be established.[2]

Our analytical and general predictions on the electronic states in low-dimensional systems and finite crystals were obtained based on a very simple model; real crystals are certainly more complicated. Nevertheless, we have clearly understood some of the most fundamental differences between the electronic states in low-dimensional systems and finite crystals and the electronic states in crystals with translational invariance by using such a simple model. The effects of those differences on properties of a low-dimensional system become more significant as the size of the system decreases.

Some of the practically most interesting and straightforward predictions of this book are as follows:

1. An ideal low-dimensional system of a cubic semiconductor may actually have a band gap smaller than the band gap of the bulk semiconductor. This is a consequence of relevant theorems such as Theorem 5.1 and the properties of the valence band maximum (VBM) of cubic semiconductors.[3, 4] Because in semiconductors the most important physical processes always happen near the band gap, an improved understanding of the band gap in low-dimensional semiconductors is essential to the physics of low-dimensional semiconductors and may have effects on possible practical applications.
2. An ideal low-dimensional system of a cubic semiconductor may even have an electrical conductivity like a metal. This is because that the boundary-like states originated from a valence band may be energetically higher than the bulk-like states originated from a conduction band. It is well known that one of the greatest successes of modern solid-state physics is that it clearly explains the significant difference between the electrical conductivity of metals and that of semiconductors (and insulators). This new prediction indicates that *such a fundamental distinction between macroscopic solids may become blurred when the size of solids becomes much smaller; thus, the effects of the boundaries of the low dimensional systems or the finite crystals have to be considered.*[5]

---

[2]Interesting numerical investigations verifying the theory can be seen, for example, in [14, 15].

[3]Even for a one-dimensional crystal, (2.96) and the analysis in Sect. 4.3 actually indicate the same consequence: An ideal finite crystal of a one-dimensional semiconductor has a band gap smaller than the band gap of the corresponding infinite crystal, as long as the boundary τ is not a zero of the VBM wave function.

[4]A recent experiment work by Tripathi et al. [16] reported that the thickness-dependent optical energy gap of ultra-thin CuO films grown by atomic layer deposition is "anomalous": the optical energy gap of CuO crystallites decreases as the size decreases.

[5]A work by Rurali and Lorenti [17] seems to support such a prediction. They studied nano Si quantum wires in the $< 100 >$ direction with density-functional calculations and found that such

We can look at these results from an even more general point of view. If one arranges various sizes of the same matter by the number of atoms in each case, the two ends of the spectrum can usually be better understood than the broad range of subjects in between. At one end is the matter consisting of a few atoms; at the other end is the crystal of infinite size, both can be easier to understand. A problem of a few atoms can be relatively easily treated in quantum mechanics and a problem on a crystal of infinite size—with translational invariance—can be essentially reduced to a problem of a few atoms in a unit cell and can also be relatively easily treated. However, the broad range of matters in between is usually harder to understand due to the mathematical difficulties in treating large systems of many atoms. In this book, it was demonstrated that in some simple cases, the electronic structure of ideal truncated periodic systems of any size—which previously were difficult to be understood due to a large number of atoms and the lack of translational invariance—now can be understood as well as the two ends. Therefore, in some sense, a route for understanding is opened, from one end of the matter with a few atoms to the other end of the crystal of infinite size, and containing a whole range of ideal truncated periodic systems of various size in between. Of them, each can be equally well understood as the two ends.

Despite all these new understandings, however, what we have understood is only the very beginning. The model used in this book is the simplest one. The low-dimensional systems and finite crystals treated in this book are also some of the simplest cases. For the little we have just understood, there is so much more we do not yet understand.

We have not touched the structures where the specific conditions such as (5.20), (6.9), or (6.20) are not satisfied. For example, we even have not understood the properties of electronic states in an (111) ideal quantum film of crystals with a simple cubic Bravais lattice in our simplest model yet, not to mention many others. There are many ways to improve the model or to investigate more different cases. Each new progress from an improved model or more investigations on various cases could further enhance our current understanding.

This book presents an investigation of electronic states in ideal low-dimensional systems and finite crystals. It is understood that the existence of the boundary-dependent electronic states in ideal low-dimensional systems or finite crystals is a fundamental distinction of the quantum confinement of Bloch waves. Nevertheless, to have a better understanding of the physical properties of and the physical processes in the low-dimensional systems or finite crystals, there is much more we need to learn. Even if we keep working with only the simplified model of ideal

(Footnote 5 continued)
Si nano wires may become strong metallic. Our treatment is a more general one, but for ideal low-dimensional systems. For quantum wires, we treated only those with rectangular cross sections. Their results might be an indication that the non-rectangular cross section of the wire and the surface reconstructions do not eliminate the metallic conductivity of Si $< 100 >$ nano wires.

low-dimensional systems or finite crystals, we need to understand issues such as how are the specific boundary locations determined for a low-dimensional system or finite crystal? What are the specific forms of those boundary-dependent electronic states in the low-dimensional system or the finite crystal with such specific boundary locations? Further, how does the existence of boundary-dependent electronic states affect the physical properties of and the physical processes in low-dimensional systems or finite crystals, such as optical transitions, scattering, transport processes, and many others? We may have reasons to expect that the physical processes between the stationary Bloch states can be understood—to a large extent—on the basis of understandings on physical processes between progressive Bloch states, as in conventional solid-state physics. We mentioned in Sect. 7.9 that the size-dependence of optical properties in semiconductor low-dimensional systems might be understood if the optical transitions between the bulk-like states contribute dominantly. However, for physical processes in which the boundary-dependent states are involved, we understand very little or almost nothing. If we go beyond the simplified model, there is even much more that we do not understand. There is still a long, long way to go before we can have a more comprehensive understanding of the physical properties of and the physical processes in low-dimensional systems or finite crystals.

8.2 Some Relevant Systems

The understandings obtained for the electronic states in ideal low-dimensional systems or finite crystals may naturally lead to insights to some related problems.

8.2.1 Other Finite Periodic Systems

Naturally, the immediate question is, do eigenmodes in other truncated periodic systems have similar interesting properties, similar to electronic states in low-dimensional systems or finite crystals?

A closely related and much-investigated problem is the properties of classical waves in periodic layered media, such as elastic waves in periodic structures of alternative elastic mediums—so-called phononic crystals—and electromagnetic waves in periodic structures of alternative dielectric or metallo-dielectric mediums—so-called photonic crystals—and so forth. In such a system, the periodic structure can be flexibly designed and shaped. In particular, the surface plane may also be located anywhere in the unit cell, therefore, the parameter τ discussed in this book can be a practically controllable quantity. The natural question is, can the conclusions obtained in this book be applied to photonic crystals and phononic crystals? If so, the boundary-dependent states might thus be tailored by specific choice of the surface location.

The modern theory of periodic Sturm–Liouville equations [2–4] summarized in Chap. 2 can treat the wave equations for some interesting one-dimensional phononic crystals or photonic crystals in the same way as the Schrödinger differential equation of a one-dimensional electronic crystal. Therefore, the question of that whether the conclusions in Part II can be applied to a one-dimensional phononic crystal or photonic crystal essentially depends on the specific boundary conditions of the relevant problem.

The theoretical investigation of one-dimensional phononic crystals in [18] and Appendix E indicates that exact and general results for an ideal one-dimensional phononic crystal of length $L = Na$ with free-surface boundaries can also be analytically obtained. For such a phononic crystal the boundary conditions are

$$p(\tau)y'(\tau) = p(\tau)y'(\tau + L) = 0. \tag{8.2}$$

There are $N - 1$ eigenmodes in each permitted band whose eigenvalues depend on L but not τ; there is one eigenmode for the forbidden range *below each permitted band* whose eigenvalue depends on τ but not L. That is, in contrast to the well-known fact that the quantum confinement effect by the boundary condition Eq. (4.6) makes the energies of the τ-dependent electronic states go higher, the confinement effect by the boundary condition Eq. (8.2) makes the τ-dependent modes go lower in a phononic crystal of finite length. The results clearly show that the boundary conditions have a strong effect as expected.

Recent theoretical and experimental investigations [18–23] on one-dimensional finite phononic crystals and photonic crystals—they may not have an eigen equation of the periodic Sturm–Liouville equation form— did show that the eigenmodes in those finite crystals behave similarly if the modes are completely confined to a finite size of N periods. There are two different types of modes in such a one-dimensional finite crystal of N periods. They are $N - 1$ eigenmodes in each permitted band whose eigenvalues depend on the N but not the boundary location and one eigenmode in a forbidden range whose eigenvalue depends on the boundary but not N. The latter mode is either a surface mode in the forbidden range or a confined band-edge mode.

Therefore, the results summarized in Eq. (8.1) might be consequences of truncated periodicity that are more general than the systems described by the periodic Sturm–Liouville equation (2.2), as long as all eigenmodes can be completely confined to a finite length of N periods.

Despite these understandings on more general one-dimensional problems, however, the issues that whether and how the corresponding theory in higher dimensions such as Theorem 5.1 and others can be extended to treat multidimensional photonic crystals and phononic crystals remain to be understood.

8.2.2 *Electronic States in Ideal Cavity Structures*

A cavity structure is a structure formed with a low-dimensional system removed from an infinite crystal. The electronic states in such a cavity structure usually were not easy to investigate theoretically. The structure does not have a translational invariance and the theoretical approaches previously used in the investigations of electronic states in low-dimensional systems—such as the effective mass approximation-based approaches or numerical calculations—are not easily and effectively used to obtain meaningful understandings. However, by using the approach in this book, understandings of electronic states in ideal cavity structures of simple crystals can be achieved. The results are presented in Appendix G.

8.3 Could a More General Theory Be Possible?

In many cases, it is the general symmetry of a system rather than the details of the specific eigen equation of a physical problem that determines the general properties of solutions. For systems with translational invariance, even though the relevant eigen equations can be very different, such as the Schrödinger differential equation with a periodic potential for the electronic states, the atomic vibrational equation for phonons in crystals, Maxwell's equations for photonic crystals, and so forth, it is the general symmetry of the system—the translational invariance and others—that determines the general properties of the solutions. The states or modes have the common property (Bloch function)

$$\phi(\mathbf{k}, \mathbf{x} + \mathbf{a}_i) = \mathrm{e}^{\mathrm{i}\mathbf{k}\cdot\mathbf{a}_i}\phi(\mathbf{k}, \mathbf{x}), \tag{8.3}$$

where $\mathbf{a}_i$ are the minimum translational units, and the wave vector $\mathbf{k}$ can be limited in a Brillouin zone, determined by the symmetry of the system. The eigenvalues of the problem are functions of the wave vector $\mathbf{k}$:

$$\lambda = \lambda_n(\mathbf{k}). \tag{8.4}$$

Although (8.3) and (8.4) can be obtained from investigations on the solutions of each different eigen equation, it is now well understood that (8.3) and (8.4) are consequences of the symmetry of the concerned system—the periodicity or the translational invariance (and other relevant symmetries) of the system—independent of the particular eigen equation(s) involved. The group theory is a powerful mathematical theory that can be used to investigate the general properties of symmetrical systems. It is the applications of the group theory in different physical problems that lead to such general understanding.

The author was quite surprised when he first obtained the result that the size effect and the boundary effect on the energies of electronic states in a simple ideal low-dimensional system or finite crystal can be separated. He did not know of any other problem that also has such an interesting behavior: Usually, when one solves a boundary eigenvalue problem of a differential equation, both the region and the boundary have an effect on all eigenvalues of the problem. He has also talked to mathematicians and so far has not met someone who knew of similar behavior in other problems.

Periodicity is one of the most fundamental and extensively investigated mathematical concepts. It is the clear understanding of the consequences of the periodicity that has played an essential role in the achievements of modern solid-state physics. Many interesting periodic systems might be truncated. The truncations of periodicity do present new issues. In comparison with investigations of periodicity, the investigations and general understandings of consequences of the truncation(s) of periodicity are much less. A comprehensive general understanding of the truncation(s) of periodicity has fundamental significance.

Preliminary understandings we learned on the truncated one-dimensional periodicity include the following:

- Depending on the original periodicity, the truncation location τ, and the truncation condition, corresponding to a forbidden range of the non-truncated periodic system, a single end truncated one-dimensional periodicity may have two possibilities:
  (i) A solution of type
  $$\psi(x+a) = \mathrm{e}^{\beta a}\psi(x) \tag{8.5}$$
  exists in the forbidden range or at a band edge, where $\beta \leq 0$ if the truncated periodicity is a right truncated periodicity in $[\tau, \infty)$, or $\beta \geq 0$ if the truncated periodicity is a left truncated periodicity in $(-\infty, \tau]$.
  (ii) No solution exists in the forbidden range or at a band edge.
- In the simplest cases, a two-end truncated one-dimensional periodicity with ends τ and $\tau + Na$—where a is the period, and N is a positive integer—has the consequences indicated in Eq. (8.1) for each permitted band. There are $N-1$ solutions of stationary Bloch wave type made of one-dimensional (8.3) in the permitted band; there is one solution of type (8.5) in the forbidden range either above or below the permitted band or at a band edge, where β is a real number.

Above are understandings summarized from existing work.

The existence of the boundary dependent confinement states is a fundamental distinction of the quantum confinement of Bloch waves. The quantum confinement effects of multidimensional Bloch waves and one-dimensional Bloch waves might be significantly different due to the differences in their band structures. Nevertheless, in the simplest cases they can both be summarized by Eq. (8.1). Whether Eq. (8.1) is a general consequence of the simplest truncated periodicity seems to be an interesting conjecture. Equations (8.3) and (8.4) can be considered as consequences of the symmetry breaking from a *continuous* translational invariance to a *discrete* translational invariance. Similarly, can Eq. (8.1) be one of the consequences of a more general mathematical theory of the symmetry-breaking concerning the truncated periodicity?

To further explore such a prospect could be interesting.

References

1. M.S.P. Eastham, *The Spectral Theory of Periodic Differential Equations* (Scottish Academic Press, Edinburgh, 1973). (and references therein)
2. J. Weidmann, *Spectral Theory of Ordinary Differential Operators* (Springer, Berlin, 1987)
3. A. Zettl, *Sturm-Liouville Theory* (The American Mathematical Society, Providence, 2005)
4. B. Malcolm Brown, M.S.P. Eastham, K.M. Schmidt, *Periodic Differential Operators*. Operator Theory: Advances and Applications, vol. 230 (Springer, Heidelberg, 2013) (and references therein)
5. G. Bethe, A. Sommerfeld, *Elektronentheorie der Metalle* (Springer, Berlin, 1967)
6. M.M. Skriganov, Sov. Math. Dokl. **20**, 956 (1979); M.M. Skriganov. Invent. Math. **80**, 107 (1985)
7. P.A. Kuchment, *Floquet Theory For Partial Differential Equations*. Operator Theory: Advances and Applications, vol. 60 (Birkhauser Verlag, Basel, 1993)
8. Y.E. Karpeshina, *Perturbation Theory for the Schrödinger Operator with a Periodic Potential*. Lecture Notes in Mathematics, vol. 1663 (Springer, Berlin, 1997)
9. O. Veliev, *Multidimensional Periodic Schrödinger Operator: Perturbation Theory and Applications*. Springer Tracts in Modern Physics, vol. 263 (Springer, Berlin, 2015)
10. P.A. Kuchment, Bull. (New Ser.) Am. Math. Soc. **53**, 343 (2016) (and references therein)
11. A. Franceschetti, A. Zunger, Appl. Phys. Lett. **68**, 3455 (1996)
12. S.B. Zhang, A. Zunger, Appl. Phys. Lett. **63**, 1399 (1993)
13. S.B. Zhang, C.-Y. Yeh, A. Zunger, Phys. Rev. B **48**, 11204 (1993)
14. A. Ajoy, S. Karmalkar, J. Phys.: Condens. Matter **22**, 435502 (2010)
15. A. Ajoy, Ph. D thesis, Indian Institute of Technology Madras (2013)
16. T.S. Tripathi, I. Terasaki, M. Karppinen, J. Phys.: Condens. Matter **28**, 475801 (2016)
17. R. Rurali, N. Lorenti, Phys. Rev. Lett. **94**, 026805 (2005)
18. S.Y. Ren, Y.C. Chang, Phys. Rev. B **75**, 212301 (2007)
19. A.-C. Hladky, G. Allan, M. de Billy, J. Appl. Phys. **98**, 054909 (2005)
20. E.H. El Boudouti, Y. El Hassouani, B. Djafari-Rouhani, H. Aynaou, Phys. Rev. E **76**, 026607 (2007)
21. Y. El Hassouani, E.H. El Boudouti, B. Djafari-Rouhani, R. Rais, J. Phys.: Conf. Ser. **92**, 012113 (2007); Y. El Hassouani, E.H. El Boudouti, B. Djafari-Rouhani, H. Aynaou, Phys. Rev. B **78**, 174306 (2008)
22. E.H. El Boudouti, B. Djafari-Rouhani, A. Akjouj, L. Dobrzynski, Surf. Sci. Rep. **64**, 471–594 (2009)
23. E.H. El Boudouti, B. Djafari-Rouhani, One-dimensional phononic crystals, in *Acoustic Metamaterials and Phononic Crystals*. Springer Series in Solid-State Sciences, vol. 173, ed. by P.A. Deymier (Springer, Berlin, 2013), p. 45

Appendix A
The Kronig–Penney Model

The Kronig–Penney model [1] played a significant and unique role in our current understanding of the electronic states in one-dimensional crystals [2–7]. Naturally, the model has also been playing an important role in various problems in solid state physics. An interesting point of the Kronig–Penney model is that not only the band structure of a Kronig–Penney crystal can be analytically obtained as well known in the solid state physics community, but also the function formalisms of all solutions, both in the permitted and forbidden energy ranges, can be analytically obtained and explicitly expressed. This point provides a significant convenience for using this model as a concrete example to illustrate the mathematical theory in Part II. This appendix is organized as follows: In Sect. A.1, we briefly describe the model. In Sect. A.2, we use the theory in Chap. 2 to obtain the band structure of the model. In Sect. A.3, the function formalisms of all solutions of a Kronig–Penney crystal in various energy ranges are obtained to illustrate the theory in Chap. 2. In Sect. A.4, a semi-infinite Kronig–Penney crystal is treated, the existence and properties of surface states are investigated based on the theory in Chap. 3 and compared with the classical results obtained by Tamm [5] and described by Seitz in his classic book [2]. Finally, in Sect. A.5 are the results for the Kronig–Penney crystal of finite length $L = Na$ and comparisons with the theory in Chap. 4.

A.1 The Model

The Schrödinger equation for an electron moving in a one-dimensional potential $U(x)$ can be written as [5]

$$y''(x) + \chi^2[\lambda - U(x)]y(x) = 0, \quad \chi^2 = \frac{8\pi^2 m}{h^2}, \tag{A.1}$$

where λ is the energy. In the Kronig–Penney model a one-dimensional crystal with a rectangular potential $U(x)$ of the period a is investigated, where two potential

S.Y. Ren, *Electronic States in Crystals of Finite Size*, Springer Tracts in Modern Physics 270, DOI 10.1007/978-981-10-4718-3

regions with length d_1 and d_2 exist in each unit cell. In the ℓth unit cell of the crystal, $a(\ell-1) \le x \le a\ell$ – here ℓ is an integer, the rectangular potential $U(x)$ can be written as

$$U(x) = \begin{cases} 0, & 0 \le x - a(\ell-1) \le d_1, \\ U_2, & d_1 \le x - a(\ell-1) \le d_1 + d_2, \end{cases} \tag{A.2}$$

where $d_1 + d_2 = a$ and U_2 is a positive real constant.

Further in the Kronig–Penney limit that

$$Lim\, d_2 = 0, \quad Lim\, d_1 = a, \quad Lim\, U_2 = \infty,$$

and keeping

$$Lim\,(U_2 d_2) = p = Const, \tag{A.3}$$

the Schrödinger equation (A.1) becomes

$$-y''(x) + \left[\sum_{n=-\infty}^{\infty} \frac{2p}{a}\delta(x-na) - \frac{\xi^2}{a^2}\right] y(x) = 0, \qquad -\infty < x < \infty \tag{A.4}$$

where

$$\xi = a\chi\sqrt{\lambda}. \tag{A.5}$$

In this Appendix, it is more convenient to discuss ξ instead of λ in many cases, similar in [5]. We will treat ξ rather than λ in most equations.

A.2 Normalized Solutions and the Discriminant

A.2.1 Normalized Solutions

We are interested in the properties of solutions of Eq. (A.4) in the interval $[-\frac{a}{2}, \frac{a}{2}]$. Two normalized solutions $\eta_1(x,\xi)$ and $\eta_2(x,\xi)$ of Eq. (A.4) are defined by the conditions that

$$\eta_1\left(-\frac{a}{2},\xi\right) = 1,\ \eta_1'\left(-\frac{a}{2},\xi\right) = 0;\ \eta_2\left(-\frac{a}{2},\xi\right) = 0,\ \eta_2'\left(-\frac{a}{2},\xi\right) = 1. \tag{A.6}$$

By integrating Eq. (A.4) from $-\varepsilon$ to ε where ε is an infinitely small positive number, for any solution y of Eq. (A.4) we have

$$y'(+\varepsilon) - y'(-\varepsilon) = 2\,\frac{p}{a}\,y(0). \tag{A.7}$$

Two normalized solutions of Eq. (A.4) can be obtained as

$$\eta_1(x,\xi) = \begin{cases} cos\frac{\xi}{2}cos\frac{\xi}{a}x - sin\frac{\xi}{2}sin\frac{\xi}{a}x, & -\frac{a}{2} \le x < 0, \\ cos\frac{\xi}{2}cos\frac{\xi}{a}x + \left(2\,\frac{p}{\xi}\,cos\frac{\xi}{2} - sin\frac{\xi}{2}\right)sin\frac{\xi}{a}x, & 0 < x \le \frac{a}{2}, \end{cases} \tag{A.8}$$

and

$$\eta_2(x,\xi) = \begin{cases} \left(\frac{\xi}{a}\right)^{-1}\left(sin\frac{\xi}{2}cos\frac{\xi}{a}x + cos\frac{\xi}{2}sin\frac{\xi}{a}x\right), & -\frac{a}{2} \le x < 0, \\ \left(\frac{\xi}{a}\right)^{-1}\left[sin\frac{\xi}{2}cos\frac{\xi}{a}x + \left(cos\frac{\xi}{2} + 2\,\frac{p}{\xi}\,sin\frac{\xi}{2}\right)sin\frac{\xi}{a}x\right], & 0 < x \le \frac{a}{2}, \end{cases} \tag{A.9}$$

where Eqs. (A.6) and (A.7) were used.

From Eqs. (A.8) and (A.9) we obtain that

$$\begin{aligned} &\eta_1\left(\tfrac{a}{2},\xi\right) = cos\,\xi + \tfrac{p}{\xi}\,sin\,\xi, && \eta_1'\left(\tfrac{a}{2},\xi\right) = -\tfrac{\xi}{a}\left(sin\,\xi - 2\,\tfrac{p}{\xi}cos^2\tfrac{\xi}{2}\right), \\ &\eta_2\left(\tfrac{a}{2},\xi\right) = \left(\tfrac{\xi}{a}\right)^{-1}\left(sin\,\xi + 2\,\tfrac{p}{\xi}\,sin^2\tfrac{\xi}{2}\right), && \eta_2'\left(\tfrac{a}{2},\xi\right) = cos\,\xi + \tfrac{p}{\xi}\,sin\,\xi. \end{aligned} \tag{A.10}$$

A.2.2 *The Discriminant $D(\xi)$*

From Eq. (A.10) we obtain the discriminant $D(\xi)$ of Eq. (A.4) as

$$D(\xi) = \eta_1\left(\frac{a}{2},\xi\right) + \eta_2'\left(\frac{a}{2},\xi\right) = 2\,cos\,\xi + 2\,\frac{p}{\xi}\,sin\,\xi. \tag{A.11}$$

In each energy range where $-2 \le D(\xi) \le 2$, we obtain Bloch wave functions $\phi_n(\pm k, x)$ as eigensolutions of (A.4):

$$\phi_n(\pm k, x+a) = e^{\pm ika}\phi_n(\pm k, x), \tag{A.12}$$

where $-\pi/a < k \le \pi/a$. The Bloch wave vector k and ξ are related by (2.74) as:

$$cos\,ka \;=\; cos\,\xi \;+\; \frac{p}{\xi}\,sin\,\xi. \tag{A.13}$$

This is the well-known band structure equation for the Kronig–Penney model [1–3]. Although this band structure of the Kronig–Penney model has been well known for many years, the approach used here is easier to be extended to treat more general and complicated cases, such as that there are more than two potential regions in one period: $a = d_1 + d_2 + d_3 \ldots$, each region has a different width and a different potential height, and so forth.

The energy ranges where $D(\xi) < -2$ are band gaps at $k = \pi/a$. Two linearly independent solutions of the Eq. (A.4) can be written as ($\beta > 0$):

$$y(\xi, \pm\beta, x + a) = -e^{\pm\beta a}\, y(\xi, \pm\beta, x), \qquad -\infty < x < +\infty, \tag{A.14}$$

where β and ξ are related by (2.82):

$$\cosh \beta a = \ - \cos \xi - \frac{p}{\xi} \sin \xi. \tag{A.15}$$

The energy ranges where $D(\xi) > 2$ are band gaps at $k = 0$. Two linearly independent solutions of the Eq. (A.4) can be written as ($\beta > 0$):

$$y(\xi, \pm\beta, x + a) = e^{\pm\beta a}\, y(\xi, \pm\beta, x), \qquad -\infty < x < +\infty, \tag{A.16}$$

where β and ξ are related by (2.78):

$$\cosh \beta a = \cos \xi + \frac{p}{\xi} \sin \xi. \tag{A.17}$$

Unlike in (A.12) where the Bloch wave vector k is a monotonic function of ξ in a specific permitted band, thus, a pair of $\pm k$ can only correspond to one unique ξ. β is not a monotonic function of ξ in a specific band gap, so one pair of $\pm\beta$ may correspond to two different ξ. Therefore, in (A.14) and (A.16) ξ is needed to specify a solution in a band gap.

A.2.3 The Band Edge Eigenvalues

The upper edge of the nth permitted band—that is the lower edge of the nth band gap, and it can be denoted as ω_n—is at

$$\xi = \omega_n = (n + 1)\pi. \qquad n = 0, 1, 2, 3, \ldots \tag{A.18}$$

The lower edge of the nth permitted band—that is the upper edge of the $n - 1$th band gap if $n > 0$—is at

$$\begin{aligned} \tfrac{\xi}{2} \tan\tfrac{\xi}{2} &= \ \ p/2, \quad n = 0, 2, 4, \ldots, \\ \tfrac{\xi}{2} \cot\tfrac{\xi}{2} &= -p/2, \quad n = 1, 3, 5, \ldots. \end{aligned} \tag{A.19}$$

Those are well-known results [1, 3, 7]. The upper edge of the nth band gap can be denoted as Ω_n, $n = 0, 1, 2, 3, \ldots$.

The band structure of a Kronig–Penney crystal depends on p in (A.4). The band edges determined by (A.18) and (A.19) are shown in Fig. A.1, as functions of p.

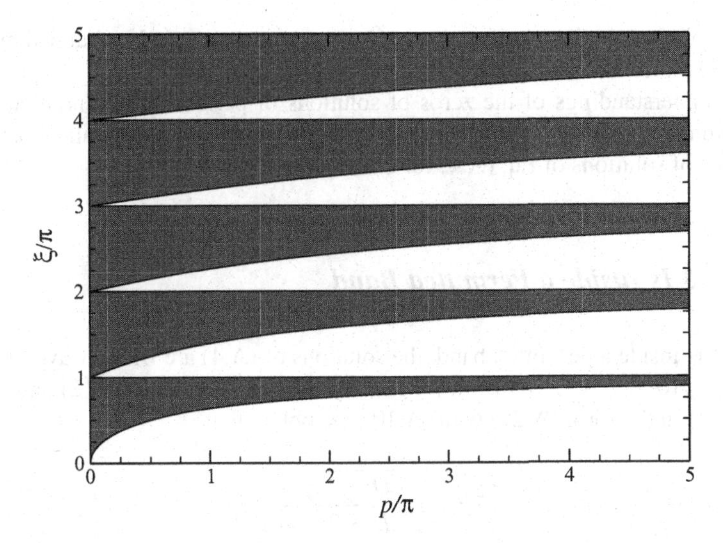

Fig. A.1 The band edges determined by (A.18) and (A.19) as functions of p in (A.4). The permitted energy band ranges are shown as shadowed regions. Note that the lower edge $\omega_n = (n+1)\pi$ of a band gap does not depend on p. As p increases, the upper edge Ω_n of each band gap increases and each permitted bandwidth decreases

In the following, in most cases $p = \frac{3\pi}{2}$ is assumed as in [1, 5]. Nevertheless, we will see that in some cases the physical results in the Kronig–Penney model might significantly even qualitatively depend on the value of p.

A.3 Solutions of the Differential Equation

In the following, we obtain the solutions of the differential equation (A.4). We consider three different cases:

(1) ξ is inside a permitted band;

(2) ξ is at a band edge: $\xi = \omega_n$ or $\xi = \Omega_n$;

(3) ξ is inside a band gap: $\omega_n < \xi < \Omega_n$.

We only need to obtain the solutions in one period $[-\frac{a}{2}, \frac{a}{2}]$. This is the reason we choose the two normalized solutions $\eta_1(x,\xi)$ and $\eta_2(x,\xi)$ of Eq. (A.4) according to (A.6). In general, a solution of Eq. (A.4) in the interval $[-\frac{a}{2}, \frac{a}{2}]$ can be written as a linear combination of two normalized solutions $\eta_1(x,\xi)$ and $\eta_2(x,\xi)$:

$$y(x,\xi) = c_1\,\eta_1(x,\xi) + c_2\,\eta_2(x,\xi), \qquad -\frac{a}{2} \le x \le \frac{a}{2}. \tag{A.20}$$

Since $\eta_1(x,\xi)$ and $\eta_2(x,\xi)$ were given in (A.8)–(A.9), the solution can be determined—with a normalization constant factor difference—if $\frac{c_2}{c_1}$ in each different case is determined.

The solutions in the whole real axis ($-\infty < x < +\infty$) can be obtained by using Eqs. (A.12), (A.14), or (A.16).

The understandings of the zeros of solutions of periodic differential equations play a fundamental role in the theory in Chap. 4. We will pay particular attention to the zeros of solutions of Eq. (A.4) for ξ in different cases.

A.3.1 ξ Is Inside a Permitted Band

When ξ is inside a permitted band, the solutions of (A.4) are Bloch wave functions $\phi_n(\pm k, x)$ $(0 < k < \pi/a)$ as in (A.12). By using $x = -\frac{a}{2}$ in (A.12) and writing $\phi_n(\pm k, x)$ in the form (A.20) with (A.10), we obtain that

$$\frac{c_2}{c_1} = \pm\sqrt{\frac{D-2}{D+2}}\frac{\frac{\xi}{a}cos\frac{\xi}{2}}{sin\frac{\xi}{2}}, \tag{A.21}$$

where D is the discriminant (A.11). Thus inside a permitted band of Eq. (A.4), the Bloch wave functions $\phi_n(\pm k, x)$ have the form

$$\phi_n(\pm k, x) = C\left[\sqrt{D+2}\,sin\frac{\xi}{2}\,\eta_1(x,\xi) \pm \sqrt{D-2}\,\frac{\xi}{a}\,cos\frac{\xi}{2}\,\eta_2(x,\xi)\right],$$
$$-\frac{a}{2} \leq x \leq \frac{a}{2}, \tag{A.22}$$

here C is a normalization constant.

Note that as a solution of (A.4) with ξ inside a permitted band, $\phi_n(\pm k, x)$ in Eq. (A.22) does not have zeros. Neither $sin\frac{\xi}{2}$ nor $cos\frac{\xi}{2}$ can be zero; $\sqrt{D+2}$ is a non-zero real number, but $\sqrt{D-2}$ is a non-zero imaginary number; and, $\eta_1(x,\xi)$ and $\eta_2(x,\xi)$ are real functions not being zero simultaneously, due to the Sturm Separation Theorem.

A.3.2 ξ Is at a Bandedge: $\xi = \omega_n$ Or $\xi = \Omega_n$

The wave function for ξ being at a band-edge can be easier obtained by noticing that $\phi_n(k_g, x)$ must be either a real symmetric function or a real antisymmetrical function due to the inversion symmetry of Eq. (A.4).

Since the potential is zero in the region except the delta-function barrier at $x = 0$, the real symmetrical solution for a specific ξ of the Eq. (A.4) is

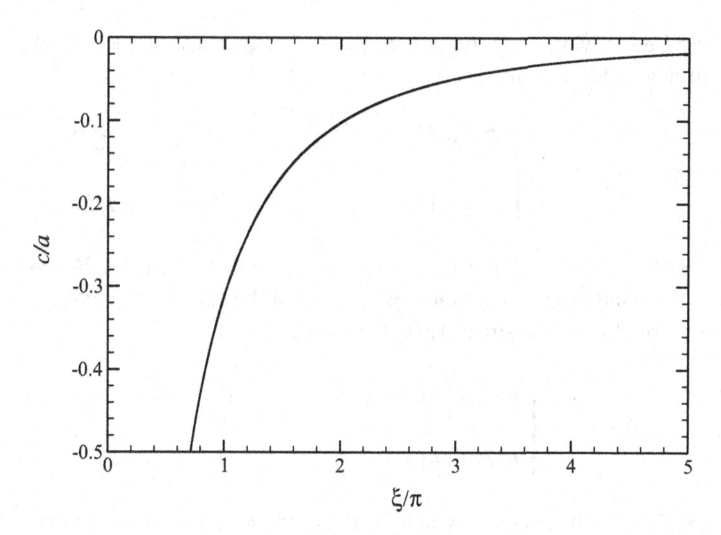

Fig. A.2 c in (A.23) as the function of ξ. $p = 3\pi/2$ is assumed

$$f_s(\xi, x) = \begin{cases} cos\left[\frac{\xi}{a}(-x + c)\right], & -a/2 \le x \le 0, \\ cos\left[\frac{\xi}{a}(x + c)\right], & 0 \le x \le a/2. \end{cases} \tag{A.23}$$

where c is a constant depending on ξ due to the existence of the delta-function barrier indicated in (A.7) and determined by

$$tan\ \frac{\xi}{a}c = -\frac{p}{\xi}. \tag{A.24}$$

In Fig. A.2 is shown c in (A.23) as the function of ξ, determined by (A.24) for a specific $p = 3\pi/2$.

The real antisymmetrical solution of Eq. (A.4) is

$$f_a(\xi, x) = sin\frac{\xi}{a}x, \qquad -a/2 \le x \le a/2. \tag{A.25}$$

Since a band-edge wave function must have $\phi_n(\pi/a, a/2) = -\phi_n(\pi/a, -a/2)$ or $\phi_n(0, a/2) = \phi_n(0, -a/2)$ by Eq. (A.12), it is easy to see that the band-edge wave function at the lower edge $\xi = \omega_n$ of each band gap is an antisymmetrical function:

$$\phi_n(k_g, x) = C\ sin\left[\frac{(n+1)\pi}{a}x\right], \qquad -a/2 \le x \le a/2. \tag{A.26}$$

where Eq. (A.18) is used.

The band-edge wave function at the upper edge $\xi = \Omega_n$ of each band gap must be an symmetrical function:

$$\phi_{n+1}(k_g, x) = \begin{cases} C\, cos\left[\frac{\xi}{a}(-x+c)\right], & -a/2 \le x \le 0, \\ C\, cos\left[\frac{\xi}{a}(x+c)\right], & 0 \le x \le a/2, \end{cases} \tag{A.27}$$

where $n = 0, 1, 2, 3, \ldots$. and $k_g = \pi/a$ or $k_g = 0$ in Eqs. (A.26) and (A.27). Similarly, the band-edge wave function $\phi_0(0, x)$ at the lowest band edge $\varepsilon_0(0)$ must be a symmetrical function since $\phi_0(0, a/2) = \phi_0(0, -a/2)$:

$$\phi_0(0, x) = \begin{cases} C\, cos\left[\frac{\xi}{a}(-x+c)\right], & -a/2 \le x \le 0, \\ C\, cos\left[\frac{\xi}{a}(x+c)\right], & 0 \le x \le a/2. \end{cases} \tag{A.28}$$

In Eqs. (A.27) and (A.28) ξ is given by Eq. (A.19). C is a normalization constant in Eqs. (A.26), (A.27) and (A.28).

In Fig. A.3 are shown the two band-edge wave functions $\phi_0(\pi/a, x)$ and $\phi_1(\pi/a, x)$ of the lowest band gap at $k{=}\pi/a$. In Figs. A.3 and A.4, we assume that $C{=}1$ for simplicity since the normalization constants do not affect the points with which we are concerned here. It can be seen that both the Bloch wave functions $\phi_0(\pi/a, x)$ and $\phi_1(\pi/a, x)$ have one zero in the interval $(-\frac{a}{2}, \frac{a}{2}]$, as in Theorem 2.7. (iii).

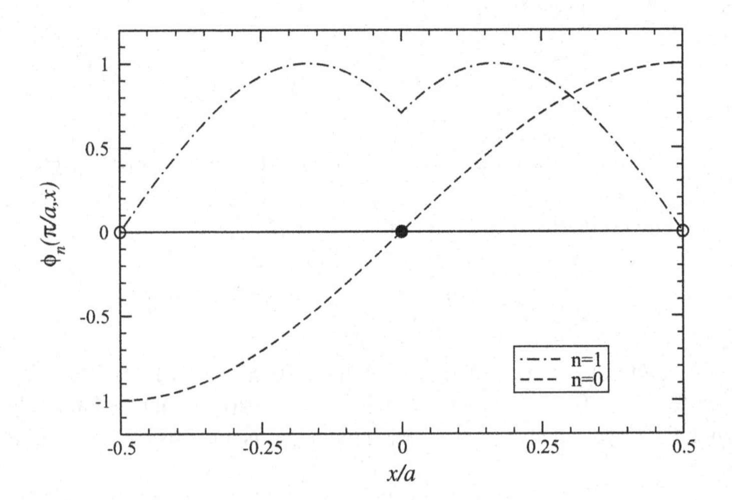

Fig. A.3 Two band-edge wave functions $\phi_0(\pi/a, x)$ and $\phi_1(\pi/a, x)$ of the lowest band gap at $k = \pi/a$. Note the discontinuity of the derivative of the upper edge wave function $\phi_1(\pi/a, x)$ at $x = 0$ due to (A.7). Note that $\phi_0(\pi/a, x)$ has one zero $x = 0$ (*solid circle*) in the interval $(-\frac{a}{2}, \frac{a}{2}]$; $\phi_1(\pi/a, x)$ has one zero $x = \frac{a}{2}$ (*open circle*) in the interval $(-\frac{a}{2}, \frac{a}{2}]$

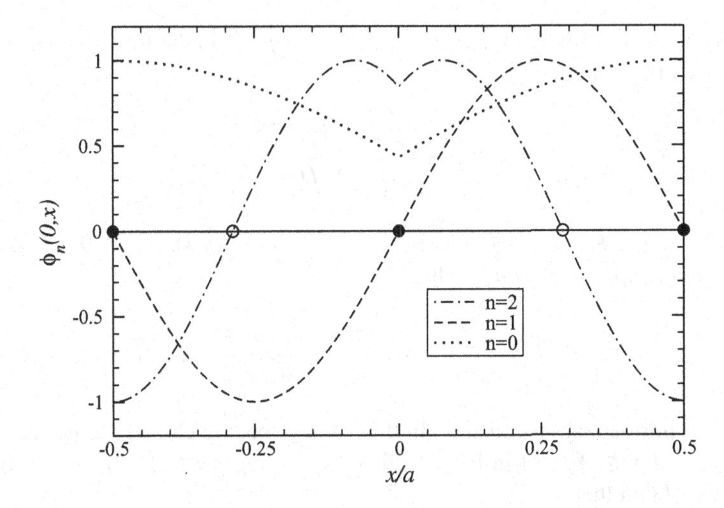

Fig. A.4 The lowest band-edge wave function $\phi_0(0, x)$ and two band-edge wave functions $\phi_1(0, x)$ and $\phi_2(0, x)$ of the lowest band gap at $k = 0$. Note the discontinuity of the derivatives of $\phi_0(0, x)$ and $\phi_2(0, x)$ at $x = 0$ due to (A.7). Note that $\phi_0(0, x)$ has no zero, $\phi_1(0, x)$ has two zeros $x = 0$ and $x = \frac{a}{2}$ (*solid circles*) in the interval $(-\frac{a}{2}, \frac{a}{2}]$; $\phi_2(0, x)$ has two symmetrical zeros (*open circles*) in the interval $(-\frac{a}{2}, \frac{a}{2}]$

In Fig. A.4 are shown the lowest band-edge wave function $\phi_0(0, x)$ and two band-edge wave functions $\phi_1(0, x)$ and $\phi_2(0, x)$ of the lowest band gap at $k = 0$. It can be seen that $\phi_0(0, x)$ has no zero in $(-\frac{a}{2}, \frac{a}{2}]$, leading to the conclusion that $\phi_0(0, x)$ has no zero in $(-\infty, +\infty)$ as in Theorem 2.7 (i). Both Bloch wave functions $\phi_1(0, x)$ and $\phi_2(0, x)$ have two zeros in the interval $(-\frac{a}{2}, \frac{a}{2}]$, as in Theorem 2.7. (ii).

From (A.26) we can see that the wave function $\phi_n(k_g, x)$ at the lower edge of the nth band gap has $n + 1$ zeros in $(-a/2, a/2]$; Eqs. (A.27) and (A.19) lead to that the wave function $\phi_{n+1}(k_g, x)$ at the upper edge of the nth band gap has $n + 1$ zeros in $(-a/2, a/2]$, in consistence with Theorem 2.7. (ii) and (iii). The lower band-edge wave function of each gap always has a zero at the potential maximum $x = 0$.

A.3.3 ξ Is Inside A Band Gap

In the following, we consider the solutions of Eq. (A.4) when ξ is inside a band gap. We need to consider two cases: ξ is inside a band gap at $k = \pi/a$ or ξ is inside a band gap at $k = 0$.

A.3.3.1 ξ Is Inside A Band Gap at $k = \pi/a$

The ξ ranges where $D(\xi) < -2$ are band gaps at $k = \pi/a$. Two linearly independent solutions of the Eq. (A.4) can be written as in the form of (A.14) $(\beta > 0)$: $y(\xi, \pm\beta, x + a) = -e^{\pm\beta a}\, y(\xi, \pm\beta, x)$, where β and ξ are related by (A.15).

By using $x = -\frac{a}{2}$ in (A.14) and writing $y(\xi, \pm\beta, x)$ in the form (A.20) with Eq. (A.10), we obtain that,

$$\frac{c_2}{c_1} = \pm\frac{\frac{\xi}{a}cos\frac{\xi}{2}}{sin\frac{\xi}{2}}\sqrt{\frac{D-2}{D+2}},$$

where $D = 2\ cos\ \xi + 2\ \frac{\beta}{\xi}\ sin\ \xi$ is the discriminant of (A.4). Since $D < -2$ for a ξ inside a band gap at $k = \pi/a$, we have

$$y(\xi, \pm\beta, x) = C\left[sin\frac{\xi}{2}\ \sqrt{-D-2}\ \eta_1(x, \xi) \pm \frac{\xi}{a}cos\frac{\xi}{2}\ \sqrt{2-D}\ \eta_2(x, \xi)\right], \quad (A.29)$$

here C is a normalization constant. In this subsection, we choose the normalization constant C of $y(\xi, \pm\beta, x)$ in Eq. (A.29) by requiring $y(\xi, \pm\beta, 0) = 1$. From Eq. (A.29) we obtain that

$$y(\xi, \pm\beta, x) = \begin{cases} \frac{2}{sin\ \xi\ (\sqrt{-D-2}\pm\sqrt{2-D})}\left[sin\frac{\xi}{2}\ \sqrt{-D-2}\ cos\left(\frac{\xi}{2}+\frac{\xi}{a}x\right)\right. \\ \qquad \left.\pm cos\frac{\xi}{2}\ \sqrt{2-D}\ sin\left(\frac{\xi}{2}+\frac{\xi}{a}x\right)\right], \quad -\frac{a}{2} \le x \le 0, \\ \frac{2}{sin\ \xi\ (\sqrt{-D-2}\mp\sqrt{2-D})}\left[sin\frac{\xi}{2}\ \sqrt{-D-2}\ cos\left(\frac{\xi}{2}-\frac{\xi}{a}x\right)\right. \\ \qquad \left.\mp cos\frac{\xi}{2}\ \sqrt{2-D}\ sin\left(\frac{\xi}{2}-\frac{\xi}{a}x\right)\right], \quad 0 \le x \le \frac{a}{2}. \end{cases} \quad (A.30)$$

The first part of Eq. (A.30) is directly from Eqs. (A.29) and (A.10), the second part comes from that due to the inversion symmetry of Eq. (A.4), $y(\xi, +\beta, x)$ in the $+x$ direction behaves exactly same as $y(\xi, -\beta, x)$ in the $-x$ direction: $y(\xi, \pm\beta, x) = y(\xi, \mp\beta, -x)$.

Zeros of Solutions:

The zeros $x = x_0$ of a solution $y(\xi, \pm\beta, x)$ in (A.30) of Eq. (A.4) for ξ in a band gap at $k = \pi/a$ is determined by

$$y(\xi, \pm\beta, x_0) = 0.$$

Thus the zeros $x = x_0$ of $y(\xi, \pm\beta, x)$ are the solutions of one of the two following equations:

$$\begin{aligned} cot\left(\frac{\xi}{2}+\frac{\xi}{a}x_0\right) &= \pm\sqrt{\frac{-sin\ \xi+\frac{\beta}{\xi}(1+cos\ \xi)}{sin\ \xi+\frac{\beta}{\xi}(1-cos\ \xi)}}, \quad -\frac{a}{2} \le x_0 \le 0 \\ cot\left(\frac{\xi}{2}-\frac{\xi}{a}x_0\right) &= \mp\sqrt{\frac{-sin\ \xi+\frac{\beta}{\xi}(1+cos\ \xi)}{sin\ \xi+\frac{\beta}{\xi}(1-cos\ \xi)}}. \quad 0 \le x_0 \le \frac{a}{2} \end{aligned} \quad (A.31)$$

A.3.3.2 ξ Is Inside A Band Gap at $k = 0$

The ξ ranges where $D(\xi) > 2$ are band gaps at $k = 0$. Two linearly independent solutions of the Eq. (A.4) can be written in the form of (A.16) ($\beta > 0$): $y(\xi, \pm\beta, x + a) = e^{\pm\beta a}\, y(\xi, \pm\beta, x)$ where β and ξ are related by (A.17). By using $x = -\frac{a}{2}$ in (A.16) and writing $y(\xi, \pm\beta, x)$ in the form (A.20) with Eq. (A.10) we obtain that

$$\frac{c_2}{c_1} = \pm \frac{\frac{\xi}{a} cos\frac{\xi}{2}}{sin\frac{\xi}{2}} \sqrt{\frac{D-2}{D+2}},$$

where $D = 2\, cos\ \xi + 2\frac{p}{\xi}\, sin\xi$ is the discriminant of (A.4). Since $D > 2$ for a ξ inside a band gap at $k = 0$, we have

$$y(\xi, \pm\beta, x) = C\left[sin\frac{\xi}{2}\sqrt{D+2}\,\eta_1(x, \xi) \pm \frac{\xi}{a} cos\frac{\xi}{2}\sqrt{D-2}\,\eta_2(x, \xi)\right], \tag{A.32}$$

where C is a normalized constant. We choose the normalization constant C in Eq. (A.32) similarly as in Sect. A.3.3.1. Thus

$$y(\xi, \pm\beta, x) = \begin{cases} \frac{2}{sin\ \xi\ (\sqrt{D+2} \pm \sqrt{D-2})}\left[sin\frac{\xi}{2}\sqrt{D+2}\, cos\left(\frac{\xi}{2} + \frac{\xi}{a}x\right)\right. \\ \left. \pm cos\frac{\xi}{2}\sqrt{D-2}\, sin\left(\frac{\xi}{2} + \frac{\xi}{a}x\right)\right], & -\frac{a}{2} \le x \le 0, \\ \frac{2}{sin\ \xi\ (\sqrt{D+2} \mp \sqrt{D-2})}\left[sin\frac{\xi}{2}\sqrt{D+2}\, cos\left(\frac{\xi}{2} - \frac{\xi}{a}x\right)\right. \\ \left. \mp cos\frac{\xi}{2}\sqrt{D-2}\, sin\left(\frac{\xi}{2} - \frac{\xi}{a}x\right)\right], & 0 \le x \le \frac{a}{2}, \end{cases} \tag{A.33}$$

by requiring $y(\xi, \pm\beta, 0) = 1$. The first part of Eq. (A.33) is directly from Eqs. (A.32) and (A.10), the second part comes from that due to the inversion symmetry of Eq. (A.4), $y(\xi, +\beta, x)$ in the $+x$ direction behaves exactly same as $y(\xi, -\beta, x)$ in the $-x$ direction: $y(\xi, \pm\beta, x) = y(\xi, \mp\beta, -x)$.

In Figs. A.5 and A.6 are shown the functions $y(\xi, \beta, x)$ and $y(\xi, -\beta, x)$ of two different ξ in the lowest band gap at $k = 0$.

Zeros of Solutions:

The zeros x_0 of a solution $y(\xi, \pm\beta, x)$ in (A.33) of Eq. (A.4) for ξ in a band gap at $k = 0$ is determined by

$$y(\xi, \pm\beta, x_0) = 0.$$

Thus the zeros $x = x_0$ of $y(\xi, \pm\beta, x)$ are the solutions of one of the following two equations:

$$\begin{aligned} cot\left(\tfrac{\xi}{2} + \tfrac{\xi}{a}x_0\right) &= \mp\sqrt{\frac{-sin\ \xi + \frac{p}{\xi}(1 + cos\ \xi)}{sin\ \xi + \frac{p}{\xi}(1 - cos\ \xi)}}, & -\tfrac{a}{2} \le x_0 \le 0 \\ cot\left(\tfrac{\xi}{2} - \tfrac{\xi}{a}x_0\right) &= \pm\sqrt{\frac{-sin\ \xi + \frac{p}{\xi}(1 + cos\ \xi)}{sin\ \xi + \frac{p}{\xi}(1 - cos\ \xi)}}. & 0 \le x_0 \le \tfrac{a}{2} \end{aligned} \tag{A.34}$$

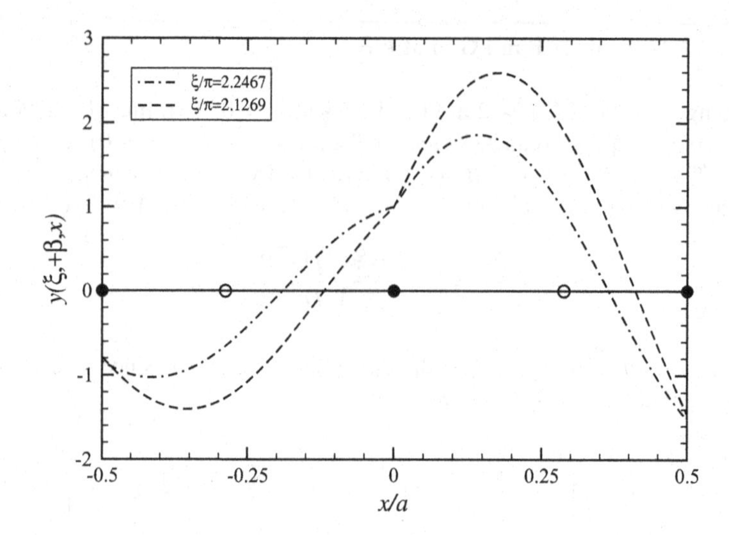

Fig. A.5 The functions $y(\xi, \beta, x)$ (A.33) with two different energies in the lowest band gap at $k = 0$. Note the discontinuity of the derivatives of the functions at $x = 0$ due to (A.7). Note that each function has two zeros in the interval $(-\frac{a}{2}, \frac{a}{2}]$ and the zeros of the functions *go left* from the zeros of $\phi_1(0, x)$ (*solid circles*) to the zeros of $\phi_2(0, x)$ (*open circles*) as ξ increases

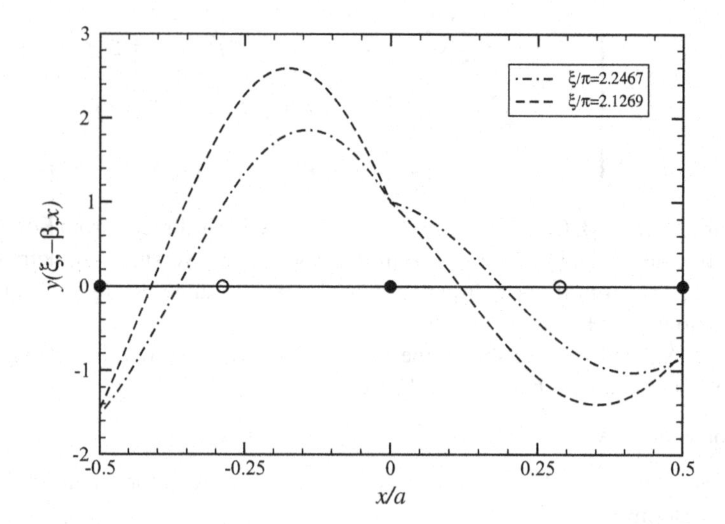

Fig. A.6 The functions $y(\xi, -\beta, x)$ (A.33) with two different energies in the lowest band gap at $k = 0$. Note the discontinuity of the derivatives of the functions at $x = 0$ due to (A.7). Note that each function has two zeros in the interval $(-\frac{a}{2}, \frac{a}{2}]$ and the zeros of the functions *go right* from the zeros of $\phi_1(0, x)$ (*solid circles*) to the zeros of $\phi_2(0, x)$ (*open circles*) as ξ increases

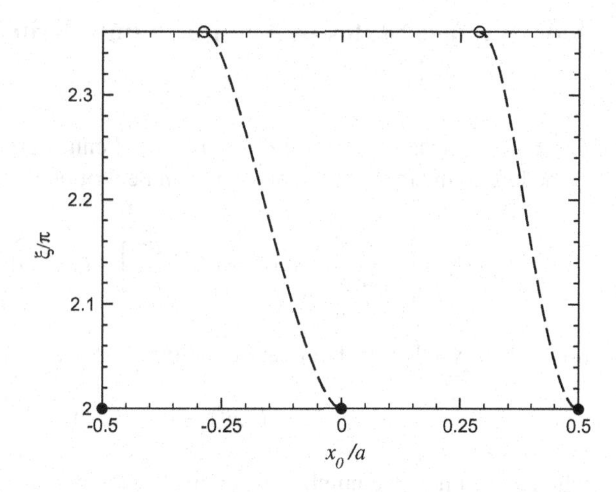

Fig. A.7 The zeros x_0 of functions $y(\xi, \beta, x)$ in the lowest band gap at $k = 0$ calculated by (A.34)

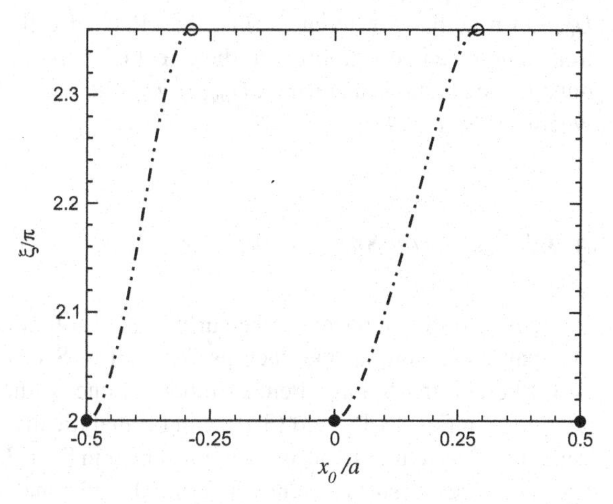

Fig. A.8 The zeros x_0 of functions $y(\xi, -\beta, x)$ in the lowest band gap at $k = 0$ calculated by (A.34)

The zeros x_0 of $y(\xi, \beta, x)$ calculated by Eq. (A.34) for different ξ in the lowest band gap at the center of the Brillouin zone $k = 0$ are shown in Fig. A.7 as dashed lines. There are two zeros of $y(\xi, \beta, x)$ in the interval $(-\frac{a}{2}, \frac{a}{2}]$, as indicated by Theorem 2.8; As ξ increases from the lower edge $\xi = \omega_1$ to the upper edge $\xi = \Omega_1$, the zeros x_0 of $y(\xi, \beta, x)$ go *left* from the zeros of $\phi_1(0, x)$ (solid circles) to the zeros of $\phi_2(0, x)$ (open circles).

The zeros x_0 of $y(\xi, -\beta, x)$ calculated by Eq. (A.34) for different ξ in the lowest band gap at the center of the Brillouin zone $k = 0$ are shown in Fig. A.8 as double-dotted chained lines. There are two zeros of $y(\xi, -\beta, x)$ in the interval $(-\frac{a}{2}, \frac{a}{2}]$, as indicated by Theorem 2.8. As ξ increases from the lower edge $\xi = \omega_1$ to the upper edge $\xi = \Omega_1$, the zeros of $y(\xi, -\beta, x)$ go *right* from the zeros of $\phi_1(0, x)$ (solid circles) to the zeros of $\phi_2(0, x)$ (open circles).

Figures A.7 and A.8 can be compared with Figs. 3.1 and 3.2 in Chap. 3.

A.4 Surface States in A Semi-infinite Kronig–Penney Crystal

For a one-dimensional semi-infinite Kronig–Penney crystal with a potential barrier at the left boundary at τ, the problem can be written as

$$-\psi''(x) + \left[\sum_{n=-\infty}^{\infty} \frac{2p}{a}\delta(x - na) - \frac{\xi^2}{a^2}\right]\psi(x) = 0, \quad \tau < x < +\infty \qquad \text{(A.35)}$$

inside the crystal and a boundary condition at τ:

$$\psi'(\tau, \xi) - \sigma\, \psi(\tau, \xi) = 0, \qquad \text{(A.36)}$$

where σ is a positive number depending on the potential barrier $U_{out}(x)$. Although $U_{out}(x)$ may have different forms, the effect of different $U_{out}(x)$ on the problem treated here can be simplified to the effect of σ. In the simplest cases, the potential outside the crystal is a constant $U_{out} = U_0$. We will only need to consider the cases where $-\frac{a}{2} \leq \tau \leq \frac{a}{2}$.

A.4.1 Surface State Solutions

We will be only interested in the surface state solutions of Eqs. (A.35) and (A.36). We may use a similar approach as we used in Sect. A.3 to obtain such solutions. Nevertheless, the Kronig–Penney model is one of the cases where $\eta_i(\tau + a)$ and $\eta_i'(\tau + a)$ in Eqs. (3.18) and (3.19) can be analytically obtained. An even more convenient way is to use the formalism presented in Sect. 3.5 to investigate the existence and properties of surface states in a one-dimensional semi-infinite Kronig–Penney crystal.

From now on the two normalized solutions $\eta_1(x, \xi)$ and $\eta_2(x, \xi)$ of Eq. (A.4) are defined by the conditions (3.14):

$$\eta_1(\tau, \xi) = 1, \ \eta_1'(\tau, \xi) = 0; \quad \eta_2(\tau, \xi) = 0, \ \eta_2'(\tau, \xi) = 1. \qquad \text{(A.37)}$$

We can consider that a unit cell of the Kronig–Penney crystal described by Eqs. (A.1) and (A.2) is composed of three regions: a region of width d_2 with a potential U_2 between a region of width d_1 on the left and a region of width d_3 on the right, both with a zero potential. Here

$$\begin{aligned} d_1 &= -\tau, \ d_3 = a + \tau, \qquad && -\tfrac{a}{2} \leq \tau \leq 0, \\ d_1 &= a - \tau, \ d_3 = \tau, \qquad && 0 \leq \tau \leq \tfrac{a}{2}. \end{aligned} \qquad \text{(A.38)}$$

In the Kronig–Penney limit

$$Lim\ d_2 = 0, \quad Lim\ (d_1 + d_3) = a, \quad Lim\ U_2 d_2 = p.$$

For such a Kronig–Penney crystal Eqs. (C.33), (C.35) and (C.36) in Appendix C give that

$$\begin{aligned} \eta_1(\tau + a, \xi) &= cos\ \xi + 2\tfrac{p}{\xi} c_{11} s_{33}, \\ \eta_2(\tau + a, \xi) &= \tfrac{a}{\xi}(sin\ \xi + 2\tfrac{p}{\xi} s_{11} s_{33}), \\ \eta_2'(\tau + a, \xi) &= cos\ \xi + 2\tfrac{p}{\xi} s_{11} c_{33}, \end{aligned} \tag{A.39}$$

where $c_{ll} = cos\ k_l d_l$, $s_{ll} = sin\ k_l d_l$ as defined in (C.11), and $k_1 = k_3 = \frac{\xi}{a}$. Note from Eq. (A.39) the discriminant $D(\xi)$ of such a Kronig–Penney crystal is

$$D(\xi) = \eta_1(\tau + a, \xi) + \eta_2'(\tau + a, \xi) = 2(cos\ \xi + \frac{p}{\xi} sin\ \xi),$$

since $d_1 + d_3 = a$. This is same as $D(\xi)$ in (A.11).

From Eqs. (3.16) and (3.17) the existence of a surface state in a band gap is determined by

$$\sigma a = \frac{p\ sin \frac{\xi}{a}(d_1 - d_3) + \frac{\xi}{2}\sqrt{D^2(\xi) - 4}}{sin\ \xi + \frac{p}{\xi}[cos \frac{\xi}{a}(d_1 - d_3) - cos\ \xi]} \tag{A.40}$$

for a band gap at the Brillouin zone boundary $k = \pi/a$; or

$$\sigma a = \frac{p\ sin \frac{\xi}{a}(d_1 - d_3) - \frac{\xi}{2}\sqrt{D^2(\xi) - 4}}{sin\ \xi + \frac{p}{\xi}[cos \frac{\xi}{a}(d_1 - d_3) - cos\ \xi]} \tag{A.41}$$

for a band gap at the Brillouin zone center $k = 0$. In (A.40) and (A.41), d_1 and d_3 are related to τ by (A.38). These two equations determine the existence of surface states in band gaps and their properties, such as how the eigenvalue $\Lambda = \xi^2/(\chi a)^2$ of each existed surface state depends on τ and σ.

For cases where the outside potential barrier is a constant $U_0 = q^2$ [2, 7], we have that

$$\sigma\ a = \sqrt{q^2 - \xi^2}. \tag{A.42}$$

In Fig. A.9 are shown the numerical results of the $\tau - \xi$ relationships for four different $U_0 = q^2$ calculated by (A.41) and (A.42), for the lowest band gap at the Brillouin zone center $k = 0$. From this figure, we can see the following: First, for each specific outside barrier potential, there are τ intervals in $[-\frac{a}{2}, \frac{a}{2}]$, no $\tau - \xi$ curves exist. That indicates that no surface state could exist in the band gap for the specific outside potential barrier $U_0 = q^2$ and a τ in those intervals; Second, those no-surface state

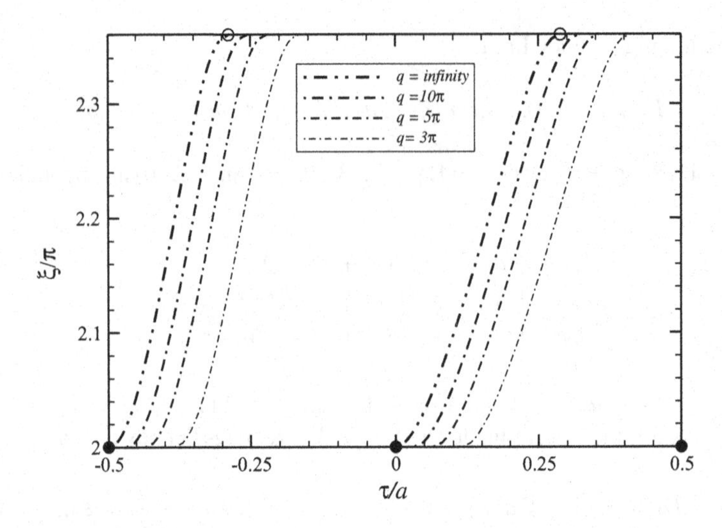

Fig. A.9 The functions $\tau - \xi$ obtained from (A.41) and (A.42) with four different constant barrier potentials q^2 outside the semi-infinite Kronig–Penney crystal in the lowest band gap at $k = 0$. The *double-dotted* chain lines corresponding to $q = \infty$ are exact the same as the *double-dotted* chained lines in Fig. A.8. The other *chained lines* of different thickness correspond to that $q = 10\pi$ (*thickest lines*), $q = 5\pi$ and $q = 3\pi$ (*thinnest lines*)

intervals move *right* as the outside barrier potential q^2 decreases, as analyzed in Chap. 3; Third, for a specific boundary τ, the energy $\Lambda = \xi^2/(\chi a)^2$ of an existing surface state increases as the outside barrier potential q^2 increases, in consistence with Eq. (3.7); Fourth, for a specific outside barrier potential q^2, the energy of an existing surface state increases as τ increases, in consistence with Eq. (3.8).

In an ideal semi-infinite Kronig–Penney crystal where the boundary condition (A.36) at τ is simplified to

$$\psi(\tau, \xi) = 0. \tag{A.43}$$

A necessary condition for existence of a surface state now is

$$\psi(\tau + a, \xi) = \psi(\tau, \xi) = 0, \tag{A.44}$$

or simpler as

$$\eta_2(\tau + a, \xi) = 0, \tag{A.45}$$

from (A.37) or (3.21).

A.4.2 Comparisons with the Tamm's Work and the Seitz's Book

As an application of Eqs. (A.40) and (A.41), we compare the results obtained here with the results obtained in the Tamm's classical work [5] and described in the Seitz's classical book [2]. In Tamm's work, a specific boundary location was used, that corresponds to $\tau \to +0$ in our more general investigation presented here.

In the limit $\tau \to +0$, both Eqs. (A.40) and (A.41) give

$$\frac{p - \sigma a}{\xi} = \sqrt{\frac{p^2}{\xi^2} + 2\,\frac{p}{\xi}\,cot\;\xi - 1}, \tag{A.46}$$

since $sin\;\xi \geq 0$ for a band gap at $k = 0$ and $sin\;\xi \leq 0$ for a band gap at $k = \pi/a$.

That is

$$\sqrt{\frac{p^2}{\xi^2} + 2\,\frac{p}{\xi}\,cot\;\xi - 1} = \frac{p - \sqrt{q^2 - \xi^2}}{\xi} \tag{A.47}$$

from (A.42). By squaring both sides of (A.47), we obtain that

$$\frac{p^2}{\xi^2} + 2\,\frac{p}{\xi}\,cot\;\xi - 1 = \frac{(q^2 - \xi^2) - 2p\sqrt{q^2 - \xi^2} + p^2}{\xi^2}$$

and thus

$$\xi\;cot\;\xi = \frac{q^2}{2p} - \sqrt{q^2 - \xi^2}. \tag{A.48}$$

This is Eq. (19) in Tamm's paper [5]. It is also Eq. (3.18) in [7] and corresponds to Eq. (6) in p. 322 in Seitz's book [2]. Equation (A.48) has one solution for each interval $(n + 1)\pi < \xi < (n + 2)\pi$ if $\sigma > 0$, i.e., $q^2 - \xi^2 > 0$. Probably it was this fact that leads to a wide and long-time belief in the solid state physics community that there is always a surface state in each band gap below the barrier height due to the termination of the periodic potential [2]. Nevertheless, as Tamm pointed out [5] that the condition $p - \sqrt{q^2 - \xi^2} > 0$ is necessary since the left side of (A.47) is positive. Consequently, *the number of solutions of* (A.47) *might be significantly less than the number of solutions of* (A.48): A termination of the periodic potential at the boundary of a semi-infinite Kronig–Penney crystal *may or may not* cause a surface state in a specific band gap below the barrier height. The widespread and longtime belief that it always does, is a misunderstanding. It might be this belief that leads to another widely accepted belief that the two terminations of the periodic potential in a simple one-dimensional crystal of finite length cause two surface states - each being associated with either end of the finite crystal - in each band gap below the barrier height.

We can further demonstrate that the number of solutions of Eq. (A.47)—that is, the solutions in the Tamm's original paper [5]—is limited, and sometimes the surface state solutions in the Tamm's original paper actually may not exist at all.

The condition $p - \sqrt{q^2 - \xi^2} > 0$ and that $\sigma > 0$ in (A.42) lead to that if a surface state - a solution of (A.47) - can exist in a band gap, the following condition must be satisfied:

$$\sqrt{q^2 - p^2} < \xi < q. \tag{A.49}$$

To understand the problem more specifically, we discuss the case where the outside barrier has a specific q. In Fig. A.10 is shown that Fig. A.1 is combined with the condition (A.49) for a specific $q = 5\pi$. There are always four band gaps (n = 0, 1, 2, 3) below the barrier; each gap may have a solution satisfying (A.48). Nevertheless, whether and how many surface state solutions can exist in these four band gaps might be quite different for different Kronig–Penney crystals.

For a Kronig–Penney crystal of $p = 3\pi/2$, we see that all the four band gaps are below the dashed line, and thus, the condition (A.49) cannot be satisfied. Accordingly, in none of the four band gaps can there be a solution satisfying (A.47) and thus a surface state, although each gap may have a solution satisfying (A.48).

However, for a Kronig–Penney crystal of $p = 4\pi$, two band gaps (n = 0, 1) are below the dashed line and thus (A.49) cannot be satisfied. In none of the two band gaps can there be a solution satisfying (A.47) and thus a surface state; The other

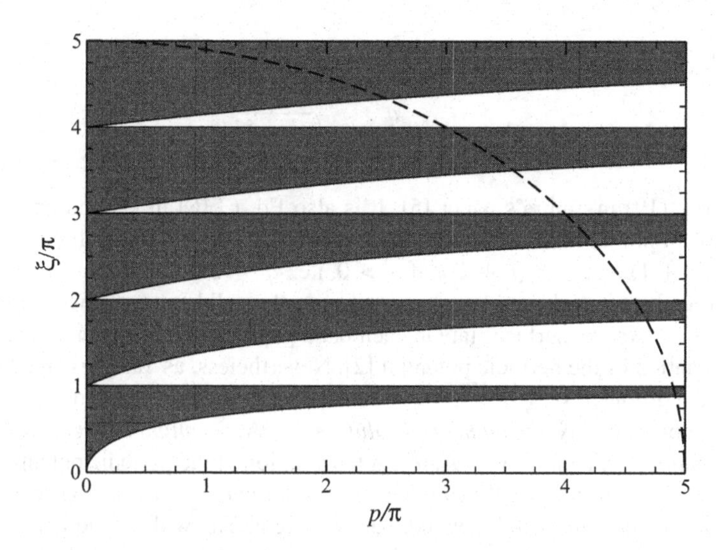

Fig. A.10 Fig. A.1 combined with the condition (A.49) for a specific barrier $q = 5\pi$. The *dashed line* corresponds to $\xi = \sqrt{q^2 - p^2}$. According to (A.49), only a ξ in a band gap *above* the *dashed line* and *below* the line $\xi = q = 5\pi$ may correspond to a surface state

two band gaps (n = 2, 3) are above the dashed line, thus (A.49) can be satisfied. Correspondingly, this Kronig–Penney crystal may have a surface state solution in each one of the two band gaps.

For a Kronig–Penney crystal of $p = 4.95\pi$, all four band gaps are above the dashed line and thus (A.49) can be satisfied. Such a Kronig–Penney crystal may have a surface state solution in each one of the four band gaps below $q = 5\pi$.

A.5 Electronic States in A Finite Kronig–Penney Crystal of Length $L = Na$

For an ideal finite Kronig–Penney crystal of length $L = Na$—where N is a positive integer—with a left boundary at τ thus a right boundary at $\tau + L$, the electronic states are the solutions of the equation

$$-\psi''(x) + \left[\sum_{n=-\infty}^{\infty} \frac{2p}{a}\delta(x-na) - \frac{\xi^2}{a^2}\right]\psi(x) = 0, \quad \tau < x < \tau + L, \quad \text{(A.50)}$$

with boundary conditions at τ and $\tau + L$ for an ideal crystal of finite length:

$$\psi(\tau) = \psi(\tau + L) = 0. \quad \text{(A.51)}$$

In general, a nontrivial solution of (A.50) with the boundary condition (A.51)—if it exists—can be expressed as

$$\psi(\xi, x) = \begin{cases} y(\xi, x), & \tau < x < \tau + L, \\ 0, & x \le \tau \text{ or } x \ge \tau + L, \end{cases}$$

where $y(\xi, x)$ is a linearly combination of the two independent solutions $y_1(\xi, x)$ and $y_2(\xi, x)$ of (A.4):

$$y(\xi, x) = c_1\, y_1(\xi, x) + c_2\, y_2(\xi, x). \quad \text{(A.52)}$$

$y_1(\xi, x)$ and $y_2(\xi, x)$ are determined by the discriminant $D(\xi)$ of Eq. (A.4). $y(\xi, x)$ in (A.52) is a general form of nontrivial solutions of Eq. (A.50). It is required to further satisfy

$$y(\xi, \tau) = y(\xi, \tau + L) = 0. \quad \text{(A.53)}$$

to give nontrivial solutions of (A.50) and (A.51).

We consider two different cases: (1) ξ is inside a permitted band and (2) ξ is not inside a permitted band. Without losing generality, we only need to consider τ in one period $[-\frac{a}{2}, \frac{a}{2}]$.

A.5.1 ξ Is Inside a Permitted Band

If ξ is inside a permitted band of Eq. (A.4), two independent solutions $y_1(\xi, x)$ and $y_2(\xi, x)$ in (A.52) are two Bloch wave solutions $\phi_n(\pm k, x)$ $(0 < k < \pi/a)$ given by (A.12):

$$y_1(\xi, x) = \phi_n(k, x), \quad y_2(\xi, x) = \phi_n(-k, x).$$

Equations (A.52) and (A.53) from Eq. (A.12) give that

$$\begin{aligned} &c_1\, \phi_n(k, \tau) + c_2\, \phi_n(-k, \tau) = 0, \\ &c_1\, e^{ikL}\phi_n(k, \tau) + c_2\, e^{-ikL}\phi_n(-k, \tau) = 0, \end{aligned} \qquad -\frac{a}{2} \leq \tau \leq \frac{a}{2}. \tag{A.54}$$

Since neither $\phi_n(k, \tau)$ nor $\phi_n(-k, \tau)$ can be zero (Sect. A.3.1), from (A.54) we obtained that

$$e^{ikL} - e^{-ikL} = 0 \tag{A.55}$$

is the condition that Eqs. (A.52) and (A.53) could have non-trivial solutions. Equation (A.55) has $N - 1$ solutions for $0 < k < \pi/a$:

$$k = \frac{j\pi}{L}, \qquad j = 1, 2, \ldots, N - 1. \tag{A.56}$$

Thus, (A.50) and (A.51) has $N - 1$ solutions in each permitted band, with ξ determined by:

$$cos\frac{j\pi}{N} = cos\ \xi + \frac{p}{\xi}\, sin\ \xi, \qquad j = 1, 2, \ldots, N - 1. \tag{A.57}$$

A.5.2 ξ Is Not Inside a Permitted Band

ξ can be either inside a band gap or at a band edge if it is not inside a permitted band of Eq. (A.4).

If ξ is inside a band gap, two independent solutions $y_1(\xi, x)$ and $y_2(\xi, x)$ of (A.4) in (A.52) can be written either as the forms of (A.14) if the band gap is at $k = \frac{\pi}{a}$ or as the forms of (A.16) if the band gap is at $k = 0$. By using a similar approach as we used to obtain (A.55), we obtain that

$$y(\tau + a, \xi) = y(\tau, \xi) = 0 \tag{A.58}$$

is a necessary condition for the existence of solutions of (A.50) and (A.51).

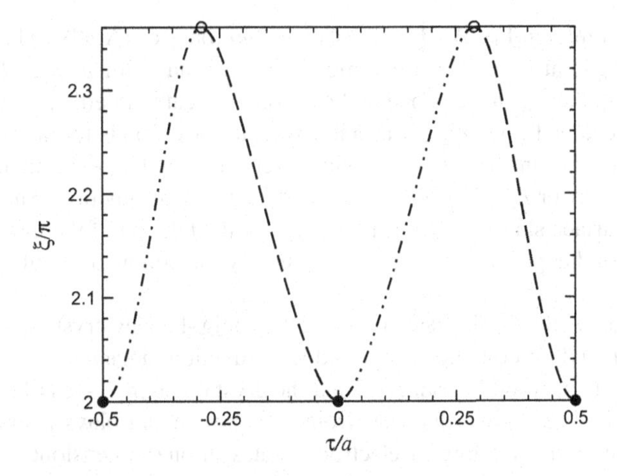

Fig. A.11 The τ-dependence of the solution ξ of (A.50) and (A.51) in the lowest band gap at $k = 0$ for a Kronig–Penney crystal with $p = 3/2\pi$. For each $-\frac{a}{2} \leq \tau \leq \frac{a}{2}$, there is always one and only one solution ξ of (A.50) and (A.51) in the band-gap, its corresponding ψ is either a surface state located near the left end of the finite crystal (*chained lines*), or a surface state located near the right end of the finite crystal (*dashed lines*), or a confined band-edge state (*solid circles* and *open circles*)

If ξ is at a band edge, we can choose $y_1(\xi, x)$ in (A.52) as a band-edge function described in Sect. A.3.2 and $y_2(\xi, x)$ as the other solution of (A.4) independent of $y_1(\xi, x)$. Based on the periodicity or semi-periodicity of the band-edge functions and the Sturm Separation Theorem, we know that only the band-edge function $y_1(\xi, x)$ can be a solution of (A.50) and (A.51) for ξ at a band edge of Eq. (A.4). We also obtain that (A.58) is a necessary condition for the existence of solutions of (A.50) and (A.51).

From (A.58) we can obtain that

$$y(\tau + \ell a, \xi) = y(\tau, \xi) = 0, \tag{A.59}$$

where $\ell = 1, 2, \ldots, N$. Therefore (A.58) is a *necessary and sufficient* condition for the existence of a solution of (A.50) with the boundary conditions (A.51) where ξ is not inside a permitted band. Theorem 2.8 indicates that corresponding to each band gap, there is always one and only one solution of (A.58).

The calculated $P = \frac{y(\tau+a,\xi)}{y(\tau,\xi)} = \frac{y'(\tau+a,\xi)}{y'(\tau,\xi)}$ from (A.58) can have three possibilities: (a) $0 < |P| < 1$; (b) $|P| = 1$; (c) $|P| > 1$. The case (a) corresponds to a solution $\psi(x, \xi)$ of Eqs. (A.50) and (A.51) oscillatory decreasing in the $+x$ direction, indicating a surface state localized near the left boundary τ; The case (b) corresponds to that $\psi(x, \xi)$ is a band-edge state of Eq. (A.4) in the finite crystal; The case (c) corresponds to an oscillatory increasing $\psi(x, \xi)$ in the $+x$ direction, thus an oscillatory decaying solution in the $-x$ direction, indicating a surface state localized near the right boundary $\tau + L$.

As an example, in Fig. A.11 is shown the solution ξ of (A.50) and (A.51) in the lowest band gap at $k = 0$ for a Kronig–Penney crystal with $p = 3/2\pi$. For each $-\frac{a}{2} \leq \tau \leq \frac{a}{2}$, there is always one and only one ξ located in either one dashed line area or one chained line area, or at one circle. It corresponds to one solution ψ of (A.50) and (A.51) in the band-gap with a zero at τ and $\tau + L$, having the form of $c_1 e^{\beta x} p_1(\xi, x)$ or $c_2 e^{-\beta x} p_2(\xi, x)$, or being a periodic function. Such a solution indicates a surface state located near the right or the left end of the finite crystal, or, a confined band edge state. Figure A.11 is simply the combination of Figs. A.7 and A.8.

Therefore, all electronic states in an ideal Kronig–Penney crystal of finite length $L = Na$—including how the energy and wavefunction of each state depend on the parameter p, the crystal boundary τ, and the crystal length L—can be analytically obtained. It is well-known that the Kronig–Penney model plays a significant role in our current understanding of electronic states in one-dimensional crystals with translational invariance. We have seen here that the model again provides a significant and analytical understanding of electronic states in an ideal one-dimensional periodic system of finite size.

References

1. R.L. Kronig, W.G. Penney, Proc. R. Soc. Lond. Ser. A. **130**, 499 (1931)
2. F. Seitz, *The Modern Theory of Solids* (McGraw-Hill, New York, 1940)
3. H. Jones, *The Theory of Brillouin Zones and Electronic States in Crystals* (North-Holland, Amsterdam, 1960)
4. R.L. Liboff, *Introductory Quantum Mechanics* 4th edn. (Addison-Wesley, Reading, 2002); E.P. O'Reilly, *Quantum Theory of Solids* (Taylor & Fransis, London, 2003); L. Kantorovich, *Quantum Theory of the Solid State: An Introduction* (Springer, Berlin, 2004); C. Kittel, *Introduction to Solid State Physics*, 8th edn. (Wiley, New York, 2005) (For example)
5. I. Tamm: Phys. Z. Sowj. **1**, 733 (1932)
6. M. Stęślicka, Prog. Surf. Sci. **5**, 157 (1974)
7. S.G. Davison, M. Stęślicka, *Basic Theory of Surface States* (Clarendon Press, Oxford, 1992)

Appendix B
Electronic States in One-Dimensional Symmetric Finite Crystals with a Finite V_{out}

The Schrödinger differential equation for a one-dimensional crystal can be written as

$$-y''(x) + [v(x) - \lambda]y(x) = 0. \tag{B.1}$$

Here, $v(x) = v(x + a)$ is the periodic potential of the crystal.

For a one-dimensional crystal of finite length $L = Na$, the eigenvalues Λ and eigenfunctions $\psi(x)$ are solutions of the equation

$$-\psi''(x) + [v(x) - \Lambda]\psi(x) = 0, \quad \tau < x < \tau + L, \tag{B.2}$$

inside the crystal with certain boundary conditions at the two boundaries τ and $\tau + L$. If the potential outside the crystal $V_{out} = +\infty$, we have the boundary conditions

$$\psi(x) = 0, \quad x = \tau \text{ or } x = \tau + L. \tag{B.3}$$

This is the case treated in Chap. 4. It is found that for each band gap, there is always one and only one state whose energy is boundary dependent but independent of the crystal length. A surface state is one of the two possibilities of such a boundary-dependent state. Therefore, there is *at most* one surface state in each band gap in an *ideal* one-dimensional finite crystal.

Many years ago, Shockley published a classic paper [1]. The paper stated that in a one-dimensional symmetric finite crystal when the potential period a is so small that the boundary curves for allowed energy bands have crossed, and the number of atoms N in the crystal is *very large*, the surface states appear *in pairs* in band gaps. To clearly understand the relationship between the results of [1] and Chap. 4, in this Appendix we investigate the cases where the electrons are not completely confined to the crystal as in [1], and the crystal length may not be very long.

Now, we need to consider the cases where V_{out} is finite. Qualitatively, the effect of a finite V_{out} can be directly obtained from a theorem in [2]: A finite V_{out} moves all energy levels lower. Quantitatively, a finite V_{out} will allow a small part of the electronic state spills out of the finite crystal and thus, make the boundary conditions be

S.Y. Ren, *Electronic States in Crystals of Finite Size*, Springer Tracts in Modern Physics 270, DOI 10.1007/978-981-10-4718-3

$$\begin{aligned}(\psi'/\psi)_{x=\tau} &= \sigma_1,\\ (\psi'/\psi)_{x=\tau+L} &= -\sigma_2,\end{aligned} \tag{B.4}$$

instead of (B.3). Here, σ_1 and σ_2 are positive numbers depending on V_{out}. Note that (B.3) corresponds to $\sigma_1 = \sigma_2 = +\infty$, and σ_1 and σ_2 will monotonically decrease as V_{out} decreases. Although V_{out} may have different forms, the effect of different V_{out} to the problem treated here can be simplified to be the effect of σ_1 and σ_2. Shockley treated one-dimensional symmetric finite crystals with finite V_{out}, where $\sigma_1 = \sigma_2 = \sigma$. His treatment provided a way to investigate how much the results obtained in Chap. 4 are dependent on V_{out} for symmetric one-dimensional finite crystals. For the convenience of comparison with his original paper, we use his approach assuming that the cell potential in the crystal is symmetric and use same notations as in [1] except that the energy is written as λ rather than E and the number of atoms in the crystal is N in this appendix. As in Shockley's paper, we also consider the two lowest band gaps: one at $k = \pi/a$ and one at $k = 0$.

Assuming $g(x)$ and $u(x)$ are two linearly independent solutions of the Schrödinger differential equation (B.1) in a unit cell, symmetric or antisymmetric to the cell center $x = 0$, Shockley obtained $g(a/2)u'(a/2)(1 - \mathrm{e}^{-ika}) = g'(a/2)u(a/2)(1 + \mathrm{e}^{ika})$ and further that $\sigma = \mu \tan(ka/2)\tan(Nka/2)$ and $\sigma = -\mu \tan(ka/2)\cot(Nka/2)$ (Eqs. (11) and (12) in [1]) give the energies of electronic states in the one-dimensional finite crystal; here, $\mu = u'(a/2)/u(a/2)$. Therefore, the effect of finite V_{out} can be found from the σ dependence of energy levels. In Fig. B.1 is shown a numerical calculation for the electronic states near the upper band-edge $\varepsilon_2(0)$ of the band gap at $k = 0$ in crystals of two different lengths, $N = 14$ and $N = 15$, with a model cell potential

$$\begin{aligned} v(x) &= -30 \quad \text{if } |x| \le 0.38 \\ &= 0 \qquad \text{if } 0.38 < |x| \le 0.5 \end{aligned}$$

and $a = 1$. It can be seen that lowering V_{out} (thus lowering σ) moves all energy levels downward. However, the energy of the state in the band gap depends on the crystal length much less than the states in the energy band: The major difference between the state corresponding to a band gap and the states corresponding to an energy band obtained in Sect. 4.2 remains.

For many physical situations, σ can be considered as sufficiently large [3]. It can be shown that for those states in Fig. B.1, in the limit of large σ (i.e., large V_{out}), the energies of the states in the energy band can approximately be given by

$$\Lambda_{2,j} = \varepsilon_2(k_j)$$

and

$$k_j = \frac{j\pi}{Na} - \frac{2}{Na}\frac{\mu}{\sigma}\tan\left(\frac{j\pi}{2N}\right), \quad j = 1, 2, \ldots, N-1, \tag{B.5}$$

where $\mu > 0$. On the other hand, the energy of the state in the gap is given approximately by $(\varepsilon_2''(0) > 0)$

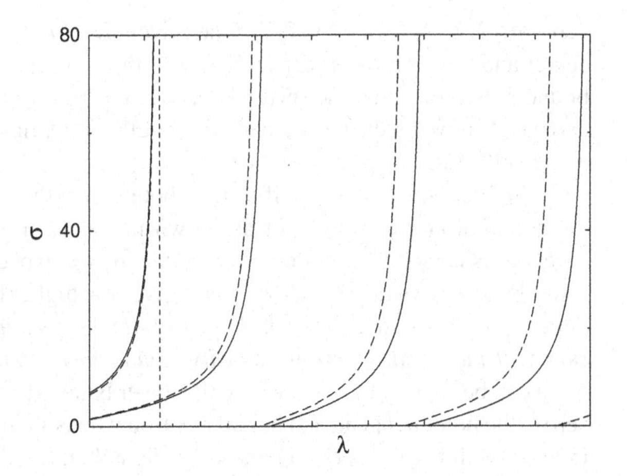

Fig. B.1 $\sigma = \mu \tan(ka/2) \tan(Nka/2)$ and $\sigma = -\mu \tan(ka/2) \cot(Nka/2)$ calculated for $N = 14$ (*solid lines*) and $N = 15$ (*long dashed lines*) near the upper band edge $\varepsilon_2(0)$ of the band gap at $k = 0$. The *short-dashed vertical line* is the band edge. Note the energy of the state in the band gap almost does not depend on the crystal length, even for a finite σ

$$\Lambda_{1,gap} = \varepsilon_2(0) - \varepsilon_2''(0)\frac{6(c-1)}{(cN^2-1)a^2};$$

here, $c = -\sigma N/\mu > 1$, $\mu < 0$, and $c \to 1$ when $\sigma \to +\infty$. Again, it can be clearly seen that lowering σ (lowering V_{out}) moves all energy levels downward, and the energy of the state in the band gap depends on the crystal length much less than the energies of the states in the energy band.

Shockley found that when (i) the potential period a is so small that the boundary curves for allowed energy bands have crossed and (ii) the number of atoms in the crystal N is *very large*, the surface states appear *in pairs* in the band gap. Now, we try to give this problem a more careful investigation and attempt to understand whether and how "two surface states" in a band gap could happen in a one-dimensional symmetric finite crystal. We also consider the cases of N is even, as in [1].

In general, inside a band gap, an electronic state as a nontrivial solution of (B.1) always has the form

$$y(x) = Ae^{\beta x} f_1(x) + Be^{-\beta x} f_2(x) \tag{B.6}$$

from (2.77) or (2.81); here, A and B are not both zero, $\beta > 0$, and $f_i(x)$ is either a periodic function ($f_i(x+a) = f_i(x)$) if the band gap is at $k = 0$ or a semi-periodic function ($f_i(x+a) = -f_i(x)$) if the band gap is at $k = \pi/a$. Equation (B.6) is more general than the simple surface states in which either A or B is zero thus the state is localized near *one* end of the crystal. Such a state (B.6) in a symmetric one-dimensional finite crystal must be either symmetric ($A = B$) or antisymmetric ($A = -B$) and, thus, is equally localized near the *both* ends of the finite crystal and can be considered as a generalized surface state. We are trying to investigate how many states of type (B.6), as solutions of (B.2) with the boundary conditions (B.4), are in a specific band gap.

For the gap at $k = \pi/a$, two band-edge states are given by either $g(a/2) = 0$ or $u'(a/2) = 0$, as in [1]. Both band-edge wavefunctions have one node in a unit cell $[-a/2, a/2)$ (Theorem 2.7). One (given by $g(a/2) = 0$) is symmetric to the cell center and has its most electron density at the cell center and zero density at the cell boundaries. The other one (given by $u'(a/2) = 0$) is antisymmetric to the cell center and has its most electronic density at the cell boundaries $x = \pm a/2$ and zero density at the cell center.

No matter how small a is, if the cell potential at the cell boundaries is higher than the potential at the cell center as shown in Fig. 1a in [1] and the form of the cell potential is reasonable and not very irregular; we expect that $g(a/2) = 0$ gives the lower band-edge state and $u'(a/2) = 0$ gives the higher band-edge state: *A state with most of its electronic density in the potential valley should have lower energy than a state with most of its electronic density around the potential peak.* As an example, this point can be seen in Fig. A.3 where the lower band-edge wavefunction $\phi_0(\pi/a, x)$ is zero at the potential peak and has all electronic density in the potential valley.[1] In fact, Levine [3] did not observe a band-crossing either. Shockley has shown that the two surface states in the gap can happen only when $g(a/2) = 0$ gives the higher band-edge state. Thus, the existence of two surface states in the lowest gap at $k = \pi/a$, as shown in Fig. 2 in [1], seems unlikely for a reasonably regular one-dimensional finite crystal. Consistent with the analysis here, many other authors did not obtain a "Shockley" surface state in the lowest gap at $k = \pi/a$ either [3, 4].

Then we consider the next band gap at $k = 0$. The two band-edge states are given by either $g'(a/2) = 0$ or $u(a/2) = 0$; which one is higher depends on the form of the cell potential. If $V_{out} = +\infty$, equations (11) and (12) in [1] for $\sigma = +\infty$ give $N - 1$ states ($k_j = j\pi/Na,\ j = 1, 2, \ldots, N - 1$) for each energy band and one confined band-edge state for each band gap. The confined band-edge state for this band gap is the band-edge state given by $u(a/2) = 0$ since its wavefunction is zero at the crystal boundaries. These results are consistent with Sect. 4.4.

If the confined band-edge state is at the *lower* band-edge $\varepsilon_1(0)$ when $V_{out} = +\infty$, an (any) finite V_{out} will move it downward into the energy band $\varepsilon_1(k)$ below and, thus, will not make a surface state. Only if the confined band-edge state is at the *upper* band-edge $\varepsilon_2(0)$ when $V_{out} = +\infty$, an (any) finite V_{out} will move it downward into the band gap and thus make a surface state. That corresponds to the case that $u(a/2) = 0$ gives the higher band-edge state.

In Fig. B.2 is shown a numerical calculation of such a situation, using the same model cell potential as in Fig. B.1, in comparison with Fig. 4 in [1]. When $V_{out} = +\infty (\sigma = +\infty)$, $u(a/2) = 0$ gives an antisymmetric confined band-edge state at the upper band-edge $\varepsilon_2(0)$. Any finite σ due to a finite V_{out} can move this state (long-dashed line) into the band gap and thus make one antisymmetric gap state. However, moving a symmetric state (solid line) crossing the higher band-edge $\varepsilon_2(0)$ into the band gap and making another surface state requires

$$\sigma < -N\gamma_u; \tag{B.7}$$

[1]Please be aware of that $\pm\frac{a}{2}$ are at a potential peak in Shockley's paper [1] whereas in a potential valley in Fig. A.3.

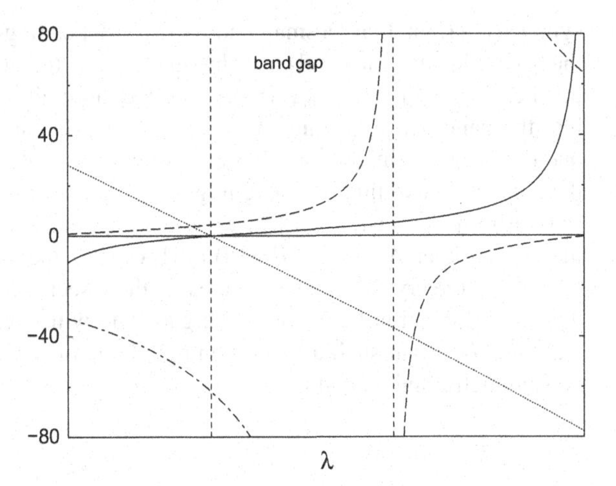

Fig. B.2 $\gamma \times 100$ (*dotted line*), μ (*chained line*), $\mu \tan(ka/2)\tan(Nka/2)$ (*solid line*) and $-\mu \tan(ka/2)\cot(Nka/2)$ (*long-dashed line*) for $N = 14$ near the band gap at $k = 0$ where the surface state(s) may exist. The *left short-dashed vertical line* corresponds to the lower band edge $\varepsilon_1(0)$, and the *right short-dashed vertical line* corresponds to the upper band edge $\varepsilon_2(0)$

here, γ_u is γ $(= g'(a/2)/g(a/2))$ at the upper band-edge, a negative number. Therefore, in principle, if there are two surface states (one antisymmetric and one symmetric) in the gap, σ (or V_{out}) needs to be small, and N needs to be large. However, a too small σ (or V_{out}) may even move the antisymmetric surface state crossing the lower band edge $\varepsilon_1(0)$ out of the band gap and into the lower energy band $\varepsilon_1(k)$. This case happens when $\sigma < -\mu_l/N$; here, μ_l is μ at the lower band-edge $\varepsilon_1(0)$, also a negative number. Note that μ_l and γ_u are determined by the cell potential, and σ is dependent on V_{out}. Figure B.2 shows the case for $N = 14$: When a small enough σ (or V_{out}) moves the symmetric state (solid line) from the upper band $\varepsilon_2(k)$ into the band gap, the antisymmetric surface state (long-dashed line) almost enters the lower band $\varepsilon_1(k)$. In fact, σ usually is quite large [3].[2] Depending on σ or V_{out}, usually a much larger N is needed to satisfy $\sigma < -N\gamma_u$. An even greater N is required if the two surface states are almost degenerated. Two degenerated gap states of (B.6) type—one symmetric and one antisymmetric in a symmetric one-dimensional finite crystal—can be linearly combined and transformed to two surface states, one at each end.

The analysis here, as in Shockley's paper [1], are for symmetric one-dimensional finite crystals. Nevertheless, we can also obtain some understanding of the surface states in general one-dimensional finite crystals. Since there is only one state corresponding to each gap for general one-dimensional finite crystals when $V_{out} = +\infty$ and a finite V_{out} always moves all energy levels downward, in any case, if there are two states in a specific band gap for a finite V_{out}, one state must come from the permitted band above that band gap and is driven downward by the finite V_{out}. This state had an energy $\varepsilon_{2m+2}(\pi/Na)$ (for a band gap at $k = 0$) or $\varepsilon_{2m+1}[(N-1)\pi/Na]$ (for a band gap at $k = \pi/a$) when $V_{out} = +\infty$, independent of whether the crystal is

[2]In almost all previously published numerical calculations, the deviations from $k_j = j\pi/(Na)$ is small; for example, see [5, 6]. Thus, from (B.5), one can obtain that usually $(2\mu/\sigma)\tan(j\pi/2N) << 1$.

symmetric or not. Only a small enough V_{out} (depending on N) can move it crossing the band-edge into the band gap. The smaller N is, the further the state is to the upper band-edge $\varepsilon_{2m+2}(0)$ or $\varepsilon_{2m+1}(\pi/a)$, and the more difficult is the state to be moved into the band gap by a finite V_{out}. Therefore, we can expect that a not very long one-dimensional finite crystal has at most one gap state in each band gap. However, the localization of this gap state might be somewhat different for a nonsymmetric finite crystal: Because when $V_{out} = +\infty$, a gap state in a nonsymmetric finite crystal may have either $A = 0$ or $B = 0$ in (B.6) (i.e., the state could be localized near one end of the crystal), as V_{out} decreases there seems no understandable reason for that the localization behavior of the gap state will have a dramatic change. Thus, we can expect that such a gap state might be mainly localized near *one* end of the nonsymmetric finite crystal.

References

1. W. Shockley, Phys. Rev. **56**, 317 (1939)
2. R. Courant, D. Hilbert, *Methods of Mathematical Physics*, vol. 1, p. 409, (Interscience, New York, 1953) (Theorem 3 and the relevant footnote)
3. J.D. Levine, Phys. Rev. **171**, 701 (1968)
4. S.G. Davison, M. Stęślicka, *Basic Theory of Surface States* (Clarendon Press, Oxford, 1992)
5. S.B. Zhang, A. Zunger, Appl. Phys. Lett. **63**, 1399 (1993)
6. S.B. Zhang, C.-Y. Yeh, A. Zunger, Phys. Rev. B **48**, 11204 (1993)

Appendix C
Layered Crystals

A layered crystal discussed in this book refers to a specific kind of periodically layered structures, in which each period is composed of two or more layers of homogeneous isotropic media. The wave equation in such a crystal can generally be written as a Sturm–Liouville equation with periodic coefficients:

$$[p(x)y'(x)]' + [\lambda w(x) - q(x)]y(x) = 0, \tag{C.1}$$

where $p(x) > 0$, $w(x) > 0$ and $q(x)$ are real constants in each layer and are all periodic with the same period a:

$$p(x + a) = p(x), \quad q(x + a) = q(x), \quad w(x + a) = w(x). \tag{C.2}$$

Each period unit of the simplest layered crystals is usually composed of two different media [1–5]. However, in some cases one might also need to deal with layered crystals in which each period is composed of three or more layers of different media [6–9].

In this Appendix, we apply the theory of periodic Sturm–Liouville equations in Chap. 2 to treat layered crystals and to obtain some general expressions for later use. We consider an (any) specific period of a layered crystal which is composed of N layers of different homogeneous isotropic media. Each layer l has a layer thickness d_l, where $l = 1, 2, \ldots, N$. The two ends of the lth layer are denoted to be x_{l-1} and x_l thus $d_l = x_l - x_{l-1}$. We choose $x_0 = 0$, thus $x_N = a$ is the period of the layered crystal. Inside the lth layer

$$p(x) = p_l, \quad q(x) = q_l, \quad w(x) = w_l, \tag{C.3}$$

are real constants and $p_l > 0$, $w_l > 0$. Equation (C.1) in each layer now can be written in the form

$$[p_l y'(x)]' + [\lambda w_l - q_l]y(x) = 0, \qquad x_{l-1} \leq x \leq x_l. \tag{C.4}$$

S.Y. Ren, *Electronic States in Crystals of Finite Size*, Springer Tracts in Modern Physics 270, DOI 10.1007/978-981-10-4718-3

Or equivalently

$$[p_l y'(x)]' + p_l k_l^2 y(x) = 0, \qquad x_{l-1} \le x \le x_l, \tag{C.5}$$

by introducing

$$p_l k_l^2 = \lambda w_l - q_l. \tag{C.6}$$

k_l is real if $\lambda w_l - q_l > 0$ or imaginary if $\lambda w_l - q_l < 0$.

According to the theory in Chap. 2, the band structure of such a layered crystal is determined by the discriminant of the Eq. (C.1):

$$D(\lambda) = \eta_1(a, \lambda) + p(a)\eta_2'(a, \lambda), \tag{C.7}$$

where $\eta_1(x, \lambda)$ and $\eta_2(x, \lambda)$ are two linearly independent normalized solutions of (C.1) that satisfy

$$\eta_1(0, \lambda) = 1, \ p(0)\eta_1'(0, \lambda) = 0; \ \ \eta_2(0, \lambda) = 0, \ p(0)\eta_2'(0, \lambda) = 1. \tag{C.8}$$

by (2.53).

From (C.5) we have in the lth layer

$$\begin{aligned} \eta_1(x, \lambda) &= A_l \cos k_l x + B_l \sin k_l x; \\ p(x)\eta_1'(x, \lambda) &= -p_l k_l \, A_l \sin k_l x + p_l k_l \, B_l \cos k_l x, \end{aligned} \qquad x_{l-1} \le x \le x_l, \tag{C.9}$$

where k_l is determined by p_l, q_l, w_l and the eigenvalue λ from (C.6). By defining

$$f_l = p_l \, k_l, \tag{C.10}$$

at $x = x_{l-1}$ we have

$$\begin{aligned} \eta_1(x_{l-1}, \lambda) &= A_l \cos k_l x_{l-1} + B_l \sin k_l x_{l-1}; \\ p_{l-1}\eta_1'(x_{l-1}, \lambda) &= -f_l \, A_l \sin k_l x_{l-1} + f_l \, B_l \cos k_l x_{l-1}. \end{aligned}$$

A_l and B_l can be obtained as,

$$\begin{aligned} A_l &= \eta_1(x_{l-1}, \lambda) \cos k_l x_{l-1} - \frac{1}{f_l} \, p_{l-1}\eta_1'(x_{l-1}, \lambda) \sin k_l x_{l-1}, \\ B_l &= \eta_1(x_{l-1}, \lambda) \sin k_l x_{l-1} + \frac{1}{f_l} \, p_{l-1}\eta_1'(x_{l-1}, \lambda) \cos k_l x_{l-1}. \end{aligned}$$

By writing

$$c_{ll} = \cos k_l d_l, \ s_{ll} = \sin k_l d_l \tag{C.11}$$

for brevity, at $x = x_l$ from (C.9) we obtain

$$\eta_1(x_l,\lambda) = \eta_1(x_{l-1},\lambda)\, c_{ll} + \frac{1}{f_l}\, p_{l-1}\eta_1'(x_{l-1},\lambda)\, s_{ll},$$
$$p_l\eta_1'(x_l,\lambda) = p_{l-1}\eta_1'(x_{l-1},\lambda)\, c_{ll} - f_l\, \eta_1(x_{l-1},\lambda)\, s_{ll}.$$

These two equations can be written as

$$\begin{pmatrix} \eta_1(x_l,\lambda) \\ p_l\eta_1'(x_l,\lambda) \end{pmatrix} = \begin{pmatrix} c_{ll} & \frac{1}{f_l}s_{ll} \\ -f_l s_{ll} & c_{ll} \end{pmatrix} \begin{pmatrix} \eta_1(x_{l-1},\lambda) \\ p_{l-1}\eta_1'(x_{l-1},\lambda) \end{pmatrix}. \tag{C.12}$$

Similarly, we have

$$\begin{pmatrix} \eta_2(x_l,\lambda) \\ p_l\eta_2'(x_l,\lambda) \end{pmatrix} = \begin{pmatrix} c_{ll} & \frac{1}{f_l}s_{ll} \\ -f_l s_{ll} & c_{ll} \end{pmatrix} \begin{pmatrix} \eta_2(x_{l-1},\lambda) \\ p_{l-1}\eta_2'(x_{l-1},\lambda) \end{pmatrix}. \tag{C.13}$$

Equations (C.12) and (C.13) can be further written as

$$\tilde{\eta}_i(x_l,\lambda) = \mathbf{M}_l\, \tilde{\eta}_i(x_{l-1},\lambda) = \prod_{j=l}^{1} \mathbf{M}_j\, \tilde{\eta}_i(x_0,\lambda) = \mathbb{M}_l\, \tilde{\eta}_i(x_0,\lambda), \quad i = 1, 2, \tag{C.14}$$

where

$$\tilde{\eta}_i(x_l,\lambda) = \begin{pmatrix} \eta_i(x_l,\lambda) \\ p_l\eta_i'(x_l,\lambda) \end{pmatrix}, \qquad l = 1, \ldots, N, \tag{C.15}$$

and

$$\mathbf{M}_j = \begin{pmatrix} c_{jj} & \frac{1}{f_j}s_{jj} \\ -f_j s_{jj} & c_{jj} \end{pmatrix}, \qquad j = 1, \ldots, N, \tag{C.16}$$

and

$$\mathbb{M}_l = \mathbf{M}_l\, \mathbf{M}_{l-1} \cdots \mathbf{M}_1 = \prod_{j=l}^{1} \mathbf{M}_j. \tag{C.17}$$

By (C.8),

$$\tilde{\eta}_1(x_0,\lambda) = \begin{pmatrix} 1 \\ 0 \end{pmatrix}, \qquad \tilde{\eta}_2(x_0,\lambda) = \begin{pmatrix} 0 \\ 1 \end{pmatrix}. \tag{C.18}$$

For a layered crystal composed of N different homogeneous isotropic media, $x_N = a$. Equation (C.14) gives that

$$\eta_1(a, \lambda) = (\mathbb{M}_N)_{11}, \tag{C.19}$$

$$p(a)\eta_1'(a, \lambda) = (\mathbb{M}_N)_{21}; \tag{C.20}$$

and

$$\eta_2(a, \lambda) = (\mathbb{M}_N)_{12}, \tag{C.21}$$

$$p(a)\eta_2'(a, \lambda) = (\mathbb{M}_N)_{22}. \tag{C.22}$$

The discriminant $D(\lambda)$ of Eq. (C.1) is

$$D(\lambda) = \eta_1(a, \lambda) + p(a)\eta_2'(a, \lambda) = \mathrm{Tr}(\mathbb{M}_N). \tag{C.23}$$

The band structure of the eigen-modes is determined by

$$cos\ ka = \frac{1}{2}D(\lambda) = \frac{1}{2}\mathrm{Tr}(\mathbb{M}_N), \tag{C.24}$$

where $-2 \leq D(\lambda) \leq 2$. Equations (C.19)–(C.24) and (C.16)–(C.17) are the basis of our treatments to layered crystals composed of different homogeneous and isotropic media, as in Appendices A, E and F.[3]

For $l = 1$, (C.19)–(C.22) give that

$$\eta_1(x_1, \lambda) = c_{11}, \quad p_1\eta_1'(x_1, \lambda) = -f_1 s_{11}, \tag{C.25}$$

$$\eta_2(x_1, \lambda) = \frac{1}{f_1}s_{11}, \quad p_1\eta_2'(x_1, \lambda) = c_{11}. \tag{C.26}$$

For a layered crystal composed of two different media, $N = 2$ and $x_2 = a$. (C.19)–(C.22) give that

$$\eta_1(a, \lambda) = c_{22}c_{11} - \frac{f_1}{f_2}s_{22}s_{11}, \tag{C.27}$$

$$p(a)\eta_1'(a, \lambda) = -f_1\, c_{22}s_{11} - f_2\, s_{22}c_{11}, \tag{C.28}$$

[3]Equations similar to (C.24), (C.17) and (C.16) were obtained as Eqs. (10), (11), (A.2) and (A.3) in [7]. The mathematical theory of periodic Sturm–Liouville equations straightforwardly gives more and general results. Equation (A.4) in [7] is simply a specific case of that the Wronskian of the solutions of a periodic Sturm–Liouville equation is a constant.

$$\eta_2(a,\lambda) = \frac{1}{f_1} c_{22}s_{11} + \frac{1}{f_2} s_{22}c_{11}, \tag{C.29}$$

and

$$p(a)\eta_2'(a,\lambda) = c_{22}c_{11} - \frac{f_2}{f_1} s_{22}s_{11}. \tag{C.30}$$

Equation (C.23) gives that

$$D(\lambda) = 2c_{22}c_{11} - \left(\frac{f_1}{f_2} + \frac{f_2}{f_1}\right) s_{22}s_{11}. \tag{C.31}$$

In the region $-2 \leq D(\lambda) \leq 2$, the band structure can be more explicitly written as

$$cos\ ka = cos\ k_2d_2\ cos\ k_1d_1 - \frac{1}{2}\left(\frac{f_1}{f_2} + \frac{f_2}{f_1}\right) sin\ k_2d_2\ sin\ k_1d_1 \tag{C.32}$$

by (C.24). This equation has been well known in the literature, see, for example, [1–5].

Further for a layered crystal of that each period is composed of three layers of different media, $N = 3$ and $x_3 = a$. Equations (C.19)–(C.22) give that

$$\eta_1(a,\lambda) = c_{33}c_{22}c_{11} - \frac{f_1}{f_2} c_{33}s_{22}s_{11} - \frac{f_1}{f_3} s_{33}c_{22}s_{11} - \frac{f_2}{f_3} s_{33}s_{22}c_{11}, \tag{C.33}$$

$$p(a)\eta_1'(a,\lambda) = -f_1\, c_{33}c_{22}s_{11} - f_2\, c_{33}s_{22}c_{11} - f_3\, s_{33}c_{22}c_{11} + \frac{f_3 f_1}{f_2} s_{33}s_{22}s_{11}, \tag{C.34}$$

$$\eta_2(a,\lambda) = \frac{1}{f_1} c_{33}c_{22}s_{11} + \frac{1}{f_2} c_{33}s_{22}c_{11} + \frac{1}{f_3} s_{33}c_{22}c_{11} - \frac{f_2}{f_1 f_3} s_{33}s_{22}s_{11}, \tag{C.35}$$

and

$$p(a)\eta_2'(a,\lambda) = c_{33}c_{22}c_{11} - \frac{f_2}{f_1} c_{33}s_{22}s_{11} - \frac{f_3}{f_1} s_{33}c_{22}s_{11} - \frac{f_3}{f_2} s_{33}s_{22}c_{11}. \tag{C.36}$$

Equation (C.23) gives that

$$D(\lambda) = 2c_{33}c_{22}c_{11} - \left(\frac{f_1}{f_2} + \frac{f_2}{f_1}\right) c_{33}s_{22}s_{11} - \left(\frac{f_1}{f_3} + \frac{f_3}{f_1}\right) s_{33}c_{22}s_{11} - \left(\frac{f_2}{f_3} + \frac{f_3}{f_2}\right) s_{33}s_{22}c_{11}. \tag{C.37}$$

In the region $-2 \leq D(\lambda) \leq 2$, by Eq. (C.24) the band structure can be more explicitly written as

$$\begin{aligned} \cos ka &= \cos k_3d_3 \cos k_2d_2 \cos k_1d_1 \\ &-\frac{1}{2}\left(\frac{f_1}{f_2}+\frac{f_2}{f_1}\right) \cos k_3d_3 \sin k_2d_2 \sin k_1d_1 \\ &-\frac{1}{2}\left(\frac{f_1}{f_3}+\frac{f_3}{f_1}\right) \sin k_3d_3 \cos k_2d_2 \sin k_1d_1 \\ &-\frac{1}{2}\left(\frac{f_2}{f_3}+\frac{f_3}{f_2}\right) \sin k_3d_3 \sin k_2d_2 \cos k_1d_1. \end{aligned} \tag{C.38}$$

Further for a layered crystal in which each period is made of four layers of different homogeneous media, $N = 4$ and $x_4 = a$. Equations (C.19)–(C.22) give that

$$\begin{aligned} \eta_1(a,\lambda) &= c_{44}c_{33}c_{22}c_{11} - \frac{f_1}{f_2}c_{44}c_{33}s_{22}s_{11} - \frac{f_1}{f_3}c_{44}s_{33}c_{22}s_{11} \\ &-\frac{f_2}{f_3}c_{44}s_{33}s_{22}c_{11} - \frac{f_1}{f_4}s_{44}c_{33}c_{22}s_{11} - \frac{f_2}{f_4}s_{44}c_{33}s_{22}c_{11} \\ &-\frac{f_3}{f_4}s_{44}s_{33}c_{22}c_{11} + \frac{f_3f_1}{f_2f_4}s_{44}s_{33}s_{22}s_{11}, \end{aligned} \tag{C.39}$$

$$\begin{aligned} p(a)\eta_1'(a,\lambda) &= -f_1c_{44}c_{33}c_{22}s_{11} - f_2c_{44}c_{33}s_{22}c_{11} - f_3c_{44}s_{33}c_{22}c_{11} \\ &-f_4s_{44}c_{33}c_{22}c_{11} + \frac{f_3f_1}{f_2}c_{44}s_{33}s_{22}s_{11} + \frac{f_1f_4}{f_2}s_{44}c_{33}s_{22}s_{11} \\ &+\frac{f_1f_4}{f_3}s_{44}s_{33}c_{22}s_{11} + \frac{f_2f_4}{f_3}s_{44}s_{33}s_{22}c_{11}, \end{aligned} \tag{C.40}$$

$$\begin{aligned} \eta_2(a,\lambda) &= \frac{1}{f_1}c_{44}c_{33}c_{22}s_{11} + \frac{1}{f_2}c_{44}c_{33}s_{22}c_{11} + \frac{1}{f_3}c_{44}s_{33}c_{22}c_{11} \\ &+\frac{1}{f_4}s_{44}c_{33}c_{22}c_{11} - \frac{f_2}{f_1f_3}c_{44}s_{33}s_{22}s_{11} - \frac{f_2}{f_1f_4}s_{44}c_{33}s_{22}s_{11} \\ &-\frac{f_3}{f_1f_4}s_{44}s_{33}c_{22}s_{11} - \frac{f_3}{f_2f_4}s_{44}s_{33}s_{22}c_{11}, \end{aligned} \tag{C.41}$$

and

$$\begin{aligned} p(a)\eta_2'(a,\lambda) &= c_{44}c_{33}c_{22}c_{11} - \frac{f_2}{f_1}c_{44}c_{33}s_{22}s_{11} - \frac{f_3}{f_1}c_{44}s_{33}c_{22}s_{11} \\ &-\frac{f_3}{f_2}c_{44}s_{33}s_{22}c_{11} - \frac{f_4}{f_1}s_{44}c_{33}c_{22}s_{11} - \frac{f_4}{f_2}s_{44}c_{33}s_{22}c_{11} \\ &-\frac{f_4}{f_3}s_{44}s_{33}c_{22}c_{11} + \frac{f_2f_4}{f_1f_3}s_{44}s_{33}s_{22}s_{11}. \end{aligned} \tag{C.42}$$

Equation (C.23) gives that,

$$D(\lambda) = 2c_{44}c_{33}c_{22}c_{11} - \left(\frac{f_1}{f_2} + \frac{f_2}{f_1}\right) c_{44}c_{33}s_{22}s_{11} - \left(\frac{f_1}{f_3} + \frac{f_3}{f_1}\right) c_{44}s_{33}c_{22}s_{11}$$
$$- \left(\frac{f_2}{f_3} + \frac{f_3}{f_2}\right) c_{44}s_{33}s_{22}c_{11} - \left(\frac{f_1}{f_4} + \frac{f_4}{f_1}\right) s_{44}c_{33}c_{22}s_{11} - \left(\frac{f_2}{f_4} + \frac{f_4}{f_2}\right) s_{44}c_{33}s_{22}c_{11}$$
$$- \left(\frac{f_3}{f_4} + \frac{f_4}{f_3}\right) s_{44}s_{33}c_{22}c_{11} + \left(\frac{f_3 f_1}{f_2 f_4} + \frac{f_2 f_4}{f_1 f_3}\right) s_{44}s_{33}s_{22}s_{11}. \quad \text{(C.43)}$$

In the region $-2 \leq D(\lambda) \leq 2$, by Eq. (C.24) the band structure can be more explicitly written as

$$\begin{aligned}
cos\ ka &= cos\ k_4 d_4\ cos\ k_3 d_3\ cos\ k_2 d_2\ cos\ k_1 d_1 \\
&- \frac{1}{2}\left(\frac{f_1}{f_2} + \frac{f_2}{f_1}\right) cos\ k_4 d_4\ cos\ k_3 d_3\ sin\ k_2 d_2\ sin\ k_1 d_1 \\
&- \frac{1}{2}\left(\frac{f_1}{f_3} + \frac{f_3}{f_1}\right) cos\ k_4 d_4\ sin\ k_3 d_3\ cos\ k_2 d_2\ sin\ k_1 d_1 \\
&- \frac{1}{2}\left(\frac{f_2}{f_3} + \frac{f_3}{f_2}\right) cos\ k_4 d_4\ sin\ k_3 d_3\ sin\ k_2 d_2\ cos\ k_1 d_1 \\
&- \frac{1}{2}\left(\frac{f_1}{f_4} + \frac{f_4}{f_1}\right) sin\ k_4 d_4\ cos\ k_3 d_3\ cos\ k_2 d_2\ sin\ k_1 d_1 \\
&- \frac{1}{2}\left(\frac{f_2}{f_4} + \frac{f_4}{f_2}\right) sin\ k_4 d_4\ cos\ k_3 d_3\ sin\ k_2 d_2\ cos\ k_1 d_1 \\
&- \frac{1}{2}\left(\frac{f_3}{f_4} + \frac{f_4}{f_3}\right) sin\ k_4 d_4\ sin\ k_3 d_3\ cos\ k_2 d_2\ cos\ k_1 d_1 \\
&+ \frac{1}{2}\left(\frac{f_3 f_1}{f_2 f_4} + \frac{f_2 f_4}{f_1 f_3}\right) sin\ k_4 d_4\ sin\ k_3 d_3\ sin\ k_2 d_2\ sin\ k_1 d_1.
\end{aligned} \quad \text{(C.44)}$$

For layered crystals of $N = 5$ or more, similar equations can be straightforwardly obtained.

Relevant expressions for the dispersion relations were obtained in the literature, see, for example [6–9]. The approach used here is more basic and simpler and provides more useful results. $\eta_i(a, \lambda)$ and $p(a)\eta_i'(a, \lambda), i = 1, 2$, play a substantial role in investigations of the existence and properties of surface states/modes in one-dimensional semi-infinite layered crystals such as in Appendices A, E, and F.

Equations (C.19)–(C.24) could be easily used in numerical investigations on layered crystals.

References

1. S.M. Rytov, Sov. Phys. JETP **2**, 466 (1956); S.M. Rytov, Sov. Phys. Acoust. **2**, 68 (1956)
2. R.E. Camley, B. Djafari-Rouhani, L. Dobrzynski, A.A. Maradudin, Phys. Rev. B **27**, 7318 (1983)
3. A. Yariv, P. Yeh, *Optical Waves in Crystals* (Wiley, New York, 1984)
4. P. Yeh, *Optical Waves in Layered Media* (Wiley, New York, 1988); (Wiley, Hoboken, 2005)
5. A.M. Kosevich, JETP Lett. **74**, 559 (2001)
6. B. Djafari-Rouhani, L. Dobrzynski, Solid State Commun. **62**, 609 (1987)
7. E.H. El Boudouti, B. Djafari-Rouhani, A. Akjouj, L. Dobrzynski, Phys. Rev. B **54**, 14728 (1996)
8. F. Szmulowicz, Phys. Lett. A **345**, 469 (2005)
9. F. Szmulowicz, Phys. Rev. B **72**, 235103 (2005)

Appendix D
Analytical Expressions of $\frac{\partial \Lambda}{\partial \tau}$ and $\frac{\partial \Lambda}{\partial \sigma}$

One of the particularly interesting recent progress in the Sturm–Liouville theory [1] is that mathematicians have found that the eigenvalues of regular Sturm–Liouville problems are differentiable functions of boundary locations, boundary conditions, as well as coefficients in the differential equation. Furthermore, the expressions of their derivatives have been obtained [2–4]. This new progress might find interesting applications in physics.

In Chap. 3 we presented a general analysis on the existence and properties of surface states caused by the termination of the periodic potential in one-dimensional semi-infinite electronic crystals. As an application of the theory of periodic Sturm–Liouville equations, in this Appendix we extend the investigations therein to further obtain equations on how the eigenvalue of *an existing surface state/mode in a one-dimensional semi-infinite crystal* depends on the boundary location and boundary condition, by following an approach developed by Kong and Zettl [3, 4]. The results obtained here are valid for the properties of an existing surface state/mode in more general one-dimensional semi-infinite crystals, including electronic crystals, photonic crystals, and phononic crystals as well.[4]

We are interested in a periodic Sturm–Liouville equation (2.2):

$$[p(x)y'(x)]' + [\lambda w(x) - q(x)]y(x) = 0, \tag{D.1}$$

where $p(x) > 0$, $w(x) > 0$ and $p(x)$, $q(x)$ and $w(x)$ are piecewise continuous real periodic functions with a finite period a:

$$p(x+a) = p(x), \ q(x+a) = q(x), \ w(x+a) = w(x).$$

We assume that Eq. (D.1) is solved, and all solutions are known. The eigenvalues are permitted bands $\varepsilon_n(k)$, and the corresponding eigenfunctions are Bloch functions

[4]Parts of the work presented in this Appendix and corresponding numerical confirmations were published in [5].

S.Y. Ren, *Electronic States in Crystals of Finite Size*, Springer Tracts in Modern Physics 270, DOI 10.1007/978-981-10-4718-3

$\phi_n(k, x)$, where $n = 0, 1, 2, \ldots$, and $-\frac{\pi}{a} < k \leq \frac{\pi}{a}$. We are mainly interested in cases where there is always a band gap between two consecutive permitted eigenvalue bands. The band gaps of Eq. (D.1) are always located either at the center of the Brillouin zone $k = 0$ or the boundary of the Brillouin zone $k = \frac{\pi}{a}$.

For a one-dimensional semi-infinite crystal, we assume that the original periodicity remains inside the semi-infinite crystal. For such a one-dimensional semi-infinite crystal with a left boundary at τ, the eigenvalues Λ, and eigenfunctions $\psi(x)$ for a specific boundary condition at the boundary τ are solutions of the differential equation

$$[p(x)\psi'(x)]' + [\Lambda w(x) - q(x)]\psi(x) = 0, \qquad \tau < x < \infty, \tag{D.2}$$

with the boundary condition at τ:

$$\sigma\psi(\tau) - p(\tau)\psi'(\tau) = 0, \tag{D.3}$$

where σ is the ratio of the quasi-derivative $p\psi'$ over the function ψ at the boundary location τ that depends on the boundary condition. For example, for a surface state in an ideal electronic crystal, the boundary condition $\psi(\tau) = 0$ leads to that $\sigma = \infty$; For a surface mode in a phononic crystal with a free boundary, the boundary condition $p(\tau)\psi'(\tau) = 0$ leads to that $\sigma = 0$. To have a localized surface state/mode within a band gap in the crystal, in most cases σ needs to be in $[0, \infty]$.

On the basis of that Eq. (D.1) is solved, the existence and the properties of the solutions of Eqs. (D.2) and (D.3) are completely determined by τ and σ in Eq. (D.3).

In general, a solution of Eqs. (D.2) and (D.3) with an eigenvalue Λ within a band gap—if it exists—has a specific form. Its eigenfunction inside the semi-infinite one-dimensional crystal always has the form

$$\psi(x, \Lambda) = \mathrm{e}^{-\beta(\Lambda)x} f(x, \Lambda), \tag{D.4}$$

where $f(x, \Lambda)$ is a periodic function $f(x + a, \Lambda) = f(x, \Lambda)$ if the band gap is at the center of the Brillouin zone $k = 0$, or a semi-periodic function $f(x + a, \Lambda) = -f(x, \Lambda)$ if the band gap is at the boundary of the Brillouin zone $k = \frac{\pi}{a}$. $\beta(\Lambda)$ is a positive real number depending on Λ. The exponential factor in Eq. (D.4) makes that $\psi(x, \Lambda)$ is localized near the left end τ of the semi-infinite crystal and thus is a surface state/mode. As Λ approaches a band edge, β decreases but remains to be finite as long as Λ does not reach the band edge. $\psi(x, \Lambda)$ in Eq. (D.4) can always be chosen as a real function and,

$$\psi(x, \Lambda) = 0, \qquad x \to \infty, \tag{D.5}$$

and can be normalized in the interval (τ, ∞):

$$\int_{\tau}^{\infty} w(x)\psi^2(x, \Lambda)\mathrm{d}x = 1. \tag{D.6}$$

In the following, we try to investigate how the eigenvalue Λ of *an existing surface state/mode* depends on τ and σ. We assume $\sigma \neq \infty$ at first. Let $\delta\sigma$ be an infinitesimal real number. Similar to that ψ is a solution of Eqs. (D.2) and (D.3), we have a real solution χ of the following equation:

$$[p(x)\chi'(x)]' + [(\Lambda + \delta\Lambda)w(x) - q(x)]\chi(x) = 0, \qquad \tau < x < \infty, \tag{D.7}$$

with the boundary condition

$$(\sigma + \delta\sigma)\chi(\tau) - p(\tau)\chi'(\tau) = 0. \tag{D.8}$$

χ will have properties similar to ψ as described in Eqs. (D.4)–(D.6). From Eqs. (D.2) and (D.7) we obtain that

$$(p\psi')'\chi - (p\chi')'\psi = [\delta\Lambda]w\psi\chi, \qquad \tau < x < \infty. \tag{D.9}$$

Since both ψ and χ have the form of (D.4), we can do integrating with both sides of Eq. (D.9) in (τ, ∞) and obtain that

$$\int_\tau^\infty [(p\psi')'\chi - (p\chi')'\psi]\mathrm{d}x = \int_\tau^\infty [\delta\Lambda]w\psi\chi\mathrm{d}x \tag{D.10}$$

The left side of Eq. (D.10)

$$\int_\tau^\infty [(p\psi')'\chi - (p\chi')'\psi]\mathrm{d}x = -[(p\psi')\chi - (p\chi')\psi]_{x=\tau}, \tag{D.11}$$

where Eq. (D.5) and thus $[(p\psi')\chi - (p\chi')\psi]_{x\to\infty} = 0$ was used.

Let $\chi = \psi + \delta\psi$, and let $\delta\sigma \to 0$, the right side of Eq. (D.10) gives

$$\int_\tau^\infty [\delta\Lambda]w\psi\chi\mathrm{d}x = \int_\tau^\infty [\delta\Lambda]w\psi\psi\mathrm{d}x, \tag{D.12}$$

since $\int_\tau^\infty [\delta\Lambda]w\psi(\delta\psi)\mathrm{d}x$ is a higher order infinitesimal number than $\int_\tau^\infty [\delta\Lambda]w\psi\psi\mathrm{d}x$ as $\delta\sigma \to 0$ and thus is negligible.

Combining Eqs. (D.11) and (D.12), and using the normalization Eq. (D.6) of ψ in Eq. (D.12), we obtain that

$$-[(p\psi')\chi - (p\chi')\psi]_{x=\tau} = \delta\Lambda. \tag{D.13}$$

For the case where $\sigma = \infty$, note that Eq. (D.3) now indicates $\psi(\tau) = 0$, the function χ can be chosen as the solution of

$$[p(x)\chi'(x)]' + [(\Lambda + \delta\Lambda)w(x) - q(x)]\chi(x) = 0, \qquad \tau + \delta\tau < x < \infty,$$

under the boundary condition $\chi(\tau + \delta\tau) = 0$, where $\delta\tau$ is an infinitesimal real number. Let $\delta\tau \to 0$, very similar arguments lead to the same Eq. (D.13).

Several quantitative relationships between the eigenvalue Λ and the boundary location τ or the boundary condition σ can be obtained from Eq. (D.13).

1. The τ-dependence of the eigenvalue with the boundary condition $\psi(\tau) = 0$.

Let the two eigenfunctions ψ and χ satisfy $\psi(\tau) = 0$, and $\chi(\tau + \delta\tau) = 0$. Equation (D.13) leads to that

$$- p(\tau)\psi'(\tau)\chi(\tau) = \delta\Lambda. \tag{D.14}$$

$-\chi(\tau)$ on the left side of Eq. (D.14) is

$$-\chi(\tau) = \chi(\tau + \delta\tau) - \chi(\tau) = \int_{\tau}^{\tau+\delta\tau} \frac{1}{p(x)}[p(x)\chi'(x)]\mathrm{d}x$$

$$= \int_{\tau}^{\tau+\delta\tau} \frac{1}{p(x)}[p(x)\psi'(x)]\mathrm{d}x + \int_{\tau}^{\tau+\delta\tau} \frac{1}{p(x)}[p(x)\delta\psi'(x)]\mathrm{d}x, \tag{D.15}$$

where $\chi(\tau + \delta\tau) = 0$, and $\chi = \psi + \delta\psi$ were used. By noting that as $\delta\tau \to 0$, the second term on the right side of Eq. (D.15) is a higher order infinitesimal quantity than the first term, and thus is negligible, we obtain that

$$p(\tau)\psi'(\tau)\left[\frac{1}{p(\tau)}p(\tau)\psi'(\tau)\right]\delta\tau = \delta\Lambda.$$

That is, as $\delta\tau \to 0$,

$$\frac{\partial \Lambda}{\partial \tau} = \frac{1}{p(\tau)}[p(\tau)\psi'(\tau)]^2. \tag{D.16}$$

2. The τ-dependence of the eigenvalue with the boundary condition $p(\tau)\psi'(\tau) = 0$.

Let the two eigenfunctions ψ and χ satisfy $p(\tau)\psi'(\tau) = 0$ and $p(\tau+\delta\tau)\chi'(\tau+\delta\tau) = 0$. Then Eq. (D.13) leads to that

$$p(\tau)\chi'(\tau)\psi(\tau) = \delta\Lambda. \tag{D.17}$$

$p(\tau)\chi'(\tau)$ on the left side of Eq. (D.17) is

$$p(\tau)\chi'(\tau) = -\int_{\tau}^{\tau+\delta\tau} [p(x)\chi'(x)]'dx = -\int_{\tau}^{\tau+\delta\tau} [q(x) - (\Lambda + \delta\Lambda)w(x)]\chi(x)dx$$

$$= -\int_{\tau}^{\tau+\delta\tau} [q(x) - (\Lambda + \delta\Lambda)w(x)]\psi(x)dx$$

$$- \int_{\tau}^{\tau+\delta\tau} [q(x) - (\Lambda + \delta\Lambda)w(x)]\delta\psi(x)dx, \tag{D.18}$$

where $p(\tau + \delta\tau)\chi'(\tau + \delta\tau) = 0$, Eq. (D.7), and $\chi = \psi + \delta\psi$ were used. By noting that in comparison with the first term on the right side of Eq. (D.18), the second term is a higher order infinitesimal quantity as $\delta\tau \to 0$ and thus is negligible, and $\delta\Lambda \to 0$, we obtain that

$$\psi(\tau)[-q(\tau)\psi(\tau) + \Lambda w(\tau)\psi(\tau)]\,\delta\tau = \delta\Lambda.$$

That is, as $\delta\tau \to 0$,

$$\frac{\partial \Lambda}{\partial \tau} = -[\psi(\tau)]^2[q(\tau) - \Lambda w(\tau)]. \tag{D.19}$$

3. The τ-dependence of the eigenvalue with the boundary condition neither $p(\tau)\psi'(\tau)$ nor $\psi(\tau)$ is zero.

For the cases where the boundary condition is that neither $p(\tau)\psi'(\tau)$ nor $\psi(\tau)$ is zero, as pointed out in [3], we can combine the results of Eqs. (D.16) and (D.19) and obtain that

$$\frac{\partial \Lambda}{\partial \tau} = \frac{1}{p(\tau)}[p(\tau)\psi'(\tau)]^2 - [\psi(\tau)]^2[q(\tau) - \Lambda w(\tau)]. \tag{D.20}$$

4. The σ-dependence of the eigenvalue of the surface state with a specific boundary location τ.

For the σ-dependence of the eigenvalue of the surface state near the neighborhood of $\sigma = 0$, note that $\sigma = 0$ means $p(\tau)\psi'(\tau) = 0$ then $\psi(\tau) \neq 0$, and thus $\chi(\tau) \neq 0$ for an infinitesimal $\delta\sigma$. By using Eqs. (D.3) and (D.8), Eq. (D.13) leads to that

$$\delta\sigma\ (\psi\chi)_{x=\tau} = \delta\Lambda.$$

Let $\delta\sigma \to 0$ we have $\chi \to \psi$ and obtain that

$$\frac{\partial \Lambda}{\partial \sigma} = [\psi(\tau)]^2. \tag{D.21}$$

For the σ-dependence of the eigenvalue of the surface state not near the neighborhood of $\sigma = 0$, then $p(\tau)\psi'(\tau) \neq 0$ and thus $p(\tau)\chi'(\tau) \neq 0$ for an infinitesimal $\delta\sigma$. By using Eqs. (D.3) and (D.8), Eq. (D.13) leads to that

$$p(\tau)\psi'(\tau)p(\tau)\chi'(\tau)\sigma^{-2}(\delta\sigma) = \delta\Lambda.$$

Let $\delta\sigma \to 0$ we have $\chi'(\tau) \to \psi'(\tau)$ and obtain that

$$\frac{\partial \Lambda}{\partial \sigma} = \sigma^{-2}[p(\tau)\psi'(\tau)]^2 = [\psi(\tau)]^2 > 0,$$

the same as in Eq. (D.21).

Equations (D.20) and (D.21) are two general equations indicating how the eigenvalue of an existing surface state/mode depends on τ and σ quantitatively. The obtained equations were verified by numerical calculations [5].

Equations (D.20) and (D.21) can be considered as further generalizations and quantifications of two equations (3.7) and (3.8) in Chap. 3. In particular, for an existing surface state in a one-dimensional semi-infinite electronic crystal:
(i) Equation (D.21) is a quantitative equation of Eq. (3.7) in Chap. 3;
(ii) Equation (D.20) is a quantitative equation of Eq. (3.8) in Chap. 3. For such a case, Eq. (D.20) can be rewritten as

$$\frac{\partial \Lambda}{\partial \tau} = [\psi(\tau)]^2 \left[\frac{1}{p(\tau)}\sigma^2 + \Lambda - q(\tau)\right].$$

$\frac{\partial \Lambda}{\partial \tau} > 0$ if $\frac{1}{p(\tau)}\sigma^2 + \Lambda > q(\tau)$, which corresponds to that the "reduced outside potential" at the surface, is higher than the "reduced periodic potential" $q(\tau)$.

For a specific σ, $\Lambda(\tau)$ is surely a periodic function of τ with a period a, similar to coefficients p, q, w in Eqs. (D.1) and (D.2). The result obtained here that $\frac{\partial \Lambda}{\partial \tau} > 0$ is true only when *Λ is the eigenvalue of an existing surface state/mode in the specific forbidden eigenvalue range* at the boundary location τ with the specific σ. A direct consequence of $\frac{\partial \Lambda}{\partial \tau} > 0$ is that the termination of the periodicity at a boundary τ in a semi-infinite periodic system *may or may not* cause a surface state/mode in a specific forbidden eigenvalue range.

The results presented here may find practical applications in relevant problems, in particular, in investigating and designing periodic man-made systems such as photonic and phononic crystals.

References

1. A. Zettl, *Sturm–Liouville Theory* (The American Mathematical Society, Providence, 2005)
2. M. Dauge, B. Helffer, J. Differ. Equs. **104**, 243 (1993); M. Dauge, B. Helffer, J. Differ. Equs. **104**, 263 (1993)
3. Q. Kong, A. Zettl, J. Differ. Equs. **126**, 389 (1996)
4. Q. Kong, A. Zettl, J. Differ. Equs. **131**, 1 (1996)
5. S.Y. Ren, Y.C. Chang, Ann. Phys. **325**, 937 (2010)

Appendix E
One-Dimensional Phononic Crystals

Investigations on classical waves in periodically arranged alternative media, such as elastic waves in periodically alternating elastic mediums—often called phononic crystals—, electromagnetic waves in periodically alternating dielectric mediums—often called photonic crystals—, and so forth, have a long history in physics. See, for example, [1–4]. These problems received increasing attention in recent decades, due to their unusual physical properties observed in these heterostructures in comparison with bulk materials as well as the tremendous technical progress made in the fabrication of these man-made heterostructures [5–15]. The most interesting property of these heterostructures is the possible existence of forbidden frequency band gaps induced by the difference in the elastic or dielectric properties of the constituents and the periodicity of these systems, which can lead to valuable practical applications. One-dimensional phononic crystals or photonic crystals are the simplest phononic crystals or photonic crystals. A clear understanding of the one-dimensional problems provides a basis for further understandings of higher-dimensional systems.

The simplest cases are that such one-dimensional crystals are made of two different homogeneous and isotropic layered media alternatively. A primary theoretical formalism widely used for investigations of such layered structures is based on the Transfer Matrix Method (TMM). In this formalism one solves the corresponding wave equation *in each layer*, matches the solutions at the interfaces of adjacent layers according to the necessary continuous conditions, and then uses the Bloch theorem to treat the periodicity [6–8].

In comparison with the extensive literature on the investigations on classical waves in periodic media, the understanding of the effects of the very existence of the boundary or boundaries in such systems is relatively less. There are interesting investigations on semi-infinite layered phononic and photonic crystals in the literature, see, for example [5–9,11,16–28] and references therein. To the author's knowledge, investigations focusing on the eigenmodes in one-dimensional phononic or photonic crystals of finite length $L = Na$—here a is the period length of the crystal, and N is a positive integer—began only recently [29–35].

S.Y. Ren, *Electronic States in Crystals of Finite Size*, Springer Tracts in Modern Physics 270, DOI 10.1007/978-981-10-4718-3

The author does not have much working experience and knows limitedly on the rich and colorful physics phenomena and their possible applications in these intensively investigated and still fast growing fields. Nevertheless, he feels that the problems studied in the major parts of this book—the effects of the boundary existence and the finite size on the electronic states in crystals—may have their correspondences in relevant problems of phononic crystals and photonic crystals. An advantage of the modern periodic Sturm–Liouville theory [36–38] is that it can straightforwardly treat wave equations of many one-dimensional phononic crystals and photonic crystals. Based on the theory summarized in Chap. 2, a theoretical formalism to treat one-dimensional phononic crystals and photonic crystals—including the layered one-dimensional phononic crystals and photonic crystals—can be developed. This theoretical formalism treats a phononic crystal or photonic crystal *as a whole* from the beginning. Major relevant equations previously obtained based on TMM can be easily re-derived from the theory as specific cases. New equations for treating more general and complicated cases can be obtained, and more comprehensive understandings of relevant problems can also be achieved.

In this Appendix, we discuss one-dimensional phononic crystals, in the order of infinite size, semi-infinite size, and finite size. In Appendix F, we will investigate one-dimensional photonic crystals. We will focus on the general formalisms and general understandings, rather than the detailed distinctions between different specific materials. Since in many cases the wave equation for a one-dimensional phononic crystal or photonic crystal can be considered as a specific form of the periodic Sturm–Liouville equations (2.2), similarly as the Schrödinger equation for a one-dimensional electronic crystal, we will see that many fundamental understandings on the one-dimensional electronic crystals presented in Chaps. 3 and 4 do have their correspondences in one-dimensional phononic crystals and photonic crystals. Scientists working in these fields might find more interesting applications of the modern theory of periodic Sturm–Liouville theory in their specific problems.

E.1 One-Dimensional Phononic Crystals of Infinite Length

We begin by considering the propagation of transverse elastic waves in a simple one-dimensional phononic crystal as illustrated in Fig. E.1. The phononic crystal is assumed to be composed of two or more isotropic media, periodic in the x direction and homogeneous in the y and z directions [30].

By choosing the displacements to be in the y direction, the dynamic equation for the transverse acoustic waves is given by [39]

$$\rho(x)\frac{\partial^2}{\partial t^2}y(x,z,t) = \frac{\partial}{\partial x}\left[\mu(x)\frac{\partial y(x,z,t)}{\partial x}\right] + \mu(x)\frac{\partial^2 y(x,z,t)}{\partial z^2},$$

where $\rho(x+a) = \rho(x)$ is the density and $\mu(x+a) = \mu(x)$ is the Lamé's coefficient, they are periodic with period a. By writing the displacements of the acoustic waves as

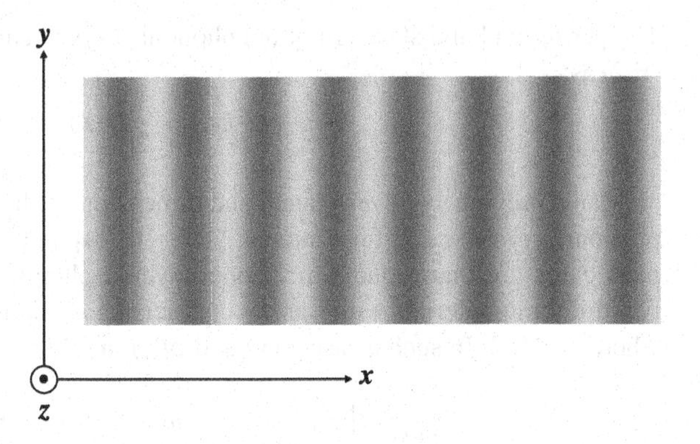

Fig. E.1 Schematic plot of a one-dimensional phononic crystal, in which the Lamé's coefficient $\mu(x)$ and the density $\rho(x)$ are periodic functions of x. Reprinted with permission from S. Y. Ren and Y. C. Chang: Phys. Rev. **B 75**, 212301 (2007). Copyright by American Physical Society

$$y(x, z, t) = y(x)e^{i(k_z z - \omega t)},$$

the wave equation of the problem now has the form

$$-[\mu(x)y'(x)]' + [\mu(x)k_z^2 - \rho(x)\lambda]y(x) = 0, \tag{E.1}$$

where $\lambda = \omega^2 \geq 0$, both $\mu(x) > 0$ and $\rho(x) > 0$. Equation (E.1) is a special case of the more general periodic Sturm–Liouville equations (2.2).[5]

Most previous investigations on one-dimensional phononic crystals focused on their permitted band structure. By (2.54), the permitted band structure of a phononic crystal described by (E.1) is determined by its discriminant $D(\lambda)$ defined as

$$D(\lambda) = \eta_1(a, \lambda) + \mu(a)\eta_2'(a, \lambda), \tag{E.2}$$

where $\eta_1(x, \lambda)$ and $\eta_2(x, \lambda)$ are two linearly independent normalized solutions of Eq. (E.1) satisfying

$$\eta_1(0, \lambda) = 1, \ \mu(0)\eta_1'(0, \lambda) = 0; \quad \eta_2(0, \lambda) = 0, \ \mu(0)\eta_2'(0, \lambda) = 1. \tag{E.3}$$

In the λ ranges where $-2 \leq D(\lambda) \leq 2$ the permitted eigenmode solutions $\phi_n(k_x, k_z, x)$ can exist as solutions of (E.1):

$$\phi_n(k_x, k_z, x + a) = e^{ik_x a}\phi_n(k_x, k_z, x).$$

The Bloch wave vector k_x in the eigenfunction $\phi_n(k_x, k_z, x)$ and the eigenvalue $\varepsilon_n(k_x, k_z)$ is limited in the Brillouin zone,

$$-\frac{\pi}{a} < k_x \leq \frac{\pi}{a}.$$

[5]The mathematical theory in this section can also be applied to the cases where the longitudinal elastic waves propagate along the x direction, which corresponds to $k_z = 0$.

The permitted band structure of the phononic crystal can be obtained from $D(\lambda)$ in (E.2) as:

$$cos\ k_x a = \frac{1}{2} D(\lambda). \tag{E.4}$$

Many previously investigated one-dimensional phononic crystals are layered phononic crystals, see, for example, [1, 6, 16–24] and references therein. The simplest cases are that a one-dimensional layered phononic crystal is constructed of two different materials with Lamé's coefficients μ_l, densities ρ_l and thicknesses d_l, where $l = 1, 2$. In such a case $\mu(x)$ and $\rho(x)$ in (E.1) can be written as

$$\mu(x) = \begin{cases} \mu_1, & na < x \leq d_1 + na, \\ \mu_2, & na + d_1 < x \leq (n+1)a, \end{cases} \tag{E.5}$$

and

$$\rho(x) = \begin{cases} \rho_1, & na < x \leq d_1 + na, \\ \rho_2, & na + d_1 < x \leq (n+1)a, \end{cases} \tag{E.6}$$

where $a = d_1 + d_2$ is the period, n is an integer and $\mu_l > 0$, $\rho_l > 0$ are real constants. Now in Eq. (E.1) $\mu(x) = \mu_l$ and $\mu(x)k_z^2 - \rho(x)\lambda$ are piecewise continuous functions and discontinuous at isolated points $x = na$ and $x = d_1 + na$.

The Eq. (E.1) now can be written in the form

$$[\mu_l y'(x)]' + \mu_l k_l^2 y(x) = 0, \tag{E.7}$$

by introducing

$$\mu_l k_l^2 = \lambda \rho_l - \mu_l k_z^2, \tag{E.8}$$

where k_l is real if $\lambda \rho_l - \mu_l k_z^2 > 0$ or imaginary if $\lambda \rho_l - \mu_l k_z^2 < 0$.

From (C.31) in Appendix C, we obtain that

$$D(\lambda) = 2\ cos\ k_2 d_2\ cos\ k_1 d_1 - \left(\frac{\mu_1 k_1}{\mu_2 k_2} + \frac{\mu_2 k_2}{\mu_1 k_1} \right) sin\ k_2 d_2\ sin\ k_1 d_1. \tag{E.9}$$

As λ increases from $-\infty$, $D(\lambda)$ changes in the way similar to what was described in Sect. 2.4.

From (E.4) and (E.9) one obtains that for such a layered phononic crystal, the permitted band structure is determined by

$$cos\ k_x a = cos\ k_2 d_2\ cos\ k_1 d_1 - \frac{1}{2} \left[\frac{\mu_1 k_1}{\mu_2 k_2} + \frac{\mu_2 k_2}{\mu_1 k_1} \right] sin\ k_2 d_2\ sin\ k_1 d_1, \tag{E.10}$$

where k_x is the Bloch wave vector. Equation (E.10) is essentially the dispersion relation frequently used in the literature, such as, in [1, 6, 10, 17, 18, 22, 24]. This approach can be easily extended to more general cases such as that each period unit contains more than two materials [20, 21] as shown in Appendix C.

For example, by (C.38), the dispersion relation of a one-dimensional layered phononic crystal composed of three different materials with Lamé's coefficients μ_1, μ_2, μ_3, densities ρ_1, ρ_2, ρ_3, and thicknesses d_1, d_2, d_3 can be written as

$$\begin{aligned} \cos k_x a = &\cos k_3 d_3 \cos k_2 d_2 \cos k_1 d_1 \\ &-\frac{1}{2}\left(\frac{\mu_1 k_1}{\mu_2 k_2} + \frac{\mu_2 k_2}{\mu_1 k_1}\right) \cos k_3 d_3 \sin k_2 d_2 \sin k_1 d_1 \\ &-\frac{1}{2}\left(\frac{\mu_1 k_1}{\mu_3 k_3} + \frac{\mu_3 k_3}{\mu_1 k_1}\right) \sin k_3 d_3 \cos k_2 d_2 \sin k_1 d_1 \\ &-\frac{1}{2}\left(\frac{\mu_2 k_2}{\mu_3 k_3} + \frac{\mu_3 k_3}{\mu_2 k_2}\right) \sin k_3 d_3 \sin k_2 d_2 \cos k_1 d_1, \end{aligned} \tag{E.11}$$

where k_x is the Bloch wave vector, and $\mu_l k_l^2 = \lambda \rho_l - \mu_l k_z^2, l = 1, 2, 3$.

E.2 Surface Modes in Semi-infinite One-Dimensional Phononic Crystals

Surface modes in semi-infinite one-dimensional phononic crystals have been investigated theoretically and experimentally for many years, see, for example, [6, 16–19, 22–24] and references therein. Since the boundary location and boundary condition of a semi-infinite one-dimensional phononic crystal can be more flexibly designed and altered than that in an electronic crystal, the theoretical and experimental investigations on how the existence and properties of surface modes depend on these factors have an apparently practical significance.

In this section, we investigate this subject based on the theory of periodic Sturm–Liouville equations in Chap. 2. We assume that (E.1) is solved. The eigenvalues of (E.1), $\varepsilon_n(k_x, k_z)$ describe the phononic band structure, and the corresponding eigenfunctions are Bloch functions $\phi_n(k_x, k_z, x)$, where $n = 0, 1, 2, \ldots$ and $-\frac{\pi}{a} < k_x \leq \frac{\pi}{a}$. For brevity, the k_z dependence of the eigenvalues and eigenfunctions will be kept implicit in the discussions below. We are mainly interested in cases where there is always a band gap between two consecutive bands of (E.1). For these cases, the band edges $\varepsilon_n(0)$ and $\varepsilon_n(\frac{\pi}{a})$ occur in the order

$$\begin{aligned} \varepsilon_0(0) < \ &\varepsilon_0\left(\tfrac{\pi}{a}\right) < \varepsilon_1\left(\tfrac{\pi}{a}\right) < \varepsilon_1(0) < \varepsilon_2(0) \\ &< \varepsilon_2\left(\tfrac{\pi}{a}\right) < \varepsilon_3\left(\tfrac{\pi}{a}\right) < \varepsilon_3(0) < \varepsilon_4(0) < \cdots. \end{aligned}$$

The band gaps are between $\varepsilon_{2m}(\frac{\pi}{a})$ and $\varepsilon_{2m+1}(\frac{\pi}{a})$ or between $\varepsilon_{2m+1}(0)$ and $\varepsilon_{2m+2}(0)$, here $m = 0, 1, 2, \ldots$.

For a semi-infinite phononic crystal with a left boundary at τ, the eigenmodes are solutions of

$$-[\mu(x)\psi'(x,\Lambda)]' + [\mu(x)k_z^2 - \rho(x)\Lambda]\psi(x,\Lambda) = 0, \qquad x \geq \tau, \tag{E.12}$$

and a boundary condition at τ. The boundary condition at the surface τ, in general, can be written as

$$\sigma\psi(\tau,\Lambda) - \mu(\tau)\psi'(\tau,\Lambda) = 0. \tag{E.13}$$

We call a semi-infinite crystal given by (E.12) and (E.13) as a right semi-infinite crystal, for the convenience of later discussions.

For surface modes in one-dimensional semi-infinite phononic crystals, we are interested in the cases where Λ is in a frequency range $[\varepsilon_{2m}(\frac{\pi}{a}), \varepsilon_{2m+1}(\frac{\pi}{a})]$ or $[\varepsilon_{2m+1}(0), \varepsilon_{2m+2}(0)]$ or $\Lambda \leq \varepsilon_0(0)$ of Eq. (E.1).[6] By the theory in Sect. 2.5, for any Λ in such a range, two linearly independent solutions of Eq. (E.1) can always be chosen in such a way that one is non-divergent, and the other one is divergent in $[\tau, +\infty)$. Only when the one non-divergent solution of (E.1) also satisfies (E.13) can we have a solution of both (E.12) and (E.13). Such a solution of (E.12) in a range of $|D(\Lambda)| \geq 2$ has the form

$$\psi(x,\Lambda) = e^{-\beta(\Lambda)x} f(x,\Lambda), \tag{E.14}$$

where $\beta \geq 0$, $f(x,\Lambda)$ is a periodic function if Λ is in $[\varepsilon_{2m+1}(0), \varepsilon_{2m+2}(0)]$ or $\Lambda \leq \varepsilon_0(0)$, or a semi-periodic function if Λ is in $[\varepsilon_{2m}(\frac{\pi}{a}), \varepsilon_{2m+1}(\frac{\pi}{a})]$.

Inside the semi-infinite crystal any solution $\psi(x,\Lambda)$ of Eq. (E.12) can be expressed as a linear combination of two normalized solutions η_1 and η_2 of Eq. (E.1):

$$\psi(x,\Lambda) = c_1\eta_1(x,\Lambda) + c_2\eta_2(x,\Lambda), \qquad x \geq \tau, \tag{E.15}$$

where $\eta_1(x,\Lambda)$ and $\eta_2(x,\Lambda)$ are two normalized solutions of Eq. (E.1) satisfying that

$$\eta_1(\tau,\Lambda) = 1,\ \mu(\tau)\eta_1'(\tau,\Lambda) = 0;\ \ \eta_2(\tau,\Lambda) = 0,\ \mu(\tau)\eta_2'(\tau,\Lambda) = 1. \tag{E.16}$$

E.2.1 Simple Cases

The simplest cases are that a one-dimensional semi-infinite phononic crystal has a stress-free boundary surface. It leads to the boundary condition at τ as

$$\mu(\tau)\psi'(\tau,\Lambda) = 0. \tag{E.17}$$

[6]For any λ inside a permitted band where $-2 < D(\lambda) < 2$, there are two non-divergent solutions of Eq. (E.1), a linear combination of two such non-divergent solutions can always satisfy Eqs. (E.12) and (E.13).

Equation (E.17) corresponds to the case $\sigma = 0$ in Eq. (E.13). Equation (E.15) becomes

$$\psi(x, \Lambda) = c_1 \eta_1(x, \Lambda), \qquad x \geq \tau, \tag{E.18}$$

by Eq. (E.16). Equations (E.14), (E.17), and (E.18) lead to that

$$\mu(\tau + a)\eta_1'(\tau + a, \Lambda) = 0 \tag{E.19}$$

is a necessary condition for the existence of a surface mode in a range $|D(\lambda)| \geq 2$ under the boundary condition (E.17).

The simplest one-dimensional phononic crystals are composed of two different isotropic homogeneous media alternatively. Without losing generality, we need only consider the cases where the boundary τ is in the first medium, i.e. $0 \leq \tau \leq d_1$. For such a case, we can treat the one-dimensional phononic crystal made of two media as a special case of that made of three media with physical parameters μ_1, μ_2, μ_1 and ρ_1, ρ_2, ρ_1 and thicknesses $d_1 - \tau, d_2, \tau$. From (C.34), Eq. (E.19) becomes

$$\begin{aligned} &-\mu_1 k_1 \, cos\, k_2 d_2 \, sin\, k_1 d_1 - \mu_2 k_2 \, sin\, k_2 d_2 \, cos\, k_1 d_1 \\ &\quad -\mu_2 k_2 \left(1 - \frac{\mu_1^2 k_1^2}{\mu_2^2 k_2^2}\right) sin\, k_2 d_2 \, sin\, k_1 (d_1 - \tau) \, sin\, k_1 \tau = 0. \end{aligned}$$

It can be written as

$$\begin{aligned} &cos\, k_2 d_2 \, sin\, k_1 d_1 + \frac{1}{2}\left(\frac{\mu_2 k_2}{\mu_1 k_1} + \frac{\mu_1 k_1}{\mu_2 k_2}\right) sin\, k_2 d_2 \, cos\, k_1 d_1 \\ &\quad + \frac{1}{2}\left(\frac{\mu_2 k_2}{\mu_1 k_1} - \frac{\mu_1 k_1}{\mu_2 k_2}\right) sin\, k_2 d_2 \, cos\, k_1 (d_1 - 2\tau) = 0, \end{aligned} \tag{E.20}$$

where k_1, k_2 are given in (E.8).

By Theorem 2.8, in each forbidden range starting from $[0, \varepsilon_0(0)]$, there is always one and only one solution $\Lambda = \nu_{\tau,n}$ of (E.19) for any specific boundary location τ. For a one-dimensional phononic crystal composed of two different isotropic homogeneous media alternatively, it is a solution of Eq. (E.20). The ratio $P = \frac{\psi(\tau+a,\Lambda)}{\psi(\tau,\Lambda)}$ for such a solution is,

$$P = \eta_1(\tau + a, \Lambda) \tag{E.21}$$

by Eqs. (E.16) and (E.18).

Using $\Lambda = \nu_{\tau,n}$ obtained from (E.20) into (C.33), we obtain the value of the surface mode at $x = \tau + a$:

$$\eta_1(\tau + a, \nu_{\tau,n}) = cos\ k_2 d_2\ cos\ k_1 d_1 - \frac{1}{2}\left[\frac{\mu_1 k_1}{\mu_2 k_2} + \frac{\mu_2 k_2}{\mu_1 k_1}\right] sin\ k_2 d_2\ sin\ k_1 d_1 \\ -\frac{1}{2}\left[\frac{\mu_1 k_1}{\mu_2 k_2} - \frac{\mu_2 k_2}{\mu_1 k_1}\right] sin\ k_2 d_2\ sin\ k_1 (d_1 - 2\tau). \quad \text{(E.22)}$$

There are three possibilities for the obtained $P = \eta_1(\tau + a, \Lambda)$ from (E.22): (a) $0 < |\eta_1(\tau + a, \nu_{\tau,n})| < 1$; (b) $|\eta_1(\tau + a, \nu_{\tau,n})| = 1$; and (c) $|\eta_1(\tau + a, \nu_{\tau,n})| > 1$. The case (a) corresponds to an oscillatory decreasing solution $\psi(x, \nu_{\tau,n}) = c_1 \eta_1(x, \nu_{\tau,n})$ of Eqs. (E.12) and (E.17), a localized surface mode having the form of (E.14) with $\beta(\nu_{\tau,n}) > 0$.[7] From Eqs. (E.20) and (E.22), one can easily obtain the expressions previously used such as in [6].

Similarly, for cases where a one-dimensional phononic crystal is composed of three different isotropic homogeneous media alternatively, we need only consider the cases where the boundary τ is in the first medium, i.e. $0 \le \tau \le d_1$. We can treat the one-dimensional phononic crystal made of three media considered here as a special case of a one-dimensional phononic crystal made of four media with physical parameters $\mu_1, \mu_2, \mu_3, \mu_1$ and $\rho_1, \rho_2, \rho_3, \rho_1$ and thicknesses $d_1 - \tau, d_2, d_3, \tau$. For such a case, equations corresponding to Eqs. (E.20) and (E.22) can be obtained from (C.39)–(C.42).

E.2.2 More General Cases

For more general cases $\sigma \neq 0$, Eqs. (E.13) and (E.14) lead to that

$$\sigma\psi(\tau + a, \Lambda) - \mu(\tau)\psi'(\tau + a, \Lambda) = \sigma\psi(\tau, \Lambda) - \mu(\tau)\psi'(\tau, \Lambda) = 0 \quad \text{(E.23)}$$

is a necessary condition for the existence of a localized surface mode with a more general boundary condition (E.13).

From Eqs. (E.23), (E.15) and (E.14), reasonings similar to Sect. 3.5 lead to that the existence and properties of a surface mode in the one-dimensional semi-infinite phononic crystal are determined by the following two equations. They are

$$\sigma = \frac{-\eta_1(\tau + a, \Lambda) + \mu(\tau)\eta_2'(\tau + a, \Lambda) - \sqrt{D^2(\Lambda) - 4}}{2\,\eta_2(\tau + a, \Lambda)}, \quad \text{(E.24)}$$

[7] The case (b) corresponds to that Eq. (E.19) or Eq. (E.20) gives a band-edge eigenvalue of Eq. (E.7), $\nu_{\tau,n} = \varepsilon_n(k_g, k_z)$. Thus, $\psi(x, \nu_{\tau,n})$ is either a periodic function or a semi-periodic function in the semi-infinite phononic crystal. The case (c) corresponds to an oscillatory increasing solution $\psi(x, \nu_{\tau,n})$ of Eq. (E.19). A surface mode can exist in the complementary left semi-infinite phononic crystal in $(-\infty, \tau]$.

for a band gap at the center of Brillouin zone $k_x = 0$; and

$$\sigma = \frac{-\eta_1(\tau + a, \Lambda) + \mu(\tau)\eta_2'(\tau + a, \Lambda) + \sqrt{D^2(\Lambda) - 4}}{2\,\eta_2(\tau + a, \Lambda)}, \tag{E.25}$$

for a band gap at the boundary of Brillouin zone $k_x = \frac{\pi}{a}$. Here $D(\lambda) = \eta_1(\tau + a, \lambda) + \mu(\tau)\eta_2'(\tau + a, \lambda)$ is the discriminant of Eq. (E.1).

Equations (E.24) and (E.25) are two equations which can be used to generally investigate the existence and properties of surface modes in a semi-infinite one-dimensional phononic crystal.

Similar to Sect. E.2.1, the simplest one-dimensional phononic crystals are those composed of two different media with Lamé's coefficients μ_l, densities ρ_l and thicknesses d_l, where $j = 1, 2$. Without losing generality, we need only consider the cases where the boundary τ is in the first medium, i.e. $0 \leq \tau \leq d_1$. We can treat the one-dimensional phononic crystal as a special case of that made of three media with physical parameters μ_1, μ_2, μ_1; ρ_1, ρ_2, ρ_1; and thicknesses $d_1 - \tau, d_2, \tau$. The existence and properties of surface modes in such a one-dimensional semi-infinite crystal can be investigated by using Eqs. (E.24) and (E.25) with $\eta_i, \mu\eta_i', i = 1, 2$ given in (C.33)–(C.36).

Similarly, the existence and properties of surface modes in a layered semi-infinite crystal composed of three homogeneous isotropic media can be investigated by using Eqs. (E.24) and (E.25) with $\eta_i, \mu\eta_i', i = 1, 2$ given in (C.39)–(C.42), etc.

If needed, surface modes in a one-dimensional phononic crystal composed of four or more different isotropic homogeneous media alternatively can be investigated similarly.

E.2.3 Brief Discussions

A one-dimensional semi-infinite phononic crystal *may or may not* have a surface mode in a specific forbidden range: (E.24) or (E.25) may or may not have a solution in such a range. Equation (E.23) is a *necessary but not sufficient condition* for having a surface mode in a specific forbidden range.

If there is a solution Λ of Eq. (E.23) existing in a specific forbidden range where $|D(\lambda) \geq 2$, then Λ can be either at a band edge or inside the forbidden range. Only when Λ is inside the forbidden range, and the solution of (E.23) has the form of (E.14), we can have a surface mode solution of (E.12) and (E.13) in the right semi-infinite phononic crystal.

If Λ is at a band edge, then the solution $\psi(x, \Lambda)$ of (E.23) will be a band-edge mode that is a periodic function or a semi-periodic function, depending on whether the band gap is at the center or the boundary of the Brillouin zone.

In the cases where $\sigma = 0$ in (E.23), if there is a surface mode $\Lambda = \nu_{\tau,n}$, the frequency of such a surface mode is below the corresponding permitted band, by Theorem 2.8.

Based on Theorem 2.8 in Chap. 2 on the zeros of solutions of (E.7), we can understand that the surface mode below the lowest bulk band has no zero in $(-\infty, +\infty)$, the surface mode in the lowest band-gap has one zero in $[0, a)$, and the surface mode in the second lowest band-gap has two zeros in $[0, a)$, ..., etc. Theorem 2.8 gives explanations on the behaviors of surface modes shown in several previous investigations. For example, the surface mode in Fig. 9a of Ref. [6] corresponds to a mode below $\varepsilon_0(0)$ thus has no zero; and the surface modes in Fig. 9b corresponds to a mode in the second lowest band-gap thus has two zeros in each unit cell. The theorem also explains the modes in Fig. 4 of Ref. [20], where each sub-figure corresponds to the surface mode in the following ranges: (a) below $\varepsilon_0(0)$; (b) in $(\varepsilon_1(0), \varepsilon_2(0))$; (c) in $(\varepsilon_0(\pi/a), \varepsilon_1(\pi/a))$; and (d) in $(\varepsilon_2(\pi/a), \varepsilon_3(\pi/a))$ respectively. Thus, each has no zero, two zeros, one zero and three zeros in a unit cell.

In this section, we have only treated surface modes in a semi-infinite crystal (without a cap layer) for simplicity. A slightly extended formalism can be developed to treat surface modes in a semi-infinite crystal with a cap layer, where the left sides in Eqs. (E.24) and (E.25) will be mathematical expressions containing parameters of the cap layer such as thickness, density, Lamé's coefficient, and so forth.

E.3 One-Dimensional Phononic Crystals of Finite Length

The mathematical theory in Chap. 2 can also be applied to treat the eigenmodes in a one-dimensional phononic crystal described by (E.1) with a finite length $L = Na$. Exact and general fundamental understandings of the eigenmodes in a one-dimensional phononic crystal of finite length with free surfaces can be analytically obtained.[8]

For a one-dimensional phononic crystal described by (E.1) with two free surfaces at $x = \tau$ and $x = \tau + L$, where $L = Na$ and N is an integer, the eigenvalues Λ and the eigenfunctions $\psi(x)$ in the finite crystal are solutions of the differential equation

$$- [\mu(x)\psi'(x, \Lambda)]' + [\mu(x)k_z^2 - \rho(x)\Lambda]\psi(x, \Lambda) = 0, \qquad \tau < x < \tau + L, \tag{E.26}$$

with the stress-free boundary condition at $x = \tau$ and $x = \tau + L$:

$$\mu(\tau)\psi'(\tau, \Lambda) = \mu(\tau)\psi'(\tau + L, \Lambda) = 0. \tag{E.27}$$

[8] Part of the results in this subsection was published in [30]. Since the author did not know then that the theory of the Hill's equation in [40] could be extended as presented in Chap. 2, $\mu'(x)$ rather than $\mu(x)$ was more restrictively assumed to be piecewise continuous.

Suppose $y_1(x, \lambda)$ and $y_2(x, \lambda)$ are two linearly independent solutions of (E.1). In general, a nontrivial solution of (E.26) and (E.27), if it exists, can be expressed as

$$\psi(x, \Lambda) = c_1\, y_1(x, \Lambda) + c_2\, y_2(x, \Lambda), \qquad \tau < x < \tau + L, \tag{E.28}$$

—in which c_1 and c_2 are not both zero—and is a nontrivial solution of (E.1) satisfying (E.27).

The forms of two linearly independent solutions $y_1(x, \lambda)$ and $y_2(x, \lambda)$ in (E.28) can be determined by the discriminant $D(\lambda)$ of (E.1). The existence and the properties of nontrivial solutions of Eqs. (E.26) and (E.27) can be obtained on this basis.

For a one-dimensional phononic crystal of finite length, both the permitted bands and the forbidden λ ranges of the infinite crystal should be considered. We need to consider the solutions of Eq. (E.27) for λ in $[0, +\infty)$. According to the theory in Sect. 2.5, there are five different cases for the linearly independent solutions $y_1(x, \lambda)$ and $y_2(x, \lambda)$ depending on $D(\lambda)$. For the problem we are interested in here, they can be classified into two different λ ranges: λ is inside a permitted band, or not inside a permitted band.

E.3.1 λ Is inside a Permitted Band

In this case, λ is inside a permitted band of (E.1), $-2 < D(\lambda) < 2$. By (2.73), two linearly independent solutions of (E.1) can be chosen as:

$$\begin{aligned} y_1(x, \lambda) &= \mathrm{e}^{\mathrm{i}k_x(\lambda)x}\, p_1(x, \lambda), \\ y_2(x, \lambda) &= \mathrm{e}^{-\mathrm{i}k_x(\lambda)x}\, p_2(x, \lambda). \end{aligned} \tag{E.29}$$

Correspondingly we have

$$\begin{aligned} \mu(x+a)y_1'(x+a, \lambda) &= \mathrm{e}^{\mathrm{i}k_x(\lambda)a}\mu(x)y_1'(x, \lambda), \\ \mu(x+a)y_2'(x+a, \lambda) &= \mathrm{e}^{-\mathrm{i}k_x(\lambda)a}\mu(x)y_2'(x, \lambda), \end{aligned} \tag{E.30}$$

where $k_x(\lambda)$ is a real number depending on λ and

$$0 < k_x(\lambda)a < \pi.$$

From (E.27) and using (E.30) for N times, we obtain that

$$\begin{aligned} c_1\, \mu(\tau)y_1'(\tau, \Lambda) + c_2\, \mu(\tau)y_2'(\tau, \Lambda) &= 0, \\ c_1\, \mathrm{e}^{\mathrm{i}k_x(\Lambda)L}\, \mu(\tau)y_1'(\tau, \Lambda) + c_2\, \mathrm{e}^{-\mathrm{i}k_x(\Lambda)L}\, \mu(\tau)y_2'(\tau, \Lambda) &= 0. \end{aligned}$$

The zeros of $\mu(x)y_1'(x,\lambda)$ are separated from the zeros of $\mu(x)y_2'(x,\lambda)$. Thus, the existence of nontrivial solutions of (E.26) and (E.27) requires that[9]

$$e^{ik_x(\Lambda)L} - e^{-ik_x(\Lambda)L} = 0. \tag{E.31}$$

Note (E.31) does not contain τ. The nontrivial solutions of the form (E.28) with (E.29) can exist if

$$k_x(\Lambda)L = j\pi,$$

where j is a positive integer. There are $N-1$ Bloch wave vector k_x satisfying

$$k_x(\Lambda_{n,j}) = \frac{j\pi}{L}, j = 1, 2, \ldots, N-1.$$

Correspondingly, for each permitted band, there are $N-1$ solutions of (E.26) and (E.27) whose eigenvalues are given by

$$\Lambda_{n,j} = \varepsilon_n\left(\frac{j\pi}{L}\right), \quad j = 1, 2, \ldots, N-1. \tag{E.32}$$

Each eigenvalue for this case is a function of L, the crystal length. However, all do not depend on the location of the crystal boundary τ or $\tau + L$. These modes $\psi(x, \Lambda_{n,j})$ are stationary Bloch modes consisting of two Bloch waves with wave vectors $k_x = \frac{j\pi}{L}$ and $-k_x = -\frac{j\pi}{L}$ in the finite crystal. For simplicity, we call these modes as L-dependent modes, although only the eigenvalue of such a mode depends only on L. The energies $\Lambda_{n,j}$ in (E.32) coincide with the band structure $\varepsilon_n(k_x)$ of the unconfined phononic crystal exactly. This case is very similar to the L-dependent confined electronic states discussed in Chap. 4. A noteworthy point is that although the boundary condition (E.27) is different from the boundary condition (4.6), the expressions of eigenvalues of the L-dependent states or modes for these two cases are the same.

E.3.2 *λ is not Inside a Permitted Band*

If λ is not inside a permitted band, it may be inside a band gap or at a band edge.

[9]Otherwise, we have

$$c_1\, \mu(\tau)y_1'(\tau, \Lambda) = 0, \text{ and } c_2\, \mu(\tau)y_2'(\tau, \Lambda) = 0.$$

Since $\mu(\tau)y_i'(\tau + a, \Lambda) = e^{\pm k_x a}\mu(\tau)y_i'(\tau, \Lambda)$, we must have $\mu(\tau)y_i'(\tau + a, \Lambda) = 0$ if we have $\mu(\tau)y_i'(\tau, \Lambda) = 0$. By Theorem 2.8, this can happen only in the intervals where $|D(\Lambda)| \geq 2$. Thus, neither $\mu(\tau)y_1'(\tau, \Lambda)$ nor $\mu(\tau)y_2'(\tau, \Lambda)$ here can be zero. Therefore, this case leads to $c_1 = c_2 = 0$ and that no nontrivial solution of (E.26) and (E.27) exists from such a case.

If λ is inside a band gap, $y_1(x, \lambda)$ and $y_2(x, \lambda)$ can be chosen as in (2.77) if the band gap is at $k_x=0$ ($D(\lambda) > 2$) or as in (2.81) if the band gap is at $k_x=\pi/a$ ($D(\lambda) < -2$). By using a similar approach as we used to obtain (E.31), we obtain that

$$\mu(\tau)\psi'(\tau + a, \Lambda) = \mu(\tau)\psi'(\tau, \Lambda) = 0 \tag{E.33}$$

is a necessary condition for the existence of a solution of (E.26) and (E.27) for λ inside a band gap, since $e^{\beta(\Lambda)L} - e^{-\beta(\Lambda)L} \neq 0$ with a real $\beta(\Lambda) \neq 0$.

If λ is at a band edge of Eq. (E.1), $y_1(\lambda, x)$ and $y_2(\lambda, x)$ can be chosen as in (2.75) if the band edge is at $k_x = 0$ ($D(\lambda) = 2$) or as in (2.79) if the band gap is at $k_x = \pi/a$ ($D(\lambda) = -2$). Simple mathematics also gives that (E.33) is a necessary condition for the existence of a solution of (E.26) and (E.27).

Therefore, an existing solution of Eqs. (E.26) and (E.27) with its Λ not inside a permitted band must have the form of $\psi(x, \Lambda) = e^{\pm\beta(\Lambda)} f(x, \Lambda)$ where $\beta(\Lambda) \geq 0$ and $f(x, \Lambda)$ is a periodic function or a semi-periodic function. (E.33) is also a sufficient condition for the existence of a solution of (E.26) and (E.27) since from (E.33) we have that

$$\mu(\tau)\psi'(\tau + \ell a, \Lambda) = \mu(\tau)\psi'(\tau, \Lambda) = 0, \tag{E.34}$$

where $\ell = 1, 2, \ldots, N$. Therefore (E.33) is a *necessary and sufficient* condition for the existence of a solution of (E.26) with the boundary condition (E.27) where Λ is not inside a permitted band.

Such eigenvalues can be written as $\Lambda_{\tau,n}$, $n = 0, 1, 2, \ldots$. The eigenvalues $\Lambda_{\tau,n}$ obtained from (E.33) is the $\nu_{\tau,n}$ given in (2.95). Equation (E.33) does not contain the crystal length L; thus, an eigenvalue $\Lambda_{\tau,n}$ of (E.26) and (E.27) obtained from (E.33) is only dependent on τ, but not on L. For simplicity, we call the corresponding eigenmodes as the τ-dependent modes.

Theorem 2.8 indicates that for any real number τ, there is always one and only one eigenvalue $\nu_{\tau,n}$ *below* or equal to the minimum of the corresponding permitted band, determined by (E.33). Such as, $\nu_{\tau,0}$ is in the interval $[0, \varepsilon_0(0)]$, $\nu_{\tau,2m+1}$ is in the interval $[\varepsilon_{2m}(\frac{\pi}{a}), \varepsilon_{2m+1}(\frac{\pi}{a})]$, $\nu_{\tau,2m+2}$ is in the interval $[\varepsilon_{2m+1}(0), \varepsilon_{2m+2}(0)]$, etc. Since in one-dimensional phononic crystals, each permitted band and each forbidden range exist alternatively, we can say that there is always one and only one τ-dependent mode corresponding to each permitted band in a finite phononic crystal of length $L = Na$. This case is similar to the τ-dependent electronic states discussed in Chap. 4.

As discussed in Sect. E.2.2, the calculated $P = \frac{\psi(\tau+a,\Lambda)}{\psi(\tau,\Lambda)}$ from (E.33) can have three possibilities: (a) $0 < |P| < 1$; (b) $|P| = 1$; (c) $|P| > 1$. The case (a) corresponds to an oscillatory decreasing solution in the $+x$ direction $\psi(x, \Lambda)$ of Eqs. (E.26) and (E.27), a surface mode localized at the left boundary τ; The case (b) corresponds to that $\psi(x, \Lambda)$ in Eq. (E.33) is a band-edge mode of Eq. (E.1); The case (c) corresponds to an oscillatory increasing solution $\psi(x, \Lambda)$ in the $+x$ direction, indicating a surface mode localized at the right boundary $\tau + L$.

A significant difference is that the τ-dependent mode here always has a lower frequency than the L-dependent modes corresponding to the same permitted band.

That is, the confinement in a one-dimensional phononic crystal of finite length due to the boundary condition (E.27) always makes the τ-dependent states go lower, rather than always making the τ-dependent states go higher in the quantum confinement of electronic states as discussed in Chap. 4. The basic reason for this significant difference is the difference between (E.27) and the boundary conditions (4.6) for electronic states. As we see in Chap. 4, the boundary conditions (4.6) leads to the requirement $y(\tau) = y(\tau + a) = 0$ for a forbidden range. Theorem 2.8 requires that the corresponding eigenvalues must be in the band gap *above* the energy band. It is this difference that makes the τ-dependent states in finite one-dimensional phononic crystals different from those in Chap. 4.

In conclusion, there are two different types of confined modes in a one dimensional phononic crystal of finite length $L = Na$. Of them, one type is that $N - 1$ modes in each permitted band whose eigenvalue depends on the crystal size L but not on the crystal boundary τ. The other type is one and only one confined mode in each forbidden range below the permitted band, whose eigenvalue depends on the crystal boundary τ but not the crystal length L.

Above results were obtained from a theory of differential equations approach, which is most suitable to treat solid-solid phononic crystals. One-dimensional finite phononic crystals of other forms were investigated in [29, 31, 33–35]. Similar results were obtained. There are two types of confined phonon modes in a finite one-dimensional phononic crystal containing N cells: Corresponding to each permitted band, there are $N - 1$ modes in the band whose frequencies depend on crystal size N; whereas there is always one and only one mode in each forbidden range whose frequency depends on the boundary location.

A.-C. Hladky, G. Allan and M. de Billy [29] investigated theoretically and experimentally the propagation of longitudinal elastic waves along a one-dimensional diatomic chain made of steel spheres of two different diameters alternatively. The vibration modes of the infinite chain have two low-frequency branches, separated by a band gap. In cases of a finite chain of N unit periods, they obtained two different types of modes: $N - 1$ vibration modes in each permitted band whose frequencies strongly depend on N and two modes whose frequency does not depend on N: one in the band gap and the other always at a zero frequency.

By using a Green's function method, El Hassouani et al. [31, 33] and El Boudouti et al. [34] did theoretical investigations on acoustic modes in finite one-dimensional structures made of N-periods, where each period is alternatively composed of a solid (such as Plexiglas) layer and a fluid (such as water) layer, and the finite phononic crystal is free of stress on both sides. They obtained two types of modes: $N - 1$ modes whose frequencies depend on N but not the boundary location in each permitted band, and, one mode with a frequency depending on the boundary location but not N in each forbidden range. This latter is either a surface mode or a confined band-edge mode.

El Boudouti and Djafari-Rouhani [35] further investigated the modes in semi-infinite and finite one-dimensional phononic crystals constructed of more general discrete or continuous media by using a Green's function approach and obtained similar results. They obtained a general expression for the Green's function of a finite crystal containing N unit cells. The unit cell can be a multilayer structure or a multi-waveguide system, etc. The expression involves the relevant matrix elements of the Green's function of the unit cell, but not the details of the unit cell. Thus, a general understanding of the eigenmodes of a finite crystal containing N unit cell can be obtained without knowing the details of the unit cell. From such an expression for a finite crystal of N unit cells, they obtained $N - 1$ confined modes in each bulk permitted band and an additional mode for each forbidden range. This additional mode is either a surface mode localized near one of the two ends if the unit cell is not symmetrical or a confined band edge mode if the unit cell is symmetrical. These extra modes are independent of N and determined by the Green's function of the unit cell.

In summary, in all different one-dimensional phononic crystals of finite size with N unit cells that have been investigated so far, the eigen modes have common general properties if the modes are completely confined to the finite size of the crystal. There are two different types of eigen modes: In each permitted band there are $N - 1$ modes whose eigenvalues depend on the crystal size N but not the boundary location τ, and the eigenvalues of those modes can be obtained from the bulk dispersion relation; There is always one and only one mode corresponding to each forbidden range whose eigenvalue depends on the boundary location τ but not the crystal size N, and this mode is either a surface mode or a confined band-edge mode.

References

1. S.M. Rytov, Akust. Zh. **2**, 71 (1956) [Sov. Phys. Acoust. **2**, 68 (1956)]; S.M. Rytov, Zh. Eksp. Teor. Fiz. **29**, 605 (1955) [Sov. Phys. JETP **2**, 466 (1956)]
2. W.M. Ewing, W.S. Jardetzky, F. Press, *Elastic Waves in Layered Media* (McGraw-Hill Book Co., Inc., New York, 1957)
3. M. Born, E. Wolf, *Principles of Optics: Electromagnetic Theory of Propagation, Interference and Diffraction of Light* (Pergamon Press, Oxford, 1964; Cambridge University Press, Cambridge, 1999)
4. T.J. Delph, G. Herrmann, R.K. Kaul, J. Appl. Mech. **45**, 343 (1972)
5. P. Yeh, A. Yariv, C.-S. Hong, J. Opt. Soc. Am. **67**, 423 (1977); P. Yeh, A. Yariv, A.Y. Cho, Appl. Phys. Lett. **32**, 104 (1978)
6. R.E. Camley, B. Djafari-Rouhani, L. Dobrzynski, A.A. Maradudin, Phys. Rev. B **27**, 7318 (1983)
7. A. Yariv, P. Yeh, *Optical Waves in Crystals* (Wiley, New York, 1984)
8. P. Yeh, *Optical Waves in Layered Media* (Wiley, New York, 1988) (Wiley, Hoboken, 2005)
9. J.D. Joannopoulos, R.D. Meade, J.N. Winn, *Photonic crystals* (Princeton University Press, Princeton, 1995)
10. S. Tamura, D.C. Hurley, J.P. Wolfe, Phys. Rev. B **38**, 1427 (1988)
11. J.D. Joannopoulos, S.G. Johnson, J.N. Winn, R.D. Meade, *Photonic crystals, Molding the Flow of Light*, 2nd edn. (Princeton University Press, Princeton, 2008)
12. J.-M. Lourtioz, H. Benisty, V. Berger, J.-M. Gérard, D. Maystre, A. Tchelnokov, D. Pagnoux, *Photonic Crystals: Towards Nanoscale Photonic Devices*, 2nd edn. (Springer, Berlin, 2008)

13. C. Sibilia, T.M. Benson, M. Marciniak, T. Szoplik (eds.), *Photonic Crystals: Physics and Technology* (Springer, Milano, 2008)
14. D.W. Prather, A. Sharkawy, S. Shi, J. Murakowski, G. Schneider, *Photonic Crystals, Theory, Applications and Fabrication* (Wiley, Hoboken, 2009)
15. A. Khelif, A. Adibi, *Phononic Crystals: Fundamentals and Applications* (Springer, Berlin, 2016)
16. B. Djafari-Rouhani, L. Dobrzynski, O. Hardouin Duparc, R.E. Camley, A. A. Maradudin, Phys. Rev. B **28**, 1711 (1983)
17. H.T. Grahn, H.J. Maris, J. Tauc, B. Abeles, Phys. Rev. B **38**, 6066 (1988)
18. E.H. El Boudouti, B. Djafari-Rouhani, E.M. Khourdifi, L. Dobrzynski, Phys. Rev. B **48**, 10987 (1993)
19. E.H. El Boudouti, B. Djafari-Rouhani, A. Nougaoni, Phys. Rev. B **51**, 13801 (1995)
20. E.H. El Boudouti, B. Djafari-Rouhani, A. Akjouj, L. Dobrzynski, Phys. Rev. B **54**, 14728 (1996)
21. B. Djafari-Rouhani, L. Dobrzynski, Solid State Commun. **62**, 609 (1987)
22. W. Chen, Y. Lu, H.J. Maris, G. Xiao, Phys. Rev. B **50**, 14506 (1994)
23. N-W. Pu, J. Bokor, Phys. Rev. Lett. **91**, 076101 (2003)
24. N-W. Pu, Phys. Rev. B **72**, 115428 (2005)
25. W.M. Robertson, G. Arjavalingam, R.D. Meade, K.D. Brommer, A.M. Rappe, J.D. Joannopoulos, Opt. Lett. **18**, 528 (1993)
26. F. Ramos-Mendieta, P. Halevi, Opt. Commun. **129**, 1 (1996)
27. F. Ramos-Mendieta, P. Halevi, J. Opt. Soc. Am. B **14**, 370 (1997)
28. A.P. Vinogradov, A.V. Dorofeenko, S.G. Erokhin, M. Inoue, A.A. Lisyansky, A.M. Merzlikin, A.B. Granovsky, Phys. Rev. B**74**, 045128 (2006)
29. A.-C. Hladky, G. Allan, M. de Billy, J. Appl. Phys. **98**, 054909 (2005)
30. S.Y. Ren, Y.C. Chang, Phys. Rev. B **75**, 212301 (2007)
31. Y. El Hassouani, E.H. El Boudouti, B. Djafari-Rouhani, R. Rais, J. Phys.: Conf. Ser. **92**, 012113 (2007)
32. E.H. El Boudouti, Y. El Hassouani, B. Djafari-Rouhani, H. Aynaou, Phys. Rev. E **76**, 026607 (2007)
33. Y. El Hassouani, E.H. El Boudouti, B. Djafari-Rouhani, H. Aynaou: Phys. Rev. B **78**, 174306 (2008)
34. E.H. El Boudouti, B. Djafari-Rouhani, A. Akjouj, L. Dobrzynski, Surf. Sci. Rep. **64**, 471 (2009)
35. E.H. El Boudouti, B. Djafari-Rouhani, One-dimensional phononic crystals, in *Acoustic Metamaterials and Phononic Crystals*. Springer Series in Solid-State Sciences, vol. 173, ed. by P.A. Deymier (Springer, Berlin, 2013), p. 45
36. J. Weidmann, *Spectral Theory of Ordinary Differential Operators* (Springer, Berlin, 1987)
37. A. Zettl, *Sturm–Liouville Theory* (American Mathematical Society, Providence, 2005) (In particular, p. 39)
38. B. Malcolm Brown, M.S.P. Eastham, K.M. Schmidt, *Periodic Differential Operators*. Operator Theory: Advances and Applications, vol. 230 (Springer, Heidelberg, 2013) (and references therein)
39. L.D. Landau, E.M. Lifshitz, *Theory of Elasticity* (Pergamon Press, New York, 1981)
40. M.S.P. Eastham, *The Spectral Theory of Periodic Differential Equations* (Scottish Academic Press, Edinburgh, 1973)

Appendix F
One-Dimensional Photonic Crystals

The interests in one-dimensional photonic crystals in the form of periodic multi-layer dielectric stacks have a long history. As early as hundreds of years ago, Hooke and Newton proposed the optical properties of stacks of thin layers [1, 2]. More recent investigations before late 1980's can be found in [3–9] and references therein. The publication of two papers [10, 11] in 1987 indicated a significant breakthrough in investigations of photonic crystals and stimulated a rapid growth of research papers in this active and fruitful field in recent years [12–24]. Nowadays, much research attentions in optics is directed at exploring effects of periodicity in photonic micro- and nanostructures [25].

The most interesting property of the photonic crystals is the possible existence of forbidden frequency band gaps originated from the difference in the electromagnetic properties of the construction media and the periodicity of the systems, which has a high potential to lead to many interesting and valuable practical applications. The Maxwell equations on photonic crystals involve two different but closely related vector fields—the electric field **E** and the magnetic field **H**—thus, the relevant physics is more plentiful and colorful in comparison with that of phononic crystals. The construction materials of photonic crystals can be more diversified; even where the medium outside a photonic crystal is a vacuum, the electromagnetic waves would not be completely confined to the photonic crystal. All of these make the physics of photonic crystals more abundant and more diversified and might lead to more practical applications.

One-dimensional photonic crystals are the simplest photonic crystals. A clear and comprehensive understanding of the physics in one-dimensional photonic crystals is a basis for further understanding of the physics in two or three-dimensional photonic crystals. So far the transfer matrix method (TMM) [5, 8, 9] has been a major theoretical tool for investigations on the simplest one-dimensional photonic crystals—a structure made of alternative layers of two different isotropic homogeneous dielectric materials. By this method, one solves the wave equations in each medium layer of a period unit, matches the relevant components of the magnetic field and electric field according to the necessary continuous conditions at the interfaces and then uses the Bloch theorem to treat the periodicity. Then the required results such as the dispersion

S.Y. Ren, *Electronic States in Crystals of Finite Size*, Springer Tracts in Modern Physics 270, DOI 10.1007/978-981-10-4718-3

relation etc. can be obtained after unknown arbitrary constants were canceled [8, 9]. Much of our current fundamental theoretical understandings on one-dimensional photonic crystals made of two different dielectric media was mainly obtained from TMM.

The modern mathematical theory of periodic Sturm–Liouville equations can treat a one-dimensional layered photonic crystal as a whole, rather than each medium layer separately, since the matching conditions at each interface required by TMM are implicitly contained in the wave equations. In this Appendix, we apply the theory summarized in Chap. 2 to treat one-dimensional photonic crystals, in the order of infinite size, semi-infinite size, and finite size, to understand some of the fundamental physics of one-dimensional photonic crystals, including the physics closely related to the very existence of the boundary and boundaries.

Many authors working on photonic crystals compare electromagnetic waves with electronic waves in periodic systems to illustrate the physics of photonic crystals, see, for example, [12, 16–18, 21, 23]. By the modern theory of periodic Sturm–Liouville equations, the wave equations for electromagnetic waves in some one-dimensional photonic crystals and the Schrödinger equation for electronic states in one-dimensional electronic crystals are special cases of the more general periodic Sturm–Liouville equations. Such a comparison now has a clear theoretical basis. We will also see that some understandings we learned on the one-dimensional electronic crystals in Chaps. 3 and 4 have their analogs or correspondences in one-dimensional photonic crystals.

F.1 Wave Equations

We treat the simplest photonic crystals. We assume that the magnetic permeability of the photonic crystal is equal to that of free space μ_0 and that the relative dielectric coefficient $\varepsilon(\mathbf{x})$ is isotropic, real, periodic with $\mathbf{x}$, and does not depend on frequency. The Maxwell equations for the propagation of light in such a photonic crystal composed of a mixed homogeneous dielectric medium with no free charges or currents lead to four equations (in CGS units) [5, 8, 9, 12, 17]:

$$\begin{array}{rcl} \frac{1}{\varepsilon(\mathbf{x})}\nabla\times\nabla\times\mathbf{E}(\mathbf{x},t) & = & -\frac{1}{c^2}\frac{\partial^2}{\partial t^2}\mathbf{E}(\mathbf{x},t), \\ \nabla\times\left[\frac{1}{\varepsilon(\mathbf{x})}\nabla\times\mathbf{H}(\mathbf{x},t)\right] & = & -\frac{1}{c^2}\frac{\partial^2}{\partial t^2}\mathbf{H}(\mathbf{x},t), \\ \nabla\cdot\varepsilon(\mathbf{x})\mathbf{E}(\mathbf{x},t) & = & 0, \\ \nabla\cdot\mathbf{H}(\mathbf{x},t) & = & 0. \end{array} \tag{F.1}$$

Here, $\mathbf{E}(\mathbf{x},t)$ and $\mathbf{H}(\mathbf{x},t)$ are the electric field and magnetic field respectively, and c is the speed of light in free space.

We are interested in the solutions of (F.1) with the form

$$\mathbf{E}(\mathbf{x}, t) = \mathbf{E}(\mathbf{x})e^{-i\omega t},$$
$$\mathbf{H}(\mathbf{x}, t) = \mathbf{H}(\mathbf{x})e^{-i\omega t},$$

where ω is the eigen-angular frequency, $\mathbf{E}(\mathbf{x})$ and $\mathbf{H}(\mathbf{x})$ are the eigenfunctions of the equations

$$\frac{1}{\varepsilon(\mathbf{x})}\nabla \times \nabla \times \mathbf{E}(\mathbf{x}) - \left(\frac{\omega}{c}\right)^2 \mathbf{E}(\mathbf{x}) = 0, \tag{F.2}$$

and

$$\nabla \times \left[\frac{1}{\varepsilon(\mathbf{x})}\nabla \times \mathbf{H}(\mathbf{x})\right] - \left(\frac{\omega}{c}\right)^2 \mathbf{H}(\mathbf{x}) = 0. \tag{F.3}$$

The electromagnetic wave modes in such a photonic crystal can be solved using (F.3) and

$$\nabla \cdot \mathbf{H}(\mathbf{x}) = 0, \tag{F.4}$$

then $\mathbf{E}(\mathbf{x})$ can be obtained from $\mathbf{H}(\mathbf{x})$ using

$$\nabla \times \mathbf{H}(\mathbf{x}) - \frac{i\omega}{c}\varepsilon(\mathbf{x})\mathbf{E}(\mathbf{x}) = 0; \tag{F.5}$$

Alternatively, they can also be solved by using (F.2) and

$$\nabla \cdot \varepsilon(\mathbf{x})\mathbf{E}(\mathbf{x}) = 0, \tag{F.6}$$

and $\mathbf{H}(\mathbf{x})$ can be obtained from $\mathbf{E}(\mathbf{x})$ using

$$\nabla \times \mathbf{E}(\mathbf{x}) + \frac{i\omega}{c}\varepsilon(\mathbf{x})\mathbf{H}(\mathbf{x}) = 0. \tag{F.7}$$

F.2 TM Modes and TE Modes

In the following, we are interested in one-dimensional photonic crystals where $\varepsilon(\mathbf{x})$ is only a function of x, and it is a periodic function of $x : \varepsilon(x + a) = \varepsilon(x)$.

We choose the plane of wave propagation as the xz plane. An electromagnetic wave propagating in the xz plane can be either a TM mode where the magnetic field $\mathbf{H}(\mathbf{x})$ is in the y direction or a TE mode where the electric field $\mathbf{E}(\mathbf{x})$ is in the y direction.

F.2.1 TM Modes

For a TM mode, the magnetic field $\mathbf{H}(\mathbf{x})$ is in the y direction. $\mathbf{H}(\mathbf{x})$ can be written as

$$H_x = 0,\ H_y = H_y(x)e^{ik_z z},\ H_z = 0.$$

Let $y(x) = H_y(x)$, from Eqs. (F.3) and (F.5) we obtain that for a TM mode

$$\left[\frac{1}{\varepsilon(x)}y'(x)\right]' + \left[\left(\frac{\omega}{c}\right)^2 - \frac{1}{\varepsilon(x)}k_z^2\right]y(x) = 0, \tag{F.8}$$

and

$$E_x\,\frac{\omega}{c}\varepsilon(x) = -k_z y(x)e^{ik_z z},\ E_y = 0,\ E_z\,\frac{i\omega}{c}\varepsilon(x) = y'(x)e^{ik_z z}. \tag{F.9}$$

Equation (F.8) is a specific form of Eq. (2.2) with $p(x)=\frac{1}{\varepsilon(x)}$, $q(x)=\frac{1}{\varepsilon(x)}k_z^2$, $w(x) = \frac{1}{c^2}$ and $\lambda = \omega^2$. In the simplest one-dimensional photonic crystals—the layered photonic crystals—, $\varepsilon(x)$ is a piecewise continuous step function rather than a continuous function. Therefore, Eq. (F.8) is not a Hill's equation in Eastham's book [26]. This is the reason that in the previous edition of this book, it was stated that the theory of the Hill's equations could not be used to treat photonic crystals. Nevertheless, the modern theory of periodic Sturm–Liouville equations summarized in Chap. 2 can be straightforward applied to treat Eq. (F.8) for layered photonic crystals.

By the mathematical theory, each solution y of Eq. (F.8) and its quasi-derivative $\frac{1}{\varepsilon(x)}y'(x)$ are continuous. Thus, the requirements that H_y and E_z are continuous for a TM mode [5, 8, 9] are implicitly contained in Eqs. (F.8) and (F.9).

According to the theory in Chap. 2, the band structure of TM modes is determined by the discriminant $D_{tm}(\lambda)$ of Eq. (F.8) defined as

$$D_{tm}(\lambda) = \eta_1(a,\lambda) + \frac{1}{\varepsilon(a)}\eta_2'(a,\lambda), \tag{F.10}$$

where $\eta_1(x,\lambda)$ and $\eta_2(x,\lambda)$ are two linearly independent normalized solutions of Eq. (F.8) satisfying that

$$\eta_1(0,\lambda) = 1,\ \frac{1}{\varepsilon(0)}\eta_1'(0,\lambda) = 0;\ \ \eta_2(0,\lambda) = 0,\ \frac{1}{\varepsilon(0)}\eta_2'(0,\lambda) = 1. \tag{F.11}$$

The permitted TM bands correspond to the λ ranges where $-2 \le D_{tm}(\lambda) \le 2$. The band structure can be obtained from $D_{tm}(\lambda)$ in these ranges by (2.74) as

$$\cos k_x a = \frac{1}{2}D_{tm}(\lambda), \tag{F.12}$$

where k_x is the Bloch wavevector: $-\frac{\pi}{a} < k_x \le \frac{\pi}{a}$.

F.2.2 TE Modes

For a TE mode, the electric field $\mathbf{E}(\mathbf{x})$ is in the y direction, and $\mathbf{E}(\mathbf{x})$ can be written as

$$E_x = 0,\ E_y = E_y(x)e^{ik_z z},\ E_z = 0.$$

Let $y = E_y(x)$, from Eqs. (F.2) and (F.7) we obtain that for a TE Mode

$$y''(x) + \left[\varepsilon(x)\left(\frac{\omega}{c}\right)^2 - k_z^2\right] y(x) = 0, \tag{F.13}$$

and

$$H_x\ \frac{\omega}{c} = k_z y(x)e^{ik_z z},\ H_y = 0,\ H_z\ \frac{i\omega}{c} = -y'(x)e^{ik_z z}. \tag{F.14}$$

Equation (F.13) is a special case of Eq. (2.2) with $p(x) = 1, q(x) = k_z^2, w(x) = \varepsilon(x)(\frac{1}{c^2})$ and $\lambda = \omega^2$. The mathematical theory in Chap. 2 can be applied to Eq. (F.13). The requirements of E_y and H_z being continuous for a TE mode [5, 8, 9] are implicitly contained in Eqs. (F.13) and (F.14), since each solution y of Eq. (F.8) and its derivative y' are continuous.

According to the theory of Chap. 2, the band structure of TE modes is determined by the discriminant $D_{te}(\lambda)$ of Eq. (F.13) defined as

$$D_{te}(\lambda) = \eta_1(a,\lambda) + \eta_2'(a,\lambda), \tag{F.15}$$

where $\eta_1(x,\lambda)$ and $\eta_2(x,\lambda)$ are two linearly independent normalized solutions of Eq. (F.13) satisfying that

$$\eta_1(0,\lambda) = 1, \eta_1'(0,\lambda) = 0;\quad \eta_2(0,\lambda) = 0, \eta_2'(0,\lambda) = 1. \tag{F.16}$$

The permitted TE bands correspond to the ranges of λ where $-2 \le D_{te}(\lambda) \le 2$. The band structure can be obtained from $D_{te}(\lambda)$ in each range by (2.74) as

$$cos\ k_x a = \frac{1}{2} D_{te}(\lambda), \tag{F.17}$$

where k_x is the Bloch wavevector: $-\frac{\pi}{a} < k_x \le \frac{\pi}{a}$.

F.2.3 Band Structures of a One-Dimensional Layered Photonic Crystal Made of Two Different Materials

The simplest cases are where a one-dimensional photonic crystal is made of two different dielectric materials with dielectric constants ε_1, ε_2 and thicknesses d_1, d_2 alternatively.

F.2.3.1 TM Modes

In such a layered photonic crystal, Eq. (F.8) can be written as

$$\left[\frac{1}{\varepsilon_l}y'(x)\right]' + \left[\frac{\lambda}{c^2} - \frac{k_z^2}{\varepsilon_l}\right]y(x) = 0, \tag{F.18}$$

where

$$\varepsilon_l = \begin{cases} \varepsilon_1, & na < x \leq d_1 + na, \\ \varepsilon_2, & na + d_1 < x \leq (n+1)a. \end{cases} \tag{F.19}$$

Here $\lambda = \omega^2 \geq 0$, $a = d_1 + d_2$ is the period, n is an integer and $\varepsilon_1 \neq \varepsilon_2$ are positive real constants. Equation (F.18) corresponds to a special case of (2.2) where $p(x) = \frac{1}{\varepsilon_l}$ and $q(x) = \frac{1}{\varepsilon_l}k_z^2$ are discontinuous at isolated points $x = na$ and $x = d_1 + na$, $w(x) = \frac{1}{c^2} > 0$.

Equation (F.18) can be rewritten as

$$\left[\frac{1}{\varepsilon_l}y'(x)\right]' + \frac{1}{\varepsilon_l}k_l^2 y(x) = 0, \tag{F.20}$$

where

$$\frac{1}{\varepsilon_l}k_l^2 = \frac{\lambda}{c^2} - \frac{k_z^2}{\varepsilon_l}. \tag{F.21}$$

Here k_l is real if $\frac{\lambda}{c^2} - \frac{k_z^2}{\varepsilon_l} > 0$ or imaginary if $\frac{\lambda}{c^2} - \frac{k_z^2}{\varepsilon_l} < 0$.

According to the theory in Chap. 2, the complex energy band structure—in both permitted bands and forbidden ranges—of the TM modes in a one-dimensional photonic crystal can be completely determined by the discriminant $D_{tm}(\lambda)$ of the Eq. (F.18). By (C.31) in Appendix C, it is

$$D_{tm}(\lambda) = 2\,cos\,k_2 d_2\,cos\,k_1 d_1 - \left[\frac{k_1\varepsilon_2}{k_2\varepsilon_1} + \frac{k_2\varepsilon_1}{k_1\varepsilon_2}\right] sin\,k_2 d_2\,sin\,k_1 d_1. \tag{F.22}$$

As λ increases, $D_{tm}(\lambda)$ changes in a way as $D(\lambda)$ described in Sect. 2.4.

Most previous theoretical investigations on TM modes in photonic crystals are of their permitted band structure where $-2 \leq D_{tm}(\lambda) \leq +2$. From (F.12) and (F.22) we obtain that for such a layered photonic crystal, the permitted band structure of TM modes is determined by

$$cos\,k_x a = cos\,k_2 d_2\,cos\,k_1 d_1 - \frac{1}{2}\left[\frac{k_1\varepsilon_2}{\varepsilon_1 k_2} + \frac{k_2\varepsilon_1}{\varepsilon_2 k_1}\right] sin\,k_2 d_2\,sin\,k_1 d_1\,. \tag{F.23}$$

Equation (F.23) essentially is the dispersion relation obtained in Sect. 6.2 in [8] and Sect. 6.2 in [9] and frequently seen in the literature, such as in [4, 27–31], etc.

The solutions in forbidden ranges where $|D_{tm}(\lambda)| > 2$ are important for semi-infinite and finite photonic crystals, their behaviors also depend on $D_{tm}(\lambda)$.

F.2.3.2 TE Modes

For TE modes in such a layered photonic crystal, Eq. (F.13) can be written as

$$y''(x) + \left[\frac{\lambda}{c^2}\,\varepsilon_l - k_z^2\right] y(x) = 0, \tag{F.24}$$

where

$$\varepsilon_l = \begin{cases} \varepsilon_1, & na < x \le d_1 + na, \\ \varepsilon_2, & na + d_1 < x \le (n+1). \end{cases} \tag{F.25}$$

Here $\lambda = \omega^2 \ge 0$, $a = d_1 + d_2$ is the period, n is an integer, and $\varepsilon_1 \ne \varepsilon_2$ are positive real constants. Equation (F.24) corresponds to a special case of Eq. (2.2) where $p(x) = 1$, $q(x) = k_z^2$, and $w(x) = \frac{\varepsilon_l}{c^2} > 0$.

Equation (F.24) can be rewritten as

$$y''(x) + k_l^2 y(x) = 0, \qquad l = 1, 2, \tag{F.26}$$

where

$$k_l^2 = \frac{\lambda}{c^2}\,\varepsilon_l - k_z^2. \tag{F.27}$$

Here k_l is real if $\frac{\lambda}{c^2}\varepsilon_l - k_z^2 > 0$ or imaginary if $\frac{\lambda}{c^2}\varepsilon_l - k_z^2 < 0$.

According to the theory in Chap. 2, the complex energy band structure—in both permitted bands and forbidden ranges—of the TE modes in the one-dimensional photonic crystal can be completely determined by the discriminant $D_{te}(\lambda)$ of Eq. (F.24). By (C.31) in Appendix C, it is

$$D_{te}(\lambda) = 2\,cos\;k_2 d_2\;cos\;k_1 d_1 - \left[\frac{k_1}{k_2} + \frac{k_2}{k_1}\right] sin\;k_2 d_2\;sin\;k_1 d_1. \tag{F.28}$$

As λ increases, $D_{te}(\lambda)$ changes in the way as $D(\lambda)$ described in Sect. 2.4.

Most previous theoretical investigations on TE modes in photonic crystals are of their permitted band structure where $-2 \le D_{te}(\lambda) \le +2$. From (F.17) and (F.28) we can obtain that for such a layered photonic crystal, the permitted band structure of TE modes is determined by

$$\cos k_x a = \cos k_2 d_2 \cos k_1 d_1 - \frac{1}{2}\left[\frac{k_1}{k_2} + \frac{k_2}{k_1}\right] \sin k_2 d_2 \sin k_1 d_1. \tag{F.29}$$

Equation (F.29) essentially is the dispersion relation obtained in Sect. 6.2 in [8] and Sect. 6.2 in [9] and frequently seen in the literature, such as in [4, 27–31], etc.

Theorem 2.7 gives the exact number of zeros of each band-edge mode. An eigenmode at the bottom or the top of the lowest band gap has exactly one zero in one period a. Those are the facts shown in Fig. 3 in Chap. 4 of [12] and Figs. 3 and 4 in Chap. 4 of [17].

The solutions in forbidden ranges where $|D_{te}(\lambda)| > 2$ are important for semi-infinite and finite photonic crystals, and their behaviors also depend on $D_{te}(\lambda)$.

F.2.4 Band Structure of a One-Dimensional Layered Photonic Crystal Made of More Different Materials

From above discussions, we have seen that an essential step for obtaining the band structures of a one-dimensional photonic crystal is to obtain the discriminants $D_{tm}(\lambda)$ and $D_{te}(\lambda)$. For a one-dimensional layered photonic crystal in which each unit cell is made of three or more different materials [32–35], the discriminant $D_{tm}(\lambda)$ of Eq. (F.8) for TM modes and the discriminant $D_{te}(\lambda)$ of Eq. (F.13) for TE modes can be obtained from relevant expressions in Appendix C.

F.2.4.1 TM Modes

For a one-dimensional layered photonic crystal in which each unit cell is made of three different homogeneous and isotropic dielectric materials with dielectric constants ε_1, ε_2, ε_3 and thicknesses d_1, d_2, d_3 alternatively, the band structure of TM modes can be obtained from (C.38) as

$$\begin{aligned}\cos k_x a = {} & \cos k_3 d_3 \cos k_2 d_2 \cos k_1 d_1 \\ & - \frac{1}{2}\left(\frac{\varepsilon_2 k_1}{\varepsilon_1 k_2} + \frac{\varepsilon_1 k_2}{\varepsilon_2 k_1}\right) \cos k_3 d_3 \sin k_2 d_2 \sin k_1 d_1 \\ & - \frac{1}{2}\left(\frac{\varepsilon_3 k_1}{\varepsilon_1 k_3} + \frac{\varepsilon_1 k_3}{\varepsilon_3 k_1}\right) \sin k_3 d_3 \cos k_2 d_2 \sin k_1 d_1 \\ & - \frac{1}{2}\left(\frac{\varepsilon_3 k_2}{\varepsilon_2 k_3} + \frac{\varepsilon_2 k_3}{\varepsilon_3 k_2}\right) \sin k_3 d_3 \sin k_2 d_2 \cos k_1 d_1.\end{aligned} \tag{F.30}$$

Here $k_l, l = 1, 2, 3$ are given by

$$\frac{1}{\varepsilon_l} k_l^2 = \frac{\lambda}{c^2} - \frac{k_z^2}{\varepsilon_l}, \tag{F.31}$$

similarly as in Eq. (F.21).

For a one-dimensional layered photonic crystal in which each unit cell is made of four different homogeneous and isotropic dielectric materials, the discriminant $D_{tm}(\lambda)$ of Eq. (F.8) can be obtained in (C.43). The band structure can be obtained accordingly from (C.44). If needed, for a one-dimensional layered photonic crystal in which each unit cell is made of five or more different homogeneous and isotropic dielectric materials, the discriminant $D_{tm}(\lambda)$ of Eq. (F.8) can be obtained similarly by using the approach in Appendix C. The band structure can be obtained accordingly.

F.2.4.2 TE Modes

For a one-dimensional layered photonic crystal in which each unit cell is made of three different homogeneous and isotropic dielectric materials, the band structure of TE modes can be obtained from (C.38) as

$$\begin{aligned} cos\ k_x a = &\ cos\ k_3 d_3\ cos\ k_2 d_2\ cos\ k_1 d_1 \\ &-\frac{1}{2}\left(\frac{k_1}{k_2}+\frac{k_2}{k_1}\right)\ cos\ k_3 d_3\ sin\ k_2 d_2\ sin\ k_1 d_1 \\ &-\frac{1}{2}\left(\frac{k_1}{k_3}+\frac{k_3}{k_1}\right)\ sin\ k_3 d_3\ cos\ k_2 d_2\ sin\ k_1 d_1 \\ &-\frac{1}{2}\left(\frac{k_2}{k_3}+\frac{k_3}{k_2}\right)\ sin\ k_3 d_3\ sin\ k_2 d_2\ cos\ k_1 d_1. \end{aligned} \tag{F.32}$$

Here $k_l, l = 1, 2, 3$ are given by

$$k_l^2 = \frac{\lambda}{c^2}\,\varepsilon_l - k_z^2, \tag{F.33}$$

similarly as in Eq. (F.27).

Similarly, for a one-dimensional layered photonic crystal in which each unit cell is made of four different homogeneous and isotropic dielectric materials, the discriminant $D_{te}(\lambda)$ of Eq. (F.13) can be obtained from (C.43). The band structure can be obtained accordingly from (C.44). If needed, for a one-dimensional layered photonic crystal in which each unit cell is made of five or more different homogeneous and isotropic dielectric materials, the discriminant $D_{tm}(\lambda)$ of Eq. (F.13) can also be obtained similarly by using the approach in Appendix C, and the band structure can be obtained accordingly.

F.3 Surface Modes in Semi-infinite One-Dimensional Photonic Crystals

A semi-infinite one-dimensional photonic crystal may have surface modes existing in band gaps and localized near the boundary surface. There have been interesting investigations on surface modes in one-dimensional photonic crystals in the literature, see, for example, [7–9, 12, 18, 27–31, 36–50] and references therein. As in cases of infinite one-dimensional photonic crystals, a previous primary theoretical tool is the transfer matrix method. Many interesting results and fundamental understandings of the surface modes in semi-infinite one-dimensional photonic crystals were obtained by this method.

In this section, we use the theory of periodic Sturm–Liouville equations in Chap. 2 to treat the problem. A general formalism of theoretical investigations on the surface modes in one-dimensional photonic crystals can be developed, and new fundamental understandings can be obtained.

The surface modes in a one-dimensional semi-infinite photonic crystal, in general, are different from the surface states in an ideal semi-infinite electronic crystal, or a semi-infinite phononic crystal with a free boundary surface: The electromagnetic waves are not completely confined to the semi-infinite photonic crystal even where the medium outside of a photonic crystal is a vacuum. Because of this, the surface modes in a one-dimensional semi-infinite photonic crystal can only exist below the light line such as shown in Fig. 13 of Chap. 4 in [12] or Fig. 14 of Chap. 4 in [17].

F.3.1 Surface TM Modes

For a simple semi-infinite one-dimensional photonic crystal with an external homogeneous medium of a dielectric constant $\varepsilon_0 > 0$ at the left boundary τ, the wave equation for the TM modes can be written as

$$\left[\frac{1}{\varepsilon(x)}\psi'(x)\right]' + \left[\left(\frac{\Omega}{c}\right)^2 - \frac{1}{\varepsilon(x)}k_z^2\right]\psi(x) = 0, \quad x > \tau, \tag{F.34}$$

in the photonic crystal and

$$\psi''(x) + \left[\varepsilon_0\left(\frac{\Omega}{c}\right)^2 - k_z^2\right]\psi(x) = 0, \quad x < \tau, \tag{F.35}$$

in the external medium, with conditions that H_y and E_z are continuous at τ.

A solution of Eq. (F.35) can be decaying in the $-x$ direction only when $\varepsilon_0(\frac{\Omega}{c})^2 - k_z^2 < 0$. That is, a localized surface mode can only exist if it is below the light line $\omega = \frac{c}{\sqrt{\varepsilon_0}}k_z$.

Equation (F.35) can be solved for a constant ε_0 as

$$\psi(x) = Ce^{\gamma x}, \qquad x < \tau, \tag{F.36}$$

where

$$\gamma = \left[k_z^2 - \varepsilon_0 \left(\frac{\Omega}{c} \right)^2 \right]^{1/2} > 0. \tag{F.37}$$

By considering the continuous conditions of H_y and E_z for TM modes at the boundary τ, the wave equation inside the photonic crystal can now be written in the form of

$$\begin{aligned} \left[\tfrac{1}{\varepsilon(x)}\psi'(x)\right]' + \left[\left(\tfrac{\Omega}{c}\right)^2 - \tfrac{1}{\varepsilon(x)}k_z^2\right]\psi(x) = 0, \quad x > \tau, \\ \sigma_{tm}\psi(x) = \tfrac{1}{\varepsilon(\tau)}\psi'(x), \qquad x = \tau + 0, \end{aligned} \tag{F.38}$$

where

$$\sigma_{tm} = \frac{1}{\varepsilon_0} \left[k_z^2 - \varepsilon_0 \left(\frac{\Omega}{c} \right)^2 \right]^{1/2} \tag{F.39}$$

is a quantity determined by Ω, k_z and ε_0.[10]

We are only interested in the solutions of Eq. (F.38) which are localized near the boundary τ. From Sect. 2.5 in Chap. 2, we know that if a solution of Eq. (F.34) localized near τ exists, it is in a band gap of Eq. (F.8). Such a solution has the form

$$\psi(x, \Lambda) = e^{-\beta(\Lambda)x} f(x, \Lambda), \tag{F.40}$$

where $\beta(\Lambda) > 0$. Here $f(x, \Lambda)$ is a periodic function $p(x + a, \Lambda) = p(x, \Lambda)$ if the band-gap is located at the center of the Brillouin zone $k_x = 0$, or a semi-periodic function $s(x + a, \Lambda) = -s(x, \Lambda)$ if the band-gap is located at the boundary of the Brillouin zone $k_x = \frac{\pi}{a}$. Consequently, the following equation is *a necessary condition* for the existence of such a surface mode:

$$\sigma_{tm}\psi(\tau + a, \Lambda) - \frac{1}{\varepsilon(\tau)}\psi'(\tau + a, \Lambda) = \sigma_{tm}\psi(\tau, \Lambda) - \frac{1}{\varepsilon(\tau)}\psi'(\tau, \Lambda) = 0. \tag{F.41}$$

Any solution ψ of Eqs. (F.34) and (F.35) or Eq. (F.38) in the semi-infinite crystal $x \geq \tau$ can be expressed as a linear combination of two normalized solutions of Eq. (F.8):

$$\psi(x, \lambda) = c_1\eta_1(x, \lambda) + c_2\eta_2(x, \lambda), \qquad x \geq \tau, \tag{F.42}$$

[10]Since for a TM mode, H_y and E_z are continuous, from Eq. (F.9) one can obtain that $\psi_{\tau-0} = \psi_{\tau+0}$ and $\frac{\psi'_{\tau-0}}{\varepsilon_0} = \frac{\psi'_{\tau+0}}{\varepsilon(\tau)}$. Thus $\sigma_{tm} = \frac{\psi'_{\tau+0}}{\varepsilon(\tau)\psi_{\tau+0}} = \frac{\psi'_{\tau-0}}{\varepsilon_0\psi_{\tau-0}} = \frac{\gamma}{\varepsilon_0}$.

where $\eta_1(x, \lambda)$ and $\eta_2(x, \lambda)$ are two linearly independent solutions of Eq. (F.8) defined by

$$\eta_1(\tau, \lambda) = 1, \ \frac{1}{\varepsilon(\tau)}\eta_1'(\tau, \lambda) = 0; \quad \eta_2(\tau, \lambda) = 0, \ \frac{1}{\varepsilon(\tau)}\eta_2'(\tau, \lambda) = 1. \quad (F.43)$$

From Eqs. (F.41), (F.42) and (F.43), a reasoning similar to Sect. 3.5 or E.2.2 leads to that the existence and properties of a TM surface mode in a one-dimensional semi-infinite photonic crystal are determined by the following two equations:

$$\frac{1}{\varepsilon_0}\left[k_z^2 - \varepsilon_0\left(\frac{\Omega}{c}\right)^2\right]^{1/2} = \frac{-\eta_1(\tau + a, \Omega^2) + \frac{1}{\varepsilon(\tau)}\eta_2'(\tau + a, \Omega^2) + \sqrt{D_{tm}^2(\Omega^2) - 4}}{2\,\eta_2(\tau + a, \Omega^2)} \quad (F.44)$$

for a band gap at $k_x = \pi/a$, or

$$\frac{1}{\varepsilon_0}\left[k_z^2 - \varepsilon_0\left(\frac{\Omega}{c}\right)^2\right]^{1/2} = \frac{-\eta_1(\tau + a, \Omega^2) + \frac{1}{\varepsilon(\tau)}\eta_2'(\tau + a, \Omega^2) - \sqrt{D_{tm}^2(\Omega^2) - 4}}{2\,\eta_2(\tau + a, \Omega^2)} \quad (F.45)$$

for a band gap at $k_x = 0$. These two equations can be used to investigate how the existence of a surface TM mode and its properties in a specific one-dimensional semi-infinite photonic crystal depend on the boundary location τ and the external medium.

More specifically, for the simple cases where a one-dimensional photonic crystal is alternatively made of two different dielectric media, without losing generality, we can consider only the cases where the surface boundary τ is in the medium 1, thus, $0 \leq \tau \leq d_1$.

The simplest cases are that the boundary layer is a full layer of medium 1, thus, $\tau = 0$. The existence and properties of a surface TM mode are determined by Eqs. (F.44) and (F.45) where

$$\begin{aligned}
\eta_1(\tau + a, \lambda) &= cos\ k_2 d_2\ cos\ k_1 d_1 - \frac{k_1 \varepsilon_2}{k_2 \varepsilon_1} sin\ k_2 d_2\ sin\ k_1 d_1, \\
\eta_2(\tau + a, \lambda) &= \frac{\varepsilon_1}{k_1} cos\ k_2 d_2\ sin\ k_1 d_1 + \frac{\varepsilon_2}{k_2} sin\ k_2 d_2\ cos\ k_1 d_1, \\
\frac{1}{\varepsilon(\tau)}\eta_2'(a, \lambda) &= cos\ k_2 d_2\ cos\ k_1 d_1 - \frac{k_2 \varepsilon_1}{k_1 \varepsilon_2} sin\ k_2 d_2\ sin\ k_1 d_1, \quad (F.46)
\end{aligned}$$

as given in (C.27), (C.29) and (C.30) and $D_{tm}(\lambda)$ is given in (F.22).

In more general cases where $\tau \neq 0$, the photonic crystal made of two media can be seen as a particular case of a photonic crystal made of three media with dielectric constants $\varepsilon_1, \varepsilon_2, \varepsilon_1$ and thicknesses of $d_1 - \tau, d_2, \tau$. We can now treat the boundary τ as the origin of the three-media photonic crystal by defining $\eta_1(x, \lambda)$ and $\eta_2(x, \lambda)$ as in (F.43). From Eqs. (C.33), (C.35) and (C.36) we can obtain that

$$\eta_1(\tau + a, \lambda) = \cos k_2 d_2 \cos k_1 d_1 - \sin k_2 d_2 \left[\frac{\varepsilon_2 k_1}{\varepsilon_1 k_2} \sin k_1 (d_1 - \tau) \cos k_1 \tau \right.$$
$$\left. + \frac{\varepsilon_1 k_2}{\varepsilon_2 k_1} \cos k_1 (d_1 - \tau) \sin k_1 \tau \right],$$
$$\eta_2(\tau + a, \lambda) = \frac{\varepsilon_1}{k_1} \cos k_2 d_2 \sin k_1 d_1 + \frac{\varepsilon_2}{k_2} \sin k_2 d_2 \cos k_1 d_1$$
$$+ \frac{\varepsilon_2}{k_2} \left(1 - \frac{\varepsilon_1^2 k_2^2}{\varepsilon_2^2 k_1^2} \right) \sin k_2 d_2 \sin k_1 (d_1 - \tau) \sin k_1 \tau,$$
$$\frac{1}{\varepsilon(\tau)} \eta_2'(\tau + a, \lambda) = \cos k_2 d_2 \cos k_1 d_1 - \sin k_2 d_2 \left[\frac{\varepsilon_1 k_2}{\varepsilon_2 k_1} \sin k_1 (d_1 - \tau) \cos k_1 \tau \right.$$
$$\left. + \frac{\varepsilon_2 k_1}{\varepsilon_1 k_2} \cos k_1 (d_1 - \tau) \sin k_1 \tau \right], \quad \text{(F.47)}$$

which can be used in Eq. (F.44) or (F.45) with $D_{tm}(\lambda)$ given in (F.22).

For cases where a one-dimensional photonic crystal is made alternatively of three different dielectric media, without losing generality, we can also consider only the cases where the boundary τ is in the medium 1 thus we have $0 \leq \tau \leq d_1$. The photonic crystal made of three media can be seen as a particular case of a photonic crystal made of four media with the dielectric constants $\varepsilon_1, \varepsilon_2, \varepsilon_3, \varepsilon_1$ and the thickness of $d_1 - \tau, d_2, d_3, \tau$. We can now treat the boundary τ as the origin of the four-media photonic crystal by defining $\eta_1(x, \lambda)$ and $\eta_2(x, \lambda)$ as in (F.43). From Eqs. (C.39), (C.41) and (C.42) we can obtain $\eta_1(\tau + a, \lambda)$, $\eta_2(\tau + a, \lambda)$, $\frac{1}{\varepsilon(\tau)} \eta_2'(\tau + a, \lambda)$ which can be used in Eqs. (F.44) or (F.45) with $D_{tm}(\lambda)$ given in (C.37).

F.3.2 Surface TE Modes

For a simple semi-infinite one-dimensional photonic crystal with an external homogeneous medium of a dielectric constant $\varepsilon_0 > 0$ at the left boundary τ, the wave equation for the TE modes can be written as

$$\psi''(x) + \left[\varepsilon(x) \left(\frac{\Omega}{c} \right)^2 - k_z^2 \right] \psi(x) = 0, \quad x > \tau, \quad \text{(F.48)}$$

and

$$\psi''(x) + \left[\varepsilon_0 \left(\frac{\Omega}{c} \right)^2 - k_z^2 \right] \psi(x) = 0, \quad x < \tau, \quad \text{(F.49)}$$

with the conditions that E_y and H_z are continuous at τ.

A solution of Eq. (F.49) can be decaying in the $-x$ only when $\varepsilon_0 \left(\frac{\Omega}{c} \right)^2 - k_z^2 < 0$. That is, a localized surface mode can only exist if it is below the light line $\omega = \frac{c}{\sqrt{\varepsilon_0}} k_z$.

Equation (F.49) can be solved for a constant ε_0 as

$$\psi(x) = Ce^{\gamma x}, \qquad x < \tau, \tag{F.50}$$

where

$$\gamma = \left[k_z^2 - \varepsilon_0\left(\frac{\Omega}{c}\right)^2\right]^{1/2} > 0. \tag{F.51}$$

By considering the continuous conditions of E_y and H_z at the boundary τ, the wave equation inside the photonic crystal can now be written in the form of

$$\begin{aligned} \psi''(x) + [\varepsilon(x)(\tfrac{\Omega}{c})^2 - k_z^2]\psi(x) = 0, \qquad x > \tau, \\ \sigma_{te}\,\psi(x) = \psi'(x), \qquad x = \tau + 0, \end{aligned} \tag{F.52}$$

where

$$\sigma_{te} = \left[k_z^2 - \varepsilon_0\left(\frac{\Omega}{c}\right)^2\right]^{1/2} \tag{F.53}$$

is a quantity determined by Ω, k_z and ε_0.[11]

We are only interested in the solutions of (F.52) which are localized near the boundary τ. From Sect. 2.5 in Chap. 2, we can see that if a solution of Eq. (F.48) localized near τ, it is in a band gap of (F.13). Such a solution has the form

$$\psi(x, \Lambda) = \mathrm{e}^{-\beta(\Lambda)x} f(x, \Lambda), \tag{F.54}$$

where $\beta(\Lambda) > 0$. Here $f(x, \Lambda)$ is either a periodic function $p(x+a, \Lambda) = p(x, \Lambda)$ if the band-gap is situated at the center of the Brillouin zone $k_x = 0$ or a semi-periodic function $s(x + a, \Lambda) = -s(x, \Lambda)$ if the band-gap is situated at the boundary of the Brillouin zone $k_x = \frac{\pi}{a}$.

Consequently, the following equation is a necessary condition for the existence of such a surface mode:

$$\sigma_{te}\psi(\tau + a, \Lambda) - \psi'(\tau + a, \Lambda) = \sigma_{te}\psi(\tau, \Lambda) - \psi'(\tau, \Lambda) = 0. \tag{F.55}$$

Any solution ψ of Eqs. (F.48) and (F.49) or Eq. (F.52) for $\tau > 0$ can be expressed as a linear combination of two normalized solutions of Eq. (F.24):

$$\psi(x, \lambda) = c_1\eta_1(x, \lambda) + c_2\eta_2(x, \lambda), \qquad x \geq \tau, \tag{F.56}$$

[11] Since for a TE mode, E_y and H_z are continuous, from Eq. (F.14) one obtains that $\psi_{\tau-0} = \psi_{\tau+0}$ and $\psi'_{\tau-0} = \psi'_{\tau+0}$. Thus $\sigma_{te} = \frac{\psi'_{\tau+0}}{\psi_{\tau+0}} = \frac{\psi'_{\tau-0}}{\psi_{\tau-0}} = \gamma$.

where $\eta_1(x,\lambda)$ and $\eta_2(x,\lambda)$ are two linearly independent solutions of Eq. (F.24) defined by

$$\eta_1(\tau,\lambda)=1,\ \eta_1'(\tau,\lambda)=0;\quad \eta_2(\tau,\lambda)=0,\ \eta_2'(\tau,\lambda)=1. \tag{F.57}$$

From Eqs. (F.55), (F.56) and (F.57), the reasonings similar to Sect. 3.5 or E.2.2 give that the existence and properties of a TE surface mode in a one-dimensional semi-infinite photonic crystal are determined by the following two equations:

$$\left[k_z^2-\varepsilon_0\left(\frac{\Omega}{c}\right)^2\right]^{1/2}=\frac{-\eta_1(\tau+a,\Omega^2)+\eta_2'(\tau+a,\Omega^2)+\sqrt{D_{te}^2(\Omega^2)-4}}{2\,\eta_2(\tau+a,\Omega^2)}, \tag{F.58}$$

for a band gap at $k_x=\pi/a$ or

$$\left[k_z^2-\varepsilon_0\left(\frac{\Omega}{c}\right)^2\right]^{1/2}=\frac{-\eta_1(\tau+a,\Omega^2)+\eta_2'(\tau+a,\Omega^2)-\sqrt{D_{te}^2(\Omega^2)-4}}{2\,\eta_2(\tau+a,\Omega^2)}, \tag{F.59}$$

for a band gap at $k_x=0$. These two equations can be used to investigate how the existence of a surface TE mode and its properties in a specific one-dimensional semi-infinite photonic crystal depend on the boundary τ and the external medium.

More specifically, for the simple cases where a one-dimensional photonic crystal is made of two different dielectric media alternatively, without losing generality, we can consider only the cases where the boundary τ is in the medium 1, thus, $0\le\tau\le d_1$.

The simplest cases are where the boundary layer is a full layer of medium 1 thus $\tau=0$, the existence and properties of a surface mode are determined by Eqs. (F.58) or (F.59) where

$$\begin{aligned}
\eta_1(\tau+a,\lambda)&=cos\ k_2d_2\ cos\ k_1d_1-\frac{k_1}{k_2}sin\ k_2d_2\ sin\ k_1d_1,\\
\eta_2(\tau+a,\lambda)&=\frac{1}{k_1}cos\ k_2d_2\ sin\ k_1d_1+\frac{1}{k_2}sin\ k_2d_2\ cos\ k_1d_1,\\
\eta_2'(\tau+a,\lambda)&=cos\ k_2d_2\ cos\ k_1d_1-\frac{k_2}{k_1}sin\ k_2d_2\ sin\ k_1d_1.
\end{aligned} \tag{F.60}$$

as given in (C.27), (C.29) and (C.30). $D_{te}(\lambda)$ is given in (F.28). The equations obtained are essentially same as equations previously obtained for the existence of TE surface modes such as Eqs. (6.9)–(5) in [8], Eqs. (11.5)–(6) in [9] and used in [30].

In more general cases where $\tau\neq 0$, the photonic crystal made of two media can be considered as a particular case of a photonic crystal made of three media with dielectric constants $\varepsilon_1,\varepsilon_2,\varepsilon_1$, and thicknesses of $d_1-\tau, d_2, \tau$. We can now treat the boundary τ as the origin of the three-media photonic crystal by defining $\eta_1(x,\lambda)$ and $\eta_2(x,\lambda)$ as (F.57). From Eqs. (C.33), (C.35) and (C.36), we obtain that

$$\eta_1(\tau + a, \lambda) = cos\ k_2 d_2\ cos\ k_1 d_1$$
$$-sin\ k_2 d_2 \left[\frac{k_1}{k_2} sin\ k_1(d_1 - \tau)\ cos\ k_1\tau + \frac{k_2}{k_1} cos\ k_1(d_1 - \tau)\ sin\ k_1\tau \right],$$
$$\eta_2(\tau + a, \lambda) = \frac{1}{k_1} cos\ k_2 d_2\ sin\ k_1 d_1 + \frac{1}{k_2} sin\ k_2 d_2\ cos\ k_1 d_1$$
$$+ \frac{1}{k_2}\left(1 - \frac{k_2^2}{k_1^2}\right) sin\ k_2 d_2\ sin\ k_1(d_1 - \tau)\ sin\ k_1\tau,$$
$$\eta_2'(\tau + a, \lambda) = cos\ k_2 d_2\ cos\ k_1 d_1$$
$$-sin\ k_2 d_2 \left[\frac{k_2}{k_1} sin\ k_1(d_1 - \tau)\ cos\ k_1\tau + \frac{k_1}{k_2} cos\ k_1(d_1 - \tau)\ sin\ k_1\tau \right], \quad \text{(F.61)}$$

which can be used in Eq. (F.58) or (F.59) with $D_{te}(\lambda)$ given in (F.28).

For cases where a one-dimensional photonic crystal is made alternatively of three different dielectric media, without losing generality, we can also consider only the cases where the surface boundary is in the medium 1, thus, $0 \le \tau \le d_1$. The photonic crystal made of three media can be considered as a particular case of a photonic crystal made of four media with dielectric constants $\varepsilon_1, \varepsilon_2, \varepsilon_3, \varepsilon_1$, and thickness $d_1 - \tau, d_2, d_3, \tau$. We can now treat the boundary τ as the origin of the four-media photonic crystal by defining $\eta_1(x, \lambda)$ and $\eta_2(x, \lambda)$ as in (F.57). From Eqs. (C.39), (C.41) and (C.42) we can obtain $\eta_1(\tau + a, \lambda)$, $\eta_2(\tau + a, \lambda)$, $\eta_2'(\tau + a, \lambda)$ which can be used in Eq. (F.58) or (F.59) with $D_{te}(\lambda)$ given in (C.37).

F.3.3 Discussions

A major theoretical formalism widely used in previous investigations on surface modes in one-dimensional photonic crystals such as [8, 9, 30] treated the simplest one-dimensional photonic crystals made of two media and the outmost layer is a whole layer of one medium. Previous theoretical investigations on surface modes with different surface termination locations used a super-cell method such as in [37, 38] or other numerical methods [12, 17]. The theoretical formalism developed in Sects. F.3.1 and F.3.2 is more general and easy to use. As a simple example, in Fig. F.1 is shown the surface band structure of a semi-infinite one-dimensional photonic crystal investigated in [12, 17], with more different surface termination locations. Note that the τ-regions where the TE surface bands exist are different for different bulk band gaps. In each bulk band gap, as the boundary goes inside—that is, as τ increases—, the surface band curve goes higher.

In this section, we only treated surface modes in semi-infinite photonic crystals (without a cap layer) for simplicity. A slightly extended formalism can be developed to treat surface modes in semi-infinite photonic crystals with a cap layer, where the left side in Eqs. (F.44) and (F.45) or Eqs. (F.58) and (F.59) will be mathematical expressions containing parameters of the cap layer such as thickness, dielectric coefficient, and so forth.

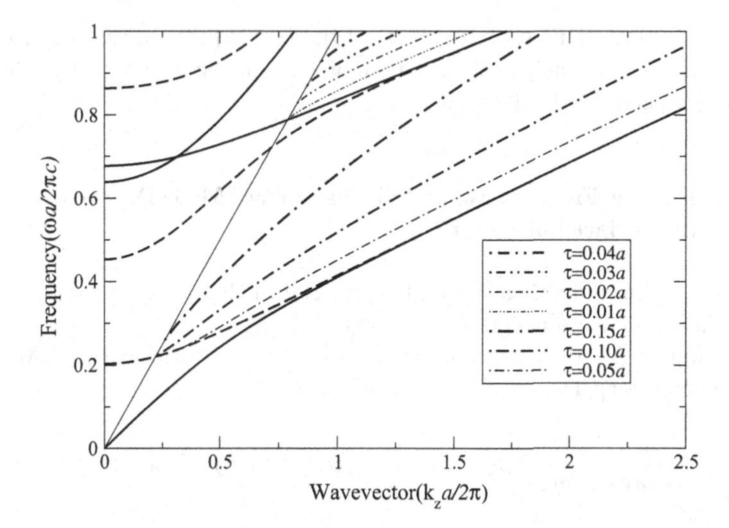

Fig. F.1 Change of TE surface band structure of a semi-infinite one-dimensional photonic crystal investigated in [12, 17] as the surface location τ changes. The one-dimensional photonic crystal is the same as the one in Fig. 13 of Chap. 4 in [12] and Fig. 14 of Chap. 4 in [17]. The outmost layer is an $\varepsilon = 13$ layer, with a layer thickness $0.2a - \tau$. The *solid* (*dashed*) *lines* are band edges at $k_x = 0$ ($k_x = \frac{\pi}{a}$) of the bulk bands. The *thin solid line* is the light line. Each one-dot chained line corresponds to a surface band in the lowest bulk band gap at $k_x = \frac{\pi}{a}$ obtained from (F.58). The *chained line* of $\tau = 0.1a$ in the figure corresponds to the surface band structure in the mentioned figures in [12, 17]. Each *two-dot chained line* corresponds to a surface band in the lowest bulk band gap at $k_x = 0$ obtained from (F.59)

The theoretical formalisms developed in Sects. F.3.1 and F.3.2 can be extended to investigate evanescent modes in more general one-dimensional photonic crystals, such as evanescent modes localized in the structure between two semi-infinite one-dimensional photonic crystals. Two photonic crystals can be of same materials, such as shown in Figs. 6 and 10 of Chap. 4 in [12], Figs. 7 and 11 of Chap. 4 in [17], or different materials, such as in Fig. 6 in [30]. The structure in-between can be merely a single interface, a layer of different thickness, or a layer of different material, and so forth.

Besides, we can also obtain some other general understandings of surface modes in one-dimensional photonic crystals from the general mathematical theory.

F.3.3.1 Zeros of the Surface Modes

Theorem 2.8 gives the numbers of zeros of an evanescent mode in each period in a one-dimensional photonic crystal. Based on the theorem, we know that any surface mode in the lowest band gap has exactly one zero in one period, any surface mode in the next lowest band gap has exactly two zeros in one period, such as shown in

Figs. 6.19, 6.20 in [8], Figs. 11.20, 11.21 in [9], Fig. 6 in [38], etc.; any surface mode in the fourth lowest band gap has exactly four zeros in one period, such as shown in Fig. 1 in [37] and Fig. 9 in [38], etc.

F.3.3.2 How the Frequency of an Existing Surface Mode Depends on the Surface Location τ?

The theory in Appendix D can be used to predict the behaviors of an existing surface mode in a one-dimensional photonic crystal.

According to (D.20), in general, the eigenvalue $\Lambda = \Omega^2$ of an existing TM surface mode or an existing TE surface mode satisfies that

$$\frac{\partial \Lambda}{\partial \tau} = \frac{1}{p(\tau)}[p(\tau)\psi'(\tau)]^2 - [\psi(\tau)]^2[q(\tau) - \Lambda w(\tau)].$$

From this equation, by using $p(x)$, $q(x)$, and $w(x)$ obtained from a comparison of Eq. (F.8) or Eq. (F.13) with Eq. (2.2), we obtain that for both an existing TM surface mode or an existing TE surface mode

$$\frac{\partial \Lambda}{\partial \tau} > 0. \tag{F.62}$$

where Eqs. (F.38) and (F.39) for a surface TM mode, or Eqs. (F.52) and (F.53) for a surface TE modes, and $\varepsilon_0 < \varepsilon(\tau)$, and that a localized surface mode can only exist if it is below the light line $\omega < \frac{c}{\sqrt{\varepsilon_0}} k_z$ were used.

Equation (F.62) indicates that if a surface TM or TE mode exists in a one-dimensional semi-infinite photonic crystal, the frequency $\Omega = \Lambda^{1/2}$ of the surface mode *always increases as the boundary τ goes inside to the photonic crystal*. This behavior was observed in previous numerical calculations such as in [38]. The calculated results in Fig. F.1 clearly shows that the frequency of an existing TE surface mode increases as the boundary τ goes inside.

$\Lambda = \Omega^2$ as a function of τ must be a periodic function with a period a. Therefore, if $\partial\Lambda/\partial\tau > 0$ in some interval(s) in any specific period, in the same period there must be some other interval(s) where this inequality can not be true. Correspondingly a surface mode—that is, a solution $\psi(x, \Lambda)$ of Eqs. (F.34) and (F.35) or Eq. (F.38) for a TM mode or a solution $\psi(x, \Lambda)$ of Eqs. (F.48) and (F.49) or Eq. (F.52) for a TE mode—could not exist in these intervals. There is a statement that "In fact, *every* periodic material has surface modes for *some* choice of termination" (p. 61) in [17]. From what we have just understood, a complementary statement that "Every periodic material can have no surface mode in any specific band gap for some choice of termination" is correct as well. Furthermore, for a specific band gap in a specific one-dimensional photonic crystal, very likely there is a way to find out the surface mode existing and non-existing termination intervals.

F.3.3.3 Extra Comments on Surface Modes in Photonic Crystals

Based on what we learned on the surface states and surface bands in electronic crystals, we may make a few extra comments on the surface modes and surface subbands in photonic crystals.

1. A Surface mode or surface subband is closely related to a bulk permitted band rather than a band gap in the physics origin. The existence of a surface mode or a surface subband originates from the existence of a bulk permitted band rather than the existence of a band gap. It is only in the one-dimensional cases that a surface mode has to be in a band gap. The "atypical case" in the statement that "In this atypical case, localization does not require a band gap." (p. 93 in [17]), in fact, is general in higher dimensional cases.

2. Correspondingly, in a multi-dimensional photonic crystal, a surface mode localized most tightly does not have to be in the mid-band-gap, or even in the band-gap at all. The statement that "as a general rule of thumb, we can localize states near the middle of the gap much more tightly than states near the gap's edge" (p. 53 in [17]) is true only in one-dimensional cases. The "exceptions" in the statement that "There are subtle exceptions to this rule, for example, with certain band structures in two and three dimensions, saddle points in the bands can lead to strong localization away from midgap" (p. 53 in [17]), in fact, can be general in high dimensional cases.

F.4 Modes in One-Dimensional Photonic Crystals of Finite Length

In principle, the mathematical theory presented in Chap. 2 can be used to treat one-dimensional layered photonic crystals of finite length $L = Na$ as well. Nevertheless, due to the continuities of H_y (E_y) and E_z (H_z) for TM (TE) waves at the interfaces between a finite photonic crystal and its external medium, such as vacuum or other dielectric homogeneous medium, in most cases neither $y = 0$ nor $py' = 0$ can be applied at the boundary of a photonic crystal. Consequently similar results—such as that presented for a one-dimensional electronic crystal in Sect. 4.2 or a one-dimensional phononic crystal in Sect. E.3—would not be obtained.

However, there is an impressive work on one-dimensional photonic crystals made of coaxial cables by El Boudouti et al. [51]. There the authors derived theoretically, confirmed by numerical calculations, and then observed experimentally the existence and behavior of two types of modes in finite size one-dimensional coaxial photonic crystals made of N cells with vanishing magnetic field on both ends.

These authors considered structures made of periodic repetitions of a specific type of one-dimensional cells—coaxial cables. They first considered the case where each unit cell is made of two different segments—say, A and B—connected in series to each other. Such a unit can be written as AB. By using a Green's function approach, they analytically obtained the dispersion relation of the infinite structure and all eigen modes of a finite structure containing N unit cells connected in series—here N is a positive integer—with the boundary conditions on both ends of the finite crystal being $H = 0$ (vanishing magnetic field). They obtained $N - 1$ confined modes in each permitted band whose frequencies depend on N and one mode for each band gap whose frequency is determined by the unit cell and does not depend on N. The latter modes are surface modes associated to one of the two surfaces surrounding the structure. These results are similar to that shown in Fig. 4.1 in Chap. 4 for the electronic states in one-dimensional crystals of finite length.

Then they considered cases where the unit cell is composed of an ABA trisegment thus is symmetrical. When connected in series, a finite structure of N unit cells is terminated by a segment A on both ends and is also symmetrical. They again obtained $N - 1$ confined modes in each permitted band and one mode determined by the single unit cell for each band gap, in an investigation of finite size coaxial cable photonic crystals made of N unit cells. Since the unit cell ABA is symmetric, the one mode associated with a band gap is a band edge mode rather than a localized surface mode. These results are similar to that shown in Fig. 4.5 in Chap. 4 for the electronic states in one-dimensional symmetrical crystals of finite length.

These results indicate once more that in simplest cases, a two-end truncation of the one-dimensional periodicity containing N unit cells leads to the consequences summarized in Eq. (8.1).

References

1. R. Hooke, Micrographia, London: The Royal Society (1665), reprinted by Palo Alto: Octavo (1998), ISBN 1891788027
2. I. Newton, *Opticks*, 4th edn. London: William Innys (1730), reprinted by New York: Dover Publications (1952), ISBN 486602052
3. F. Abeles, Ann. Phys. (Paris) **5**, 596 (1950); **5**, 706 (1950)
4. S.M. Rytov, Zh. Eksp. Teor. Fiz. **29**, 605 (1955) [Sov. Phys. JETP **2**, 466 (1956)]
5. M. Born, E. Wolf, *Principles of Optics: Electromagnetic Theory of Propagation, Interference and Diffraction of Light* (Pergamon Press, Oxford, 1964; Cambridge University Press, Cambridge, 1999)
6. L.M. Brekhovskikh, *Waves in Layered Media* (Academic, New York, 1967, 1981)
7. P. Yeh, A. Yariv, C.-S. Hong, J. Opt. Soc. Am. **67**, 423 (1977); P. Yeh, A. Yariv, A.Y. Cho, Appl. Phys. Lett. **32**, 104 (1978)
8. A. Yariv, P. Yeh, *Optical Waves in Crystals* (Wiley, New York, 1984)
9. P. Yeh, *Optical Waves in Layered Media* (Wiley, New York, 1988) (Wiley, Hoboken, 2005)
10. E. Yablonovitch, Phys. Rev. Lett. **58**, 2059 (1987)
11. S. John, Phys. Rev. Lett. **58**, 2486 (1987)
12. J.D. Joannopoulos, R.D. Meade, J.N. Winn, *Photonic Crystals* (Princeton University Press, Princeton, 1995)
13. S.G. Johnson, J.D. Joannopoulos, *Photonic Crystals: The Road from Theory to Practice* (Kluwer, Boston, 2002)
14. K. Sakoda, *Optical Properties of Photonic Crystals* (Springer, Berlin, 2001)
15. K. Inoue, K. Ohtaka (eds.), *Photonic Crystals* (Springer, New York, 2004)

16. J. Lourtioz, H. Benisty, V. Berger, J. Gerard, D. Maystre, A. Tchelnokov, *Photonic Crystals: Towards Nanoscale Photonic Devices* (Springer, Berlin, 2005)
17. J.D. Joannopoulos, S.G. Johnson, J.N. Winn, R.D. Meade, *Photonic Crystals, Molding the Flow of Light*, 2nd edn. (Princeton University Press, Princeton, 2008)
18. J.-M. Lourtioz, H. Benisty, V. Berger, J.-M. Gérard, D. Maystre, A. Tchelnokov, D. Pagnoux, *Photonic Crystals: Towards Nanoscale Photonic Devices*, 2nd edn. (Springer, Berlin, 2008)
19. C. Sibilia, T.M. Benson, M. Marciniak, T. Szoplik, *Photonic Crystals: Physics and Technology* (Springer, Milano, 2008)
20. I.A. Sukhoivanov, I.V. Guryev, *Photonic Crystals: Physics and Practical Modeling* (Springer, Heidelberg, 2009)
21. D.W. Prather, A. Sharkawy, S. Shi, J. Murakowski, G. Schneider, *Photonic Crystals, Theory, Applications and Fabrication* (Wiley, Hoboken, 2009)
22. A. Massaro (ed.), *Photonic Crystals - Introduction, Applications and Theory* (open access book) (InTech 2012)
23. Q. Gong, X. Hu, *Photonic Crystals: Principles and Applications* (CRC Press, Boca Eaton, 2013)
24. B. Goodwin, *Photonic Crystals: Characteristics, Performance and Applications* (Nova Science Pub Inc., 2016)
25. I.L. Garanovich, S. Longhi, A.A. Sukhorukov, Y.S. Kivshar, Phys. Rep. **518**, 1 (2012)
26. M.S.P. Eastham, *The Spectral Theory of Periodic Differential Equations* (Scottish Academic Press, Edinburgh, 1973) (and references therein)
27. X.I. Saldana, G.G. de la Cruz, J. Opt. Soc. Am. A **8**, 36 (1991)
28. M.L. Bah, A. Akjouj, E.H. El Boudouti, B. Djafari-Rouhani, L. Dobrzynski, J. Phys. Condens. Matter **8** (1996) 4171 (1996)
29. A.M. Kosevich, JETP Lett. **74**, 559 (2001)
30. A.P. Vinogradov, A.V. Dorofeenko, S.G. Erokhin, M. Inoue, A.A. Lisyansky, A.M. Merzlikin, A.B. Granovsky, Phys. Rev. B **74**, 045128 (2006)
31. O. El Abouti, E.H. El Boudouti, Y. El Hassouani, A. Noual, B. Djafari-Rouhani, Phys. Plasmas **23**, 082115 (2016)
32. A.N. Naumov, A.M. Zheltikov, Laser Phys. **11**, 879 (2001)
33. B. Park, M.N. Kim, S.W. Kim, J.H. Park, Opt. Express **16**, 14524 (2008)
34. S. Prasad, V. Singh, A.K. Singh, Optik **121**, 1520 (2010)
35. F. Scotognella, Opt. Mater. **34**, 1610 (2012)
36. W.M. Robertson, G. Arjavalingam, R.D. Meade, K.D. Brommer, A.M. Rappe, J.D. Joannopoulos, Opt. Lett. **18**, 528, (1993)
37. F. Ramos-Mendieta, P. Halevi, Opt. Commun. **129**, 1 (1996)
38. F. Ramos-Mendieta, P. Halevi, J. Opt. Soc. Am. B **14**, 370 (1997)
39. W.M. Robertson, J. Lightwave Technol. **17**, 2013 (1999)
40. W.M. Robertson, M.S. May, Appl. Phys. Lett. **74**, 1800 (1999)
41. F. Villa, J.A. Gaspar-Armenta, Opt. Commun. **223**, 109 (2003)
42. J.A. Gaspar-Armenta, F. Villa, J. Opt. Soc. Am. B **21**, 405 (2004)
43. J. Martorell, D.W.L. Sprung, G.V. Morozov, J. Opt. A: Pure Appl. Opt. **8**, 630 (2006)
44. J. Bravo-Abad, M. Ibanescu, J.D. Joannopolous, M. Solačić, Phys. Rev. A **74**, 053619 (2006)
45. M.E. Sasin, R.P. Seisyan, M.A. Kalitteevski, S. Brand, R.A. Abram, J.M. Chamberlain, A.Yu. Egorov, A.P. Vasilev, V.S. Mikhrin, A.V. Kavokin, Appl. Phys. Lett. **92**, 251112 (2008)
46. T.B. Wang, C.-P. Yin, W.-Y. Liang, J.-W. Dong, H.-Z. Wang, J. Opt. Soc. Am. B **26**, 1635 (2009)
47. M. Ballarini, F. Frascella, F. Michelotti, G. Digregorio, P. Rivolo, V. Paeder, V. Musi, F. Giorgis, E. Descrovi, Appl. Phys. Lett. **99**, 043302 (2011)
48. R. Brückner, M. Sudzius, S.I. Hintschich, H. Frob, V.G. Lyssenko, M.A. Kaliteevski, I. Iorsh, R.A. Abram, A.V. Kavokin, K. Leo, Appl. Phys. Lett. **100**, 062101 (2012)

49. A. Angelini, E. Enrico, N. De Leo, P. Munzert, L. Boarino, F. Michelotti, F. Giorgis, E. Descrovi, New J. Phys. **15**, 073002 (2013)
50. R. Das, T. Srivastava, R. Jha, Opt. Lett. **39**, 896 (2014)
51. E.H. El Boudouti, Y. El Hassouani, B. Djafari-Rouhani, H. Aynaou, Phys. Rev. E **76**, 026607 (2007)

Appendix G
Electronic States in Ideal Cavity Structures

In this Appendix, we investigate the electronic states in cavity structures where a low-dimensional system such as treated in Chaps. 4–7 is removed from an infinite crystal.

For the electronic states in ideal cavity structures treated in this Appendix, we assume that (i) the potential $v(x)$ or $v(\mathbf{x})$ *outside* the cavity is the same as in (4.1) or (5.1) and (ii) the electronic states are completely confined outside the cavity.

G.1 Electronic States in An Ideal Cavity Structure of A One-Dimensional Crystal

An ideal cavity structure of a one-dimensional crystal is a structure formed when a one-dimensional finite crystal bounded at τ and $\tau + L$ was removed from an infinite one-dimensional crystal with a potential period a. Here, $L = Na$ and N is a positive integer.

The eigenvalues Λ and eigenfunctions $\psi(x)$ of the electronic states in such an ideal cavity structure are solutions of the following two equations:

$$\begin{cases} -\psi''(x) + [v(x) - \Lambda]\psi(x) = 0, & x < \tau \text{ or } x > \tau + L, \\ \psi(x) = 0, & \tau \leq x \leq \tau + L, \end{cases} \tag{G.1}$$

Equation (G.1) can be considered as the equations of electronic states in two separate ideal semi-infinite one-dimensional crystals: one left semi-infinite crystal in the range of $(-\infty, \tau)$ and one right semi-infinite crystal in the range of $(\tau + L, +\infty)$. The two ideal semi-infinite crystals are not independent of each other, since $L = Na$, and N is a positive integer. For those two semi-infinity crystals, as in Sect. 3.1, we are interested only in the electronic states whose energy depends on the boundary τ or $\tau + L$. The properties and energies of those boundary-dependent electronic states

S.Y. Ren, *Electronic States in Crystals of Finite Size*, Springer Tracts in Modern Physics 270, DOI 10.1007/978-981-10-4718-3

in the cavity structure can be easily obtained as long as the τ-dependent electronic states in the ideal finite crystal are obtained.

We have presented an analysis of the τ-dependent states in ideal one-dimensional finite crystals in Sect. 4.3. The boundary-dependent electronic states in an ideal cavity structure of a one-dimensional crystal can be easily obtained from that analysis.

We also take a band gap at $k = 0$ as an example. For a specific band gap index n, the boundary τ could be in one of three cases.

1. If τ is in the set $L(n)$, in the finite crystal bounded at τ and $\tau + L$, there is an electronic state with a form of $e^{-\beta x} p(x, \Lambda)$ $(\beta > 0)$ in the band gap. It is a surface state with an energy Λ located near the left boundary τ of the finite crystal. Correspondingly, $\tau + L$ is also in the set $L(n)$; therefore, there is a surface state with the same form of $e^{-\beta x} p(x, \Lambda)$ and energy Λ near the left boundary $\tau + L$ of the right semi-infinite crystal $(\tau + L, +\infty)$, whereas no τ-dependent state exists in the left semi-infinite crystal $(-\infty, \tau)$.
2. If τ is in the set $R(n)$, in the finite crystal bounded at τ and $\tau + L$, there is an electronic state with a form of $e^{\beta x} p(x, \Lambda)$ $(\beta > 0)$ in the band gap. It is a surface state with an energy Λ located near the right boundary $\tau + L$ of the finite crystal. Correspondingly, there is a surface state with the same form of $e^{\beta x} p(x, \Lambda)$ and energy Λ near the right boundary τ of the left semi-infinite crystal $(-\infty, \tau)$, whereas no τ-dependent state exists in the right semi-infinite crystal $(\tau + L, +\infty)$.
3. If τ is in the set $M(n)$, a band edge state with a form of $p(x, \Lambda)$ and the band edge energy exists in the finite crystal bounded at τ and $\tau + L$, indicating a confined band edge state periodically distributed in the finite crystal. Correspondingly, a band edge state with the same form of $p(x, \Lambda)$ and the same energy Λ exists in both the right semi-infinite crystal $(\tau + L, +\infty)$ and the left semi-finite crystal $(-\infty, \tau)$.

Band gaps at $k = \pi/a$ can be similarly analyzed; only a semi-periodic functions $s(x, \Lambda)$ should be used instead of the periodic functions $p(x, \Lambda)$.

Therefore, the τ-dependent states in such a cavity structure can be obtained similarly to the τ-dependent states in the finite crystal removed.

G.2 Electronic States in an Ideal Two-Dimensional Cavity Structure of A Three-Dimensional Crystal

A two-dimensional cavity structure in an infinite three-dimensional crystal is a structure formed when a film of a specific orientation and a specific thickness was removed from an infinite crystal. In this section, we are only interested in such cavity structures where an ideal quantum film as investigated in Chap. 5 was removed from an infinite crystal. As in Chap. 5, we assume that the film plane is defined by two primitive lattice vectors $\mathbf{a}_1$ and $\mathbf{a}_2$, $x_3 = \tau_3$ defines the bottom of the removed film and N_3 is a positive integer indicating the thickness of the removed film. Such a cavity structure has two separate parts: an upper semi-infinite crystal part and a lower semi-infinite crystal part.

The electronic states $\hat{\psi}(\hat{\mathbf{k}}, \mathbf{x})$ in a two-dimensional cavity are solutions of the following two equations:

$$\begin{cases} -\nabla^2\hat{\psi}(\hat{\mathbf{k}}, \mathbf{x}) + [v(\mathbf{x}) - \hat{\Lambda}]\hat{\psi}(\hat{\mathbf{k}}, \mathbf{x}) = 0, & x_3 < \tau_3 \text{ or } x_3 > \tau_3 + N_3, \\ \hat{\psi}(\hat{\mathbf{k}}, \mathbf{x}) = 0, & \tau_3 \leq x_3 \leq \tau_3 + N_3. \end{cases} \tag{G.2}$$

The electronic states $\hat{\psi}(\hat{\mathbf{k}}, \mathbf{x})$ in such a cavity structure are two-dimensional Bloch waves with a wave vector $\hat{\mathbf{k}}$ in the film plane.

As in Sect. G.1, we are only interested in the boundary-dependent electronic states in such a cavity structure. These states in such a cavity structure can be similarly obtained as the boundary-dependent states in the removed film treated in Chap. 5: For each bulk energy band n and each wave vector $\hat{\mathbf{k}}$ in the film plane, there is one such electronic state in the cavity structure, which can be obtained from (5.11) by assigning a non-divergent $\hat{\phi}_n(\hat{\mathbf{k}}, \mathbf{x}; \tau_3)$ in the cavity structure:

$$\hat{\psi}_n(\hat{\mathbf{k}}, \mathbf{x}; \tau_3) = \begin{cases} c\,\hat{\phi}_n(\hat{\mathbf{k}}, \mathbf{x}; \tau_3), & x_3 < \tau_3 \text{ or } x_3 > \tau_3 + N_3, \\ 0, & \tau_3 \leq x_3 \leq \tau_3 + N_3. \end{cases} \tag{G.3}$$

where c is a normalization constant. Unlike in (5.29), c in (G.3) does not depend on the thickness N_3 of the removed film. The divergent part of $\hat{\phi}_n(\hat{\mathbf{k}}, \mathbf{x}; \tau_3)$ in (G.3) should be abandoned. Correspondingly, the energy of such a state is given by

$$\hat{\Lambda}_n(\hat{\mathbf{k}}; \tau_3) = \hat{\lambda}_n(\hat{\mathbf{k}}; \tau_3), \tag{G.4}$$

as in (5.30). There is one solution (G.3) of (G.2) for each energy band n and each $\hat{\mathbf{k}}$. Each $\hat{\psi}_n(\hat{\mathbf{k}}, \mathbf{x}; \tau_3)$ defined in (G.3) is an electronic state in the cavity structure whose energy $\hat{\Lambda}_n(\hat{\mathbf{k}}; \tau_3)$ in (G.4) depends on the cavity boundary τ_3 but not on the cavity size N_3. By Theorem 5.1, $\hat{\Lambda}_n(\hat{\mathbf{k}}; \tau_3)$ is either above or at the energy maximum of $\varepsilon_n(\mathbf{k})$ with that n and that $\hat{\mathbf{k}}$.

In the special cases where $\hat{\phi}_n(\hat{\mathbf{k}}, \mathbf{x}; \tau_3)$ in (G.3) is a Bloch function,

$$\hat{\phi}_n(\hat{\mathbf{k}}, \mathbf{x}; \tau_3) = \phi_{n'}(\mathbf{k}, \mathbf{x}), \quad n \leq n', \tag{G.5}$$

the corresponding Bloch function $\phi_{n'}(\mathbf{k}, \mathbf{x})$ has a nodal surface at $x_3 = \tau_3$ and thus has nodal surfaces at $x_3 = \tau_3 + \ell$, where $\ell = 1, 2, \ldots, N_3$. The wave function $\hat{\phi}_n(\hat{\mathbf{k}}, \mathbf{x}; \tau_3)$ in (G.3) exists in both the upper semi-infinite crystal part and the lower semi-infinite crystal part of the cavity structure.

In most cases, $\hat{\phi}_n(\hat{\mathbf{k}}, \mathbf{x}; \tau_3)$ in (G.3) is not a Bloch function. In such a situation there is a nonzero imaginary part of k_3 in (5.11), indicating that $\hat{\psi}_n(\hat{\mathbf{k}}, \mathbf{x}; \tau_3)$ in (G.3) is a surface state located near either the top surface of the lower semi-infinite crystal part (if the imaginary part of k_3 in (5.11) is negative) or the bottom surface of the upper semi-infinite crystal part (if the imaginary part of k_3 in (5.11) is positive) of the cavity structure: It exists in one of the two semi-infinite crystal parts of the cavity structure: for some $\hat{\mathbf{k}}$ it is in the upper part above the cavity, whereas for some other $\hat{\mathbf{k}}$ it is in the lower part below the cavity. Correspondingly, the energy of such a state

$$\hat{\Lambda}_n(\hat{\mathbf{k}}; \tau_3) > \varepsilon_n(\mathbf{k}), \quad \text{for } (\mathbf{k} - \hat{\mathbf{k}}) \cdot \mathbf{a}_i = 0, \;\; i = 1, 2, \tag{G.6}$$

by Theorem 5.1. However, there is no reason to expect that $\hat{\Lambda}_n(\hat{\mathbf{k}}; \tau_3)$ has to be in a band gap.

Therefore, for each bulk energy band n, there is one surface-like subband $\hat{\Lambda}_n(\hat{\mathbf{k}}; \tau_3)$ in (G.4) in such an ideal cavity structure.

Those results should be correct for cavity structures of crystals with a sc, tetr, or an ortho Bravais lattice for which an ideal (001) film was removed. More generally, they should also be correct for ideal cavity structures of crystals with an fcc or a bcc Bravais lattice for which an ideal (001) or (110) film was removed.

We have seen in Sects. G.1 and G.2 that the boundary-dependent electronic states in a cavity structure actually can be obtained similarly to the boundary-dependent electronic states in the removed low-dimensional systems. This is due to the mere fact that the ideal cavity structure and the ideal low-dimensional system removed have the same boundary. The same idea can be applied to obtain the boundary-dependent electronic states in an ideal one-dimensional or zero-dimensional cavity structure in a three-dimensional crystal.

G.3 Electronic States in an Ideal One-Dimensional Cavity Structure of a Three-Dimensional Crystal

A one-dimensional cavity structure in a three-dimensional crystal is a structure formed when a quantum wire was removed from an infinite crystal. In this section, we are only interested in such cavity structures where an ideal rectangular quantum wire investigated in Chap. 6 was removed from an infinite crystal.

As in Chap. 6, we choose the primitive vector $\mathbf{a}_1$ in the wire cavity direction. Such a rectangular wire cavity can be defined by a bottom face $x_3 = \tau_3$, a top face $x_3 = \tau_3 + N_3$, a front face perpendicularly intersecting the $\mathbf{a}_2$ axis at $\tau_2\mathbf{a}_2$, and a rear face perpendicularly intersecting it at $(\tau_2 + N_2)\mathbf{a}_2$. Here, τ_2 and τ_3 define the boundary faces of the wire cavity, and N_2 and N_3 are two positive integers indicating the size and shape of the wire cavity.

For the electronic states in such an ideal cavity structure, we look for the eigenvalues $\bar{\Lambda}$ and eigenfunctions $\bar{\psi}(\bar{\mathbf{k}}, \mathbf{x})$ of the following two equations:

$$\begin{cases} -\nabla^2\bar{\psi}(\bar{\mathbf{k}}, \mathbf{x}) + [v(\mathbf{x}) - \bar{\Lambda}]\bar{\psi}(\bar{\mathbf{k}}, \mathbf{x}) = 0, & \mathbf{x} \notin \text{the cavity}, \\ \bar{\psi}(\bar{\mathbf{k}}, \mathbf{x}) = 0, & \mathbf{x} \in \text{the cavity}. \end{cases} \tag{G.7}$$

The solutions $\bar{\psi}(\bar{\mathbf{k}}, \mathbf{x})$ of (G.7) are one-dimensional Bloch waves with a wave vector $\bar{\mathbf{k}}$ in the wire direction $\mathbf{a}_1$.

There are different types of electronic state solutions of these two equations. As in Sects. G.1 and G.2, in this section, we are only interested in the solutions of (G.7) whose energies are dependent on the cavity boundary locations τ_2 and/or τ_3. Based on similar arguments we had in Chaps. 5–6 and in Sects. G.1 and G.2, we can understand that the electronic states whose energies are dependent on the cavity boundary location τ_2 *or* τ_3 are surface-like states in the cavity structure. They are located near the opposite surface of the cavity structure in comparison with the corresponding surface-like states in the removed quantum wire: If there is a surface-like state located near the top surface of the removed quantum wire, then there is a corresponding surface-like state located near the bottom surface of the cavity and vice versa. If there is a surface-like state located near the front surface of the removed quantum wire, then there is a corresponding surface-like state located near the rear surface of the cavity and vice versa. Similarly, the electronic states whose energies are dependent on the cavity boundary locations τ_2 *and* τ_3 are edge-like states in the cavity structure; they are located near the opposite edge of the cavity in comparison with the corresponding edge-like states in the removed quantum wire.

G.3.1 Wire Cavities in Crystals with A sc, tetr, or ortho Bravais Lattice

For an ideal one-dimensional cavity structure of a crystal with a sc, tetr, or ortho Bravais lattice, we consider the case where the rectangular quantum wire removed has two boundary faces at $x_3 = \tau_3$ and $x_3 = \tau_3 + N_3$ in the $\mathbf{a}_3$ direction, two other boundary faces at $x_2 = \tau_2$ and $x_2 = \tau_2 + N_2$ in the $\mathbf{a}_2$ direction. For each bulk energy band n in the cavity structure, there are
$(N_3 - 1)$ surface-like subbands with energies

$$\bar{\Lambda}_{n,j_3}(\bar{\mathbf{k}};\tau_2) = \hat{\Lambda}_n\left(\bar{\mathbf{k}} + \frac{j_3\pi}{N_3}\mathbf{b}_3;\tau_2\right); \tag{G.8}$$

$(N_2 - 1)$ surface-like subbands with energies

$$\bar{\Lambda}_{n,j_2}(\bar{\mathbf{k}};\tau_3) = \hat{\Lambda}_n\left(\bar{\mathbf{k}} + \frac{j_2\pi}{N_2}\mathbf{b}_2;\tau_3\right); \tag{G.9}$$

one edge-like subband with energy $\bar{\Lambda}_n(\bar{\mathbf{k}};\tau_2,\tau_3)$ depending on both τ_2 and τ_3, similar to (6.29), (6.30), and (6.31) in Sect. 6.4.

Here, $j_2 = 1, 2, \ldots, N_2 - 1$ and $j_3 = 1, 2, \ldots, N_3 - 1$. $\hat{\Lambda}_n(\hat{\mathbf{k}};\tau_3)$ is the surface-like band structure in a quantum film with the film plane oriented in the $\mathbf{a}_3$ direction. $\hat{\Lambda}_n(\hat{\mathbf{k}};\tau_2)$ is the surface-like band structure in a quantum film with the film plane oriented in the $\mathbf{a}_2$ direction.

However, probably the practically more interesting cases are cavity structures of crystals with an fcc or bcc Bravais lattice. In the following, we give predictions on the electronic states in several such one-dimensional cavity structures.

G.3.2 *Wire Cavities with (001) and (110) Surfaces in an fcc Crystal*

A cavity structure with (001) and (110) surfaces in an fcc crystal is a structure formed when an $[1\bar{1}0]$ quantum wire was removed from an infinite crystal with an fcc Bravais lattice. The removed quantum wire has (001) and (110) surfaces and has a rectangular cross section $N_{110}a/\sqrt{2} \times N_{001}a$, where N_{110} and N_{001} are two positive integers.

For each bulk energy band n, there are $(N_{001} - 1) + (N_{110} - 1)$ surface-like subbands in such a cavity structure. They are $(N_{001} - 1)$ subbands with energies

$$\bar{\Lambda}^{sf,a_1}_{n,j_{001}}(\bar{\mathbf{k}}; \tau_{110}) = \hat{\Lambda}_n \left[\bar{\mathbf{k}} + \frac{j_{001}\pi}{N_{001}a}(0, 0, 1); \tau_{110} \right] \tag{G.10}$$

and $(N_{110} - 1)$ subbands with energies

$$\bar{\Lambda}^{sf,a_2}_{n,j_{110}}(\bar{\mathbf{k}}; \tau_{001}) = \hat{\Lambda}_n \left[\bar{\mathbf{k}} + \frac{j_{110}\pi}{N_{110}a}(1, 1, 0); \tau_{001} \right], \tag{G.11}$$

similar to (6.51) and (6.52). Here, τ_{001} or τ_{110} define the boundary faces of the cavity in the [001] or [110] direction, $j_{001} = 1, 2, \ldots, N_{001} - 1$, and $j_{110} = 1, 2, \ldots, N_{110} - 1$. $\hat{\Lambda}_n(\hat{\mathbf{k}}; \tau_{001})$ is the surface-like band structure in a quantum film with the film plane oriented in the [001] direction. $\hat{\Lambda}_n(\hat{\mathbf{k}}; \tau_{110})$ is the surface-like band structure in a quantum film with the film plane oriented in the [110] direction.

For each bulk energy band n, there is one edge-like subband in the cavity structure with energy $\bar{\Lambda}^{eg}_n(\bar{\mathbf{k}}; \tau_{001}, \tau_{110})$ depending on both τ_{001} and τ_{110}, similar to (6.38) or (6.45).

G.3.3 *Wire Cavities with (110) and (1$\bar{1}$0) Surfaces in an fcc Crystal*

A cavity structure with (110) and $(1\bar{1}0)$ surfaces in an fcc crystal is a structure formed when a [001] quantum wire was removed from an infinite crystal with an fcc Bravais lattice. The removed quantum wire has (110) and $(1\bar{1}0)$ surfaces and has a rectangular cross section $N_{110}a/\sqrt{2} \times N_{1\bar{1}0}a/\sqrt{2}$, where N_{110} and $N_{1\bar{1}0}$ are two positive integers.

For each bulk energy band n, there are $(N_{1\bar{1}0} - 1) + (N_{110} - 1)$ surface-like subbands in the cavity structure. They are $(N_{1\bar{1}0} - 1)$ subbands with energies

$$\bar{\Lambda}^{sf,a_1}_{n,j_{1\bar{1}0}}(\bar{\mathbf{k}};\tau_{110}) = \hat{\Lambda}_n\left[\bar{\mathbf{k}} + \frac{j_{1\bar{1}0}\pi}{N_{1\bar{1}0}a}(1,-1,0);\tau_{110}\right] \tag{G.12}$$

and $(N_{110} - 1)$ subbands with energies

$$\bar{\Lambda}^{sf,a_2}_{n,j_{110}}(\bar{\mathbf{k}};\tau_{1\bar{1}0}) = \hat{\Lambda}_n\left[\bar{\mathbf{k}} + \frac{j_{110}\pi}{N_{110}a}(1,1,0);\tau_{1\bar{1}0}\right], \tag{G.13}$$

similar to (6.61) and (6.62). Here, τ_{110} or $\tau_{1\bar{1}0}$ define the boundary faces of the cavity in the [110] or [1$\bar{1}$0] direction, $j_{110} = 1, 2, \ldots, N_{110}-1$, and $j_{1\bar{1}0} = 1, 2, \ldots, N_{1\bar{1}0}-1$. $\hat{\Lambda}_n(\hat{\mathbf{k}};\tau_{110})$ is the surface-like band structure in a quantum film with the film plane oriented in the [110] direction. $\hat{\Lambda}_n(\hat{\mathbf{k}};\tau_{1\bar{1}0})$ is the surface-like band structure in a quantum film with the film plane oriented in the [1$\bar{1}$0] direction.

For each bulk energy band n, there is one edge-like subband in the cavity structure with energy $\bar{\Lambda}^{eg}_n(\bar{\mathbf{k}};\tau_{110},\tau_{1\bar{1}0})$ depending on both τ_{110} and $\tau_{1\bar{1}0}$, similar to (6.63).

G.3.4 Wire Cavities with (010) and (001) Surfaces in a bcc Crystal

A cavity structure with (010) and (001) surfaces in a bcc crystal is a structure formed when a [100] quantum wire was removed from an infinite crystal with a bcc Bravais lattice. The removed quantum wire has (010) and (001) surfaces and has a rectangular cross section $N_{010}a \times N_{001}a$, where N_{010} and N_{001} are two positive integers.

For each bulk energy band n, there are $(N_{001} - 1) + (N_{010} - 1)$ surface-like subbands in the cavity structure. They are $(N_{001} - 1)$ subbands with energies

$$\bar{\Lambda}^{sf,a_1}_{n,j_{001}}(\bar{\mathbf{k}};\tau_{010}) = \hat{\Lambda}_n\left[\bar{\mathbf{k}} + \frac{j_{001}\pi}{N_{001}a}(0,0,1);\tau_{010}\right] \tag{G.14}$$

and $(N_{010} - 1)$ subbands with energies

$$\bar{\Lambda}^{sf,a_2}_{n,j_{010}}(\bar{\mathbf{k}};\tau_{001}) = \hat{\Lambda}_n\left[\bar{\mathbf{k}} + \frac{j_{010}\pi}{N_{010}a}(0,1,0);\tau_{001}\right], \tag{G.15}$$

similar to (6.69) and (6.70). Here, τ_{010} or τ_{001} define the boundary faces of the cavity in the [010] or [001] direction, $j_{001} = 1, 2, \ldots, N_{001}-1$, and $j_{010} = 1, 2, \ldots, N_{010}-1$. $\hat{\Lambda}_n(\hat{\mathbf{k}};\tau_{001})$ is the surface-like band structure in a quantum film with the film plane oriented in the [001] direction. $\hat{\Lambda}_n(\hat{\mathbf{k}};\tau_{010})$ is the surface-like band structure in a quantum film with the film plane oriented in the [010] direction.

For each bulk energy band n, there is one edge-like subband in the cavity structure with energy $\bar{\Lambda}^{eg}_n(\bar{\mathbf{k}};\tau_{001},\tau_{010})$ depending on both τ_{001} and τ_{010}, similar to (6.71).

G.4 Electronic States in an Ideal Zero-Dimensional Cavity Structure of a Three-Dimensional Crystal

A zero-dimensional cavity structure in an infinite three-dimensional crystal is a structure formed when a quantum dot was removed from the infinite crystal. In this section, we are only interested in ideal cavity structures where a quantum dot of rectangular cuboid shape investigated in Chap. 7 was removed from an infinite crystal.

Such a cavity can be defined by a bottom and top face at $x_3 = \tau_3$ and $x_3 = \tau_3 + N_3$, a front and rear face perpendicularly intersecting the $\mathbf{a}_2$ axis at $x_2 = \tau_2$ and $x_2 = \tau_2 + N_2$, and a left and right face perpendicularly intersecting the $\mathbf{a}_1$ axis at $x_1 = \tau_1$ and $x_1 = \tau_1 + N_1$. Here, τ_1, τ_2, and τ_3 define the boundary faces of the cavity and N_1, N_2, and N_3 are three positive integers indicating the cavity size and/or shape. We look for the eigenvalues Λ and eigenfunctions $\psi(\mathbf{x})$ of the following two equations:

$$\begin{cases} -\nabla^2\psi(\mathbf{x}) + [v(\mathbf{x}) - \Lambda]\psi(\mathbf{x}) = 0, & \mathbf{x} \notin \text{ the cavity}, \\ \psi(\mathbf{x}) = 0, & \mathbf{x} \in \text{ the cavity}. \end{cases} \tag{G.16}$$

There are different types of electronic state solutions of (G.16). As in Sects. G.1–G.3, in this section, we are only interested in the solutions of (G.16) whose energies are dependent on the cavity boundary τ_1, τ_2, and/or τ_3. Based on similar arguments we had in Chaps. 5–7 and in Sects. G.1–G.3, we can understand that the electronic states whose energies are dependent on *one* of the cavity boundary locations τ_1, τ_2, or τ_3 are surface-like states in the cavity structure; they are located near the opposite surface of the cavity structure in comparison with the corresponding surface-like states in the removed quantum dot: If there is a surface-like state located near one specific surface in the removed quantum dot, then there is a corresponding surface-like state located near the opposite surface of the cavity. The electronic states whose energies are dependent on *two* of the cavity boundary locations τ_1, τ_2 or τ_3 are edge-like states in the cavity structure; they are located near the opposite edge of the cavity in comparison with the corresponding edge-like states in the removed quantum dot. The electronic states whose energies are dependent on all *three* cavity boundary locations τ_1, τ_2, and τ_3 are vertex-like states in the cavity structure; they are located near the opposite vertex of the cavity in comparison with the corresponding vertex-like states in the removed quantum dot.

G.4.1 *Dot Cavities In Crystals with a sc, tetr, or ortho Bravais Lattice*

For such a cavity structure with sizes N_1a_1, N_2a_2 and N_3a_3 in the $\mathbf{a}_1$, $\mathbf{a}_2$ and $\mathbf{a}_3$ directions in a crystal with a sc, tetr, or ortho Bravais lattice, for each bulk energy band there are $(N_1 - 1)(N_2 - 1) + (N_2 - 1)(N_3 - 1) + (N_3 - 1)(N_1 - 1)$ surface-like states, $(N_1 - 1) + (N_2 - 1) + (N_3 - 1)$ edge-like states, and one vertex-like state in the cavity structure. They are as follows:

$(N_1 - 1)(N_2 - 1)$ surface-like states with energies

$$\Lambda_{n,j_1,j_2}(\tau_3) = \hat{\Lambda}_n \left[\frac{j_1\pi}{N_1}\mathbf{b}_1 + \frac{j_2\pi}{N_2}\mathbf{b}_2; \tau_3 \right]; \tag{G.17}$$

$(N_2 - 1)(N_3 - 1)$ surface-like states with energies

$$\Lambda_{n,j_2,j_3}(\tau_1) = \hat{\Lambda}_n \left[\frac{j_2\pi}{N_2}\mathbf{b}_2 + \frac{j_3\pi}{N_3}\mathbf{b}_3; \tau_1 \right]; \tag{G.18}$$

$(N_3 - 1)(N_1 - 1)$ surface-like states with energies

$$\Lambda_{n,j_3,j_1}(\tau_2) = \hat{\Lambda}_n \left[\frac{j_3\pi}{N_3}\mathbf{b}_3 + \frac{j_1\pi}{N_1}\mathbf{b}_1; \tau_2 \right]; \tag{G.19}$$

$(N_1 - 1)$ edge-like states with energies

$$\Lambda_{n,j_1}(\tau_2, \tau_3) = \bar{\Lambda}_n \left[\frac{j_1\pi}{N_1}\mathbf{b}_1; \tau_2, \tau_3 \right]; \tag{G.20}$$

$(N_2 - 1)$ edge-like states with energies

$$\Lambda_{n,j_2}(\tau_3, \tau_1) = \bar{\Lambda}_n \left[\frac{j_2\pi}{N_2}\mathbf{b}_2; \tau_3, \tau_1 \right]; \tag{G.21}$$

$(N_3 - 1)$ edge-states with energies

$$\Lambda_{n,j_3}(\tau_1, \tau_2) = \bar{\Lambda}_n \left[\frac{j_3\pi}{N_3}\mathbf{b}_3; \tau_1, \tau_2 \right]; \tag{G.22}$$

and one vertex-like state with energy $\Lambda_n(\tau_1, \tau_2, \tau_3)$ depending all three τ_1, τ_2, and τ_3, similar to (7.39)–(7.45).
Here $j_1 = 1, 2, \ldots, N_1 - 1$, $j_2 = 1, 2, \ldots, N_2 - 1$, and $j_3 = 1, 2, \ldots, N_3 - 1$. τ_1, τ_2, and τ_3 define the boundary faces of the cavity in the $\mathbf{a}_1$, $\mathbf{a}_2$, and $\mathbf{a}_3$ directions. $\hat{\Lambda}_n[\hat{\mathbf{k}}; \tau_l]$ is the surface-like band structure of a quantum film with the film plane oriented in the $\mathbf{a}_l$ direction. $\bar{\Lambda}_n[\bar{\mathbf{k}}; \tau_l, \tau_m]$ is the edge-like band structure of a rectangular quantum wire with the wire faces oriented in the $\mathbf{a}_l$ or the $\mathbf{a}_m$ direction.

Probably the practically more interesting cases are the cavity structures in crystals with an fcc or bcc Bravais lattice. Similar to Sects. G.1–G.3, electronic states in those cavity structures can be obtained.

G.4.2 Dot Cavities with $(1\bar{1}0)$, (110), and (001) Surfaces in An fcc Crystal

For a cavity structure in a crystal with an fcc Bravais lattice, if the cavity has (001), (110), and $(1\bar{1}0)$ surfaces and a rectangular cuboid size $N_{001}a \times N_{110}a/\sqrt{2} \times N_{1\bar{1}0}a/\sqrt{2}$, the boundary-dependent electronic states in such a cavity structure can be obtained similarly to the boundary-dependent electronic states in an ideal quantum dot obtained in Sect. 7.7.

For each bulk energy band n, there are $(N_{001}-1)(N_{1\bar{1}0}-1)+(N_{110}-1)(N_{001}-1)+(N_{1\bar{1}0}-1)(N_{110}-1)$ surface-like states in the cavity structure. They are as follows:
$(N_{001}-1)(N_{1\bar{1}0}-1)$ states with energies

$$\Lambda^{sf,a_1}_{n,j_{001},j_{1\bar{1}0}}(\tau_{110}) = \hat{\Lambda}_n\left[\frac{j_{001}\pi}{N_{001}a}(0,0,1)+\frac{j_{1\bar{1}0}\pi}{N_{1\bar{1}0}a}(1,-1,0);\tau_{110}\right]; \tag{G.23}$$

$(N_{110}-1)(N_{001}-1)$ states with energies

$$\Lambda^{sf,a_2}_{n,j_{110},j_{001}}(\tau_{1\bar{1}0}) = \hat{\Lambda}_n\left[\frac{j_{110}\pi}{N_{110}a}(1,1,0)+\frac{j_{001}\pi}{N_{001}a}(0,0,1);\tau_{1\bar{1}0}\right]; \tag{G.24}$$

$(N_{1\bar{1}0}-1)(N_{110}-1)$ states with energies

$$\Lambda^{sf,a_3}_{n,j_{1\bar{1}0},j_{110}}(\tau_{001}) = \hat{\Lambda}_n\left[\frac{j_{1\bar{1}0}\pi}{N_{1\bar{1}0}a}(1,-1,0)+\frac{j_{110}\pi}{N_{110}a}(1,1,0);\tau_{001}\right], \tag{G.25}$$

similar to (7.56)–(7.58).

Here, $j_{001} = 1, 2, \ldots, N_{001}-1$, $j_{1\bar{1}0} = 1, 2, \ldots, N_{1\bar{1}0}-1$, and $j_{110} = 1, 2, \ldots, N_{110}-1$. τ_{110}, $\tau_{1\bar{1}0}$, or τ_{001} define the boundary faces of the cavity in the [110], $[1\bar{1}0]$, or [001] direction, $\hat{\Lambda}_n[\hat{\mathbf{k}};\tau_l]$ is the surface-like band structure of a quantum film with the film plane oriented in the $[l]$ direction. l can be either one of 110, $1\bar{1}0$, or 001.

For each energy band n, there are $(N_{001}-1)+(N_{110}-1)+(N_{1\bar{1}0}-1)$ edge-like states in the cavity structure. They are as follows:
$(N_{001}-1)$ states with energies

$$\Lambda^{eg,a_1}_{n,j_{001}}(\tau_{1\bar{1}0},\tau_{110}) = \bar{\Lambda}_n\left[\frac{j_{001}\pi}{N_{001}a}(0,0,1);\tau_{1\bar{1}0},\tau_{110}\right]; \tag{G.26}$$

$(N_{110}-1)$ states with energies

$$\Lambda^{eg,a_2}_{n,j_{110}}(\tau_{1\bar{1}0},\tau_{001}) = \bar{\Lambda}_n\left[\frac{j_{110}\pi}{N_{110}a}(1,1,0);\tau_{1\bar{1}0},\tau_{001}\right]; \tag{G.27}$$

$(N_{1\bar{1}0} - 1)$ states with energies

$$\Lambda^{eg,a_3}_{n,j_{1\bar{1}0}}(\tau_{001}, \tau_{110}) = \bar{\Lambda}_n \left[\frac{j_{1\bar{1}0}\pi}{N_{1\bar{1}0}a}(1, -1, 0); \tau_{001}, \tau_{110} \right], \tag{G.28}$$

similar to (7.59)–(7.61).

Here, $\bar{\Lambda}_n[\bar{\mathbf{k}}; \tau_l, \tau_m]$ is the edge-like band structure of a rectangular quantum wire with the wire faces oriented in the $[l]$ or the $[m]$ direction. l and m can be two of 001, 110, and $1\bar{1}0$.

For each bulk energy band n, there is one vertex-like state in the cavity structure with energy $\Lambda^{vt}_n(\tau_{001}, \tau_{1\bar{1}0}, \tau_{110})$ depending all three τ_{001}, $\tau_{1\bar{1}0}$, and τ_{110}, similar to (7.62).

G.4.3 Dot Cavities with (100), (010), and (001) Surfaces in bcc Crystals

In a crystal with a bcc Bravais lattice, the boundary-dependent electronic states in a cavity structure with (100), (010) and (001) surfaces and a rectangular cuboid size $N_{100}a \times N_{010}a \times N_{001}a$ can be obtained similarly to the boundary-dependent electronic states in an ideal quantum dot obtained in Sect. 7.8.

For each bulk energy band n, there are $(N_{100} - 1)(N_{010} - 1) + (N_{010} - 1)(N_{001} - 1) + (N_{001} - 1)(N_{100} - 1)$ surface-like states in the cavity structure. They are as follows:

$(N_{010} - 1)(N_{001} - 1)$ states with energies

$$\Lambda^{sf,a_1}_{n,j_{010},j_{001}}(\tau_{100}) = \hat{\Lambda}_n \left[\frac{j_{010}\pi}{N_{010}a}(0, 1, 0) + \frac{j_{001}\pi}{N_{001}a}(0, 0, 1); \tau_{100} \right]; \tag{G.29}$$

$(N_{001} - 1)(N_{100} - 1)$ states with energies

$$\Lambda^{sf,a_2}_{n,j_{001},j_{100}}(\tau_{010}) = \hat{\Lambda}_n \left[\frac{j_{001}\pi}{N_{001}a}(0, 0, 1) + \frac{j_{100}\pi}{N_{100}a}(1, 0, 0); \tau_{010} \right]; \tag{G.30}$$

$(N_{100} - 1)(N_{010} - 1)$ states with energies

$$\Lambda^{sf,a_3}_{n,j_{100},j_{010}}(\tau_{001}) = \hat{\Lambda}_n \left[\frac{j_{100}\pi}{N_{100}a}(1, 0, 0) + \frac{j_{010}\pi}{N_{010}a}(0, 1, 0); \tau_{001} \right], \tag{G.31}$$

similar to (7.72)–(7.74).

Here, $j_{100} = 1, 2, \ldots, N_{100} - 1$, $j_{010} = 1, 2, \ldots, N_{010} - 1$, and $j_{001} = 1, 2, \ldots, N_{001} - 1$. τ_{100}, τ_{010}, or τ_{001} define the boundary faces of the cavity in the [100], [010], or [001] direction, $\hat{\Lambda}_n[\hat{\mathbf{k}}; \tau_l]$ is the surface-like band structure of a quantum film with the film plane oriented in the $[l]$ direction. l can be either one of 100, 010, or 001.

For each bulk energy band n, there are $(N_{100} - 1) + (N_{010} - 1) + (N_{001} - 1)$ edge-like states in the cavity structure. They are as follows:
$(N_{100} - 1)$ states with energies

$$\Lambda^{eg,a_1}_{n,j_{100}}(\tau_{010}, \tau_{001}) = \bar{\Lambda}_n \left[\frac{j_{100}\pi}{N_{100}a}(1, 0, 0); \tau_{010}, \tau_{001} \right]; \tag{G.32}$$

$(N_{010} - 1)$ states with energies

$$\Lambda^{eg,a_2}_{n,j_{010}}(\tau_{001}, \tau_{100}) = \bar{\Lambda}_n \left[\frac{j_{010}\pi}{N_{010}a}(0, 1, 0); \tau_{001}, \tau_{100} \right]; \tag{G.33}$$

$(N_{001} - 1)$ states with energies

$$\Lambda^{eg,a_3}_{n,j_{001}}(\tau_{100}, \tau_{010}) = \bar{\Lambda}_n \left[\frac{j_{001}\pi}{N_{001}a}(0, 0, 1); \tau_{100}, \tau_{010} \right], \tag{G.34}$$

similar to (7.75)–(7.77).

Here, $\bar{\Lambda}_n[\bar{\mathbf{k}}; \tau_l, \tau_m]$ is the edge-like band structure of a rectangular quantum wire with the wire faces oriented in the $[l]$ or the $[m]$ direction. l and m can be two of 100, 010, and 001.

For each bulk energy band n, there is one vertex-like state in the cavity structure with energy $\Lambda^{vt}_n(\tau_{100}, \tau_{010}, \tau_{001})$ depending all three τ_{100}, τ_{010}, and τ_{001}, similar to (7.78).

Index

S.Y. Ren, *Electronic States in Crystals of Finite Size*, Springer Tracts in Modern Physics 270, DOI 10.1007/978-981-10-4718-3

Printed in the United States
By Bookmasters